Structural Depth Reference Manual
for the Civil PE Exam

Third Edition

Alan Williams, PhD, SE, FICE, C Eng

Professional Publications, Inc. • Belmont, California

Benefit by Registering This Book with PPI

- Get book updates and corrections.
- Hear the latest exam news.
- Obtain exclusive exam tips and strategies.
- Receive special discounts.

Register your book at **www.ppi2pass.com/register**.

Report Errors and View Corrections for This Book

PPI is grateful to every reader who notifies us of a possible error. Your feedback allows us to improve the quality and accuracy of our products. You can report errata and view corrections at **www.ppi2pass.com/errata**.

STRUCTURAL DEPTH REFERENCE MANUAL FOR THE CIVIL PE EXAM
Third Edition

Current printing of this edition: 2

Printing History

edition number	printing number	update
2	4	Minor changes. Copyright update.
3	1	New edition. Copyright update.
3	2	Minor changes.

Copyright © 2012 by Professional Publications, Inc. (PPI). All rights reserved. No part of this publication may be reproduced, stored in a retrieval system, or transmitted, in any form or by any means, electronic, mechanical, photocopying, recording, or otherwise, without the prior written permission of the publisher.

Printed in the United States of America.

PPI
1250 Fifth Avenue, Belmont, CA 94002
(650) 593-9119
www.ppi2pass.com

ISBN: 978-1-59126-392-0

Library of Congress Control Number: 2008936268

Table of Contents

Preface and Acknowledgments............ v

Introduction................................ vii

**References and Codes
Used to Prepare This Book**................ ix

Chapter 1: Reinforced Concrete Design
1. Strut-and-Tie Models...................... 1-1
2. Corbels................................... 1-5
3. Design for Torsion........................ 1-7
   Practice Problems......................... 1-11
   Solutions................................. 1-12

Chapter 2: Foundations
1. Eccentrically Loaded Column Bases......... 2-1
2. Combined Footings......................... 2-10
3. Strap Footings............................ 2-16
   Practice Problems......................... 2-21
   Solutions................................. 2-22

Chapter 3: Prestressed Concrete Design
1. Strength Design of Flexural Members....... 3-1
2. Design for Shear and Torsion.............. 3-7
3. Prestress Losses.......................... 3-15
4. Composite Construction.................... 3-19
5. Load Balancing Procedure.................. 3-24
6. Concordant Cable Profile.................. 3-26
   Practice Problems......................... 3-28
   Solutions................................. 3-29

Chapter 4: Structural Steel Design
1. Plastic Design............................ 4-1
2. Eccentrically Loaded Bolt Groups.......... 4-11
3. Eccentrically Loaded Weld Groups.......... 4-19
4. Composite Beams........................... 4-27
   Practice Problems......................... 4-33
   Solutions................................. 4-34

Chapter 5: Design of Wood Structures
1. Design Principles......................... 5-1
2. Design for Flexure........................ 5-6
3. Design for Compression.................... 5-10
4. Design for Tension........................ 5-15
5. Design for Shear.......................... 5-18
6. Design of Connections..................... 5-22
   Practice Problems......................... 5-32
   Solutions................................. 5-33

Chapter 6: Design of Reinforced Masonry
1. Design Principles......................... 6-1
2. Design for Flexure........................ 6-4
3. Design for Shear.......................... 6-10
4. Design of Masonry Columns................. 6-12
5. Design of Masonry Shear Walls............. 6-14
6. Wall Design for Out-of-Plane Loads........ 6-18
7. Design of Anchor Bolts.................... 6-24
8. Design of Prestressed Masonry............. 6-26
   Practice Problems......................... 6-33
   Solutions................................. 6-34

Index...................................... I-1

Index of Codes............................. IC-1

Preface and Acknowledgments

The structural depth section of the civil PE exam requires detailed understanding of engineering codes and their application. My goal in writing this book has been to provide a guide to the relevant codes, and to demonstrate their use in calculations for structures as necessary to help you pass the civil PE exam.

Different from the previous editions, this third edition has been revised and expanded to reflect the most current NCEES design standards. (For a complete list of updated codes, see References and Codes Used to Prepare This Book.) Nomenclature, equations, examples, and practice problems have been checked and updated so that they are congruent with current codes. New text has also been added to sections previously dependent on older editions of exam-referenced codes.

I would like to say thank you to C. Dale Buckner, PhD, PE, for performing the initial technical review of this book, and to the PPI product development and implementation staff who worked on the third edition, including Sarah Hubbard, director of product development and implementation; Cathy Schrott, director of production services; Megan Synnestvedt, Jenny Lindeburg King, and Julia White, editorial project managers; Bill Bergstrom, Lisa Devoto Farrell, Tyler Hayes, Chelsea Logan, Scott Marley, Magnolia Molcan, and Bonnie Thomas, copy editors; Kate Hayes, production associate; Andrew Chan and Todd Fisher, calculation checkers; Tom Bergstrom, technical illustrator; and Amy Schwertman La Russa, cover designer.

Alan Williams, PhD, SE, C Eng, FICE

Introduction

This book addresses the structural depth section of the civil PE examination administered by the National Council of Examiners for Engineering and Surveying (NCEES). The civil PE examination provides the qualifying test for candidates seeking registration as civil engineers. The structural section of the examination is intended to assess the candidate's depth of knowledge of structural design principles and practice. The problems in the examination are intended to be representative of the process of designing portions of real structures.

This book is written with the exam in mind. The national codes are referenced, and the appropriate sections of the codes are explained and analyzed in concise and simple manners. Illustrative examples are also presented. Each example focuses on one specific code issue. This book provides a comprehensive guide and reference for self-study of the subject, and it supplies a rapid and concise solution technique for any particular problem type. In addition to providing clarification and interpretation of the applicable sections of the codes, extensive reference publications are cited to reflect current design procedures.

The text is organized into six chapters, corresponding to the subject areas covered on the NCEES civil PE structural depth exam.

- reinforced concrete design
- foundations and retaining structures
- prestressed concrete design
- structural steel design
- timber design
- masonry design

The American Concrete Institute has published *Building Code Requirements for Structural Concrete, ACI 318-08*, and this edition of the code has been adopted by the NCEES for their examinations. The first three chapters of this book conform to the 2008 edition of the ACI code.

For questions involving structural steel design, the candidate may choose to use either the allowable stress design (ASD) method or the load and resistance factor (LRFD) method. In Ch. 4 of this book, both methods have been used. The NCEES has adopted the *Steel Construction Manual*, 13th ed. 2005, published by the American Institute of Steel Construction (AISC). Chapter 4 of this book conforms to the 2005 edition of the *AISC Steel Construction Manual*.

The American Forest and Paper Association has published the *National Design Specification for Wood Construction, ANSI/NF & PA NDS-2005*, and this edition of the code has been adopted by the NCEES for their examinations. Chapter 5 of this book conforms to the 2005 edition of the NDS code.

The American Concrete Institute has published the *Building Code Requirements for Masonry Structures: ACI 530-08/ASCE 5-08/TMS 402-08*, and this edition of the code has been adopted by the NCEES for their examinations. Chapter 6 of this book conforms to the 2008 edition of the ACI code.

The NCEES has also adopted the *International Building Code, IBC-2009*, published by the International Code Council, and *Minimum Design Loads for Buildings and Other Structures, ASCE 7-05*, published by the American Society of Civil Engineers, and all chapters of this book conform to these codes.

Abbreviations are used throughout this book to refer to common reference sources. These sources are listed in the References and Codes section, with their abbreviations in brackets. This book also cites other publications that discuss current design procedures. These other publications are also listed in the References and Codes section.

References and Codes Used to Prepare This Book

The information that was used to write and update this book was based on the exam specifications at the time of publication. However, as with engineering practice itself, the PE examination is not always based on the most current codes or cutting-edge technology. Similarly, codes, standards, and regulations adopted by state and local agencies often lag issuance by several years. It is likely that the codes that are most current, the codes that you use in practice, and the codes that are the basis of your exam will all be different. PPI lists on its website the dates and editions of the codes, standards, and regulations on which NCEES has announced the PE exams are based. It is your responsibility to find out which codes are relevant to your exam. In the meantime, the following codes have been incorporated into this edition. The list of references are those that were used in the writing of this book, and may prove helpful in your own exam preparation.

CODES

American Concrete Institute. *Building Code Requirements for Structural Concrete, (ACI 318-08)*. 2008. Farmington Hills, MI. [ACI]

American Concrete Institute. *Building Code Requirements for Masonry Structures, (ACI 530-08/ASCE 5-08/TMS 402-08)*. 2008. Farmington Hills, MI. [MSJC]

American Forest and Paper Association. *National Design Specification for Wood Construction, with Commentary and Supplement, (ANSI/AF&PA NDS-2005)*. 2005. Washington, DC. [NDS]

American Institute of Steel Construction. *Steel Construction Manual*, 13th ed. 2005. Chicago, IL. [AISC]

American Society of Civil Engineers. *Minimum Design Loads for Buildings and Other Structures, (ASCE 7-05)*. 2005. Reston, VA. [ASCE]

International Code Council. 2009 *International Building Code without Supplements*. 2009. Falls Church, VA. [IBC]

REFERENCES

American Plywood Association. *Glued Laminated Beam Design Tables*. 2007. Tacoma, WA.

Buckner, C. D. *246 Solved Structural Engineering Problems*, 3rd ed. PPI. 2003. Belmont, CA.

Building Seismic Safety Council. *NEHRP Recommended Seismic Provisions for New Buildings and Other Structures*. 2009 ed. Washington, DC.

Durning, T. A. "Prestressed Masonry." *Structural Engineer*. June 2000. Atlanta, GA.

Ekwueme, C. G. "Design of Anchor Bolts in Concrete Masonry." *Masonry Chronicles*. Winter 2009-10. Concrete Masonry Association of California and Nevada. Citrus Heights, CA.

Ekwueme, C. G. "Out-of-Plane Design of Masonry Walls." *Structural Engineer*. October 2003. Atlanta, GA.

Freyermuth, C. L., and Schoolbred, R. A. *Post-Tensioned Prestressed Concrete*. Portland Cement Association. 1967. Skokie, IL.

Horne, M. R., and Morris, L. J. *Plastic Design of Low-Rise Frames*. MIT Press. 1982. Cambridge, MA.

Kubischta, M. "Comparison of the 1997 UBC and the 2002 MSJC Code." *Masonry Chronicles*. Spring 2003. Concrete Masonry Association of California and Nevada. Citrus Heights, CA.

Kubischta, M. "In-Plane Loads on Masonry Walls." *Masonry Chronicles*. Fall 2003. Concrete Masonry Association of California and Nevada. Citrus Heights, CA.

Lin, T. Y. "Load Balancing Method for Design and Analysis of Prestressed Concrete Structures." Proceedings American Concrete Association. 60: 719–742, 1963.

Masonry Society. *Masonry Designers' Guide*, 6th ed. 2010. Boulder, CO.

Neal, B. G. *Plastic Methods of Structural Analysis*. Chapman and Hall. 1970. London.

Portland Cement Association. *Notes on ACI 318-08: Building Code Requirements for Reinforced Concrete*. 2008. Skokie, IL.

Prestressed Concrete Institute. *PCI Design Handbook, Precast and Prestressed Concrete*, 6th ed. 2004. Chicago, IL.

Western Wood Products Association. *Western Lumber Span Tables*. 1992. Portland, OR.

Williams, A. *Steel Structures Design*. McGraw Hill/International Code Council. 2007. New York/Falls Church, VA.

Williams, A. *Structural Engineering Reference Manual*, 6th ed. PPI. 2011. Belmont, CA.

Zia, P., et al. "Estimating Prestress Losses." *Concrete International: Design and Construction.* (1)6: 32–38, June 1979.

Reinforced Concrete Design

1. Strut-and-Tie Models 1-1
2. Corbels 1-5
3. Design for Torsion 1-7
   Practice Problems 1-11
   Solutions 1-12

1. STRUT-AND-TIE MODELS

Nomenclature

a	shear span, distance between concentrated load and face of support	ft
A_{cs}	effective cross-sectional area of a strut in a strut-and-tie model, taken perpendicular to the axis of the strut	in^2
A_{nz}	effective cross-sectional area of the face of a nodal zone	in^2
A_{si}	area of reinforcement in the ith layer of reinforcement crossing the strut	in^2
A_{ts}	total area of nonprestressed reinforcement in a tie	in^2
A_v	area of shear reinforcement perpendicular to tension reinforcement within a distance s	in^2
A_{vh}	area of shear reinforcement parallel to tension reinforcement within a distance s	in^2
b	width of member	in
b_w	web width	in
c	clear cover to reinforcement	in
C	compressive force acting on a nodal zone	kips
d	distance from extreme compression fiber to centroid of longitudinal tension reinforcement	in
d_b	nominal diameter of bar, wire, or prestressing strand	in
f'_c	compressive strength of concrete	lbf/in^2
f_{ce}	effective compressive strength of the concrete in a strut	kips/in^2
F_{ns}	nominal compressive strength of a strut	kips
F_{nt}	nominal tensile strength of a tie	kips
h	depth of member	in
l_a	anchorage length of a reinforcing bar	in
l_b	width of bearing plate	in
l_{dh}	development length in tension of a hooked bar	in
R	support reaction acting on a nodal zone	kips
s	stirrup spacing	in
s_i	spacing of the ith layer of reinforcement	in
T	tension force acting on a nodal zone	kips
w_s	effective width of strut perpendicular to the axis of the strut	in
w_t	effective width of concrete concentric with a tie	in

Symbols

β_n	factor to account for the effect of the anchorage of ties on the effective compressive strength of a nodal zone	–
β_s	factor to account for the effect of cracking and confining reinforcement on the effective compressive strength of the concrete in a strut	–
α_i	angle between the strut and the bars in the ith layer of reinforcement crossing the strut	deg
θ	angle between the axis of a strut and a tension chord	deg
λ	correction factor related to unit weight of concrete	–
ρ	reinforcement ratio	–
ϕ	strength reduction factors	–

A strut-and-tie model consists of the application of an analogous truss model to regions of a member in which the stress distribution is nonuniform and the normal beam theory does not apply. An example of this situation is a deep beam, which may be designed using nonlinear analysis methods or by the strut-and-tie analogy method given in ACI 318. A deep beam is defined in ACI Sec. 11.7.1 as a beam in which the ratio of clear span to overall depth does not exceed four, or in which the shear span to depth ratio does not exceed two.

As shown in Fig. 1.1, a discontinuity, or D region, occurs at a change in the geometry of a member and at concentrated loads and reactions. The D region extends a distance equal to the overall depth of the member from the location of the change in geometry or the point of application of a load or reaction. Outside of the D region, beam theory is applicable, and this region is known as a beam, or B, region.

Typical examples of strut-and-tie models are shown in Fig. 1.2. Several possible solutions may exist for a given example. Compressive forces are resisted by concrete struts and tensile forces are resisted by steel ties. Struts, ties, and external forces meet at nodes. Tensile stresses in the concrete are neglected, and struts should be oriented parallel to the directions of anticipated cracking. Forces in struts and ties are uniaxial, and equilibrium must be maintained at the nodes and in the truss

Figure 1.1 Discontinuity Regions

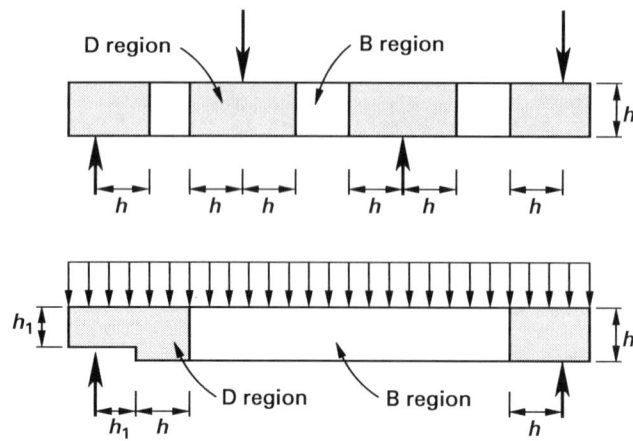

Figure 1.2 Typical Strut-and-Tie Model Examples

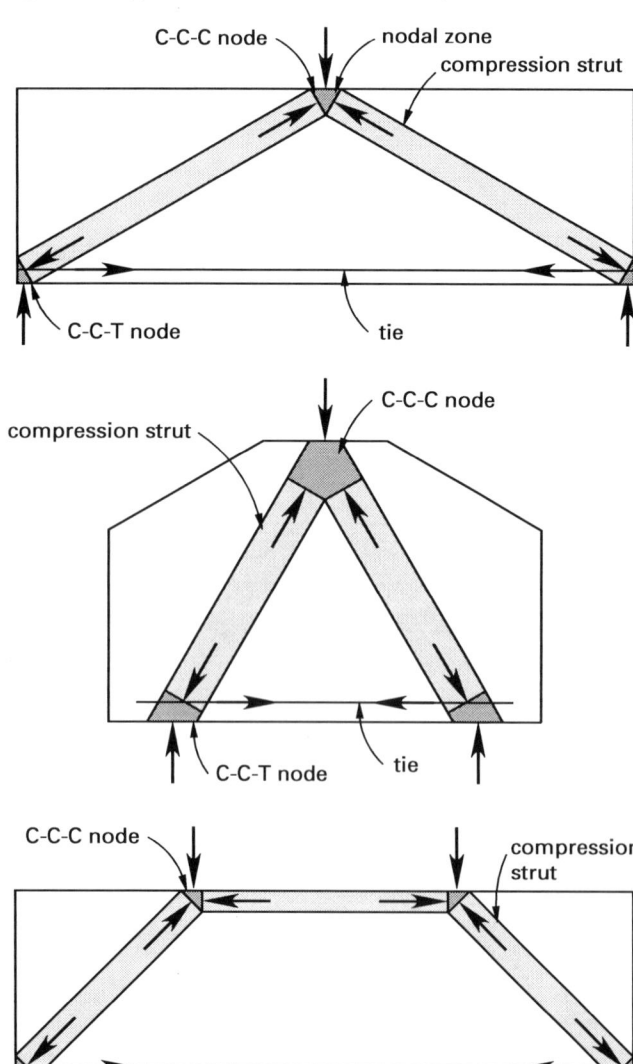

as a whole. In setting up the strut-and-tie model, the following procedure may be followed.

- Determine the reactions on the model.
- Select the location of the members by aligning the direction of struts in the direction of the anticipated cracking.
- Determine the areas of struts, ties, and nodes necessary to provide the required strength.
- Provide anchorage for the ties.
- Provide crack control reinforcement.

The two types of concrete struts (see Fig. 1.3) that may form in the model are the prism strut and the bottle-shaped strut. The prism strut has a uniform cross section and forms in the compression zone of a beam. A bottle-shaped strut expands and contracts along its length. The strength of this strut is governed by the transverse tension developed by the lateral spread of the applied compression force. As specified in ACI Sec. A.3.3, using two orthogonal layers of confinement reinforcement near each face to resist the transverse tension increases the strength of the strut. The required reinforcement ratio is

$$\sum \frac{A_{si}}{bs_i} \sin \alpha_i \geq 0.003 \quad \text{[ACI A-4]}$$

Figure 1.3 Types of Concrete Struts

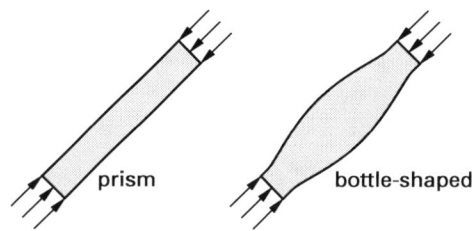

The effective compressive strength of the concrete in a strut is specified in ACI Sec. A.3.2 as

$$f_{ce} = 0.85\beta_s f'_c \quad \text{[ACI A-3]}$$

β_s is the factor, given in Table 1.1, that accounts for the effect of cracking and confining reinforcement on the effective compressive strength of the concrete in a strut.

In accordance with ACI Sec. A.3.1, the nominal compressive strength of a strut is

$$F_{ns} = f_{ce}A_{cs} \quad \text{[ACI A-2]}$$
$$= f_{ce}w_s b$$

Ties are the tension members of the strut-and-tie model that are formed by reinforcing or prestressing steel. They may consist of longitudinal chord reinforcement

Table 1.1 Values of β_s and λ

β_s	type of strut
1.0	strut of uniform cross section
0.75	bottle-shaped strut, with reinforcement as specified in ACI Sec. A.3.3
0.60λ	unreinforced, bottle-shaped strut
0.40	strut in a tension member or the tension flange of a member
0.60	all other cases

λ	weight of concrete
1.0	normal
0.85	sand-lightweight
0.75	all lightweight

or vertical stirrups. Adequate anchorage must be provided to the reinforcement by means of end plates or hooks, or by a straight development length. In a reinforced concrete beam, the nominal strength of a reinforcing bar acting as a tie is given by ACI Sec. A.4.1 as

$$F_{nt} = A_{ts}f_y$$

If the bars in a tie are in one layer, as shown in Fig. 1.4, the width of the tie may be taken as the diameter of the bars, plus twice the cover to the surface of the bars.

A node is the location at which struts, ties, and external loads meet. As shown in Fig. 1.2, a node may be classified as C-C-T, with two of the members acting on the node in compression and the third member in tension. Similarly, when all three members acting on the node are in compression, the node is classified as C-C-C.

A nodal zone, as defined in ACI Sec. A.1 and shown in Fig. 1.4, is the volume of concrete surrounding a node that is assumed to transfer strut-and-tie forces through the node. The effective compressive strength of the concrete in a nodal zone is specified in ACI Sec. A.5.2 as

$$f_{ce} = 0.85\beta_n f'_c \quad \text{[ACI A-8]}$$

β_n is the factor, given in Table 1.2, that accounts for the effect of anchoring ties on the effective compressive strength of a nodal zone.

Table 1.2 Values of β_n

β_n	type of nodal zone
1.0	bounded on all sides by struts, or bearing areas, or both
0.80	anchoring one tie
0.60	anchoring two or more ties

In accordance with ACI Sec. A.5.1, the nominal compressive strength of a nodal zone is

$$F_{nn} = f_{ce}A_{nz} \quad \text{[ACI A-7]}$$
$$= f_{ce}w_s b$$

The faces of the nodal zone shown in Fig. 1.4 are perpendicular to the axes of the strut, tie, and bearing plate. The lengths of the faces are in direct proportion to the forces acting. Hence, the node has equal stresses on all faces and is termed a *hydrostatic nodal zone*. The effective width of the strut shown in Fig. 1.4 is

$$w_s = w_t \cos\theta + l_b \sin\theta$$

The extended nodal zone shown in Fig. 1.4 is that portion of the member bounded by the intersection of the effective strut width and the effective tie width. As specified in ACI Sec. A.4.3.2, the anchorage length of the reinforcement is measured from the point of intersection of the bar and the extended nodal zone. The reinforcement may be anchored by a plate, by hooks, or by a straight development length.

To control cracking in a deep beam, ACI Secs. 11.7.4 and 11.7.5 require the provision of two orthogonal layers of confinement reinforcement near each face. The maximum reinforcement spacing in each layer is

$$s \leq \frac{d}{5} \leq 12 \text{ in}$$

The area of horizontal reinforcement must not be less than

$$A_{vh} = 0.0015 b_w s$$

The area of vertical reinforcement must not be less than

$$A_v = 0.0025 b_w s$$

Figure 1.4 Nodal Zones

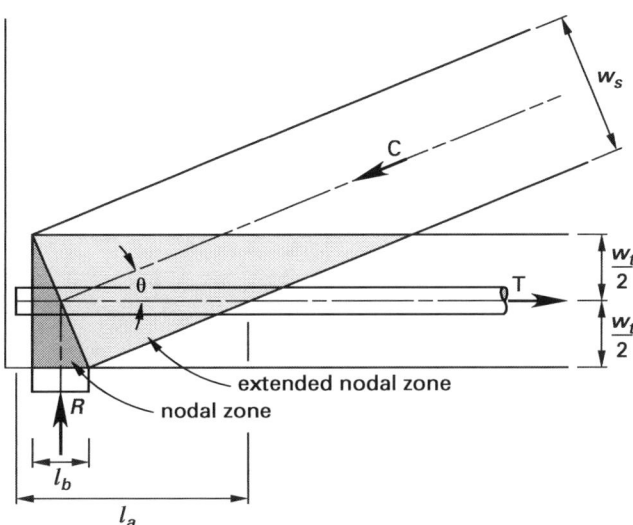

Alternatively, the reinforcement specified in ACI Sec. A.3.3 may be provided.

Example 1.1

A reinforced concrete beam with a clear span of 6 ft, an effective depth of 21 in, and a width of 12 in, as shown in the illustration, has a concrete compressive strength of 5000 lbf/in². The factored applied force of 100 kips includes an allowance for the self-weight of the beam. Determine the area of grade 60 tension reinforcement required, and check that the equivalent concrete strut and nodal zone at the left support comply with the requirements of ACI App. A.

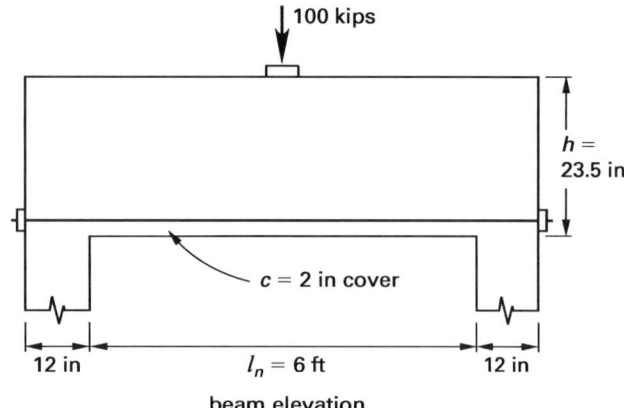

beam elevation

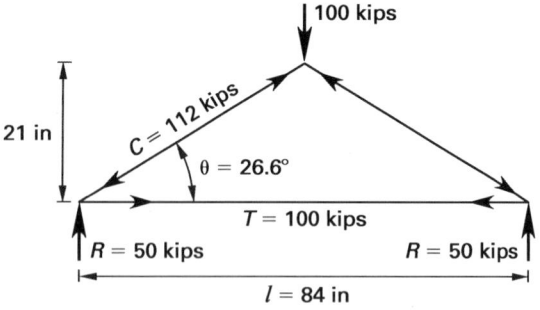

strut-and-tie model

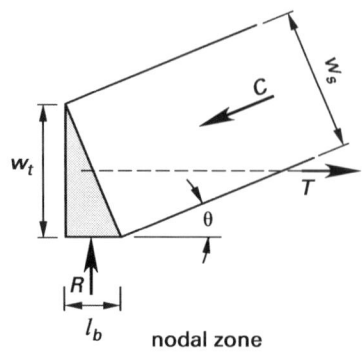

nodal zone

(not to scale)

Solution

The clear span to depth ratio is

$$\frac{l_n}{h} = \frac{(6 \text{ ft})\left(12\ \frac{\text{in}}{\text{ft}}\right)}{23.5 \text{ in}} = 3.1$$
$$< 4 \quad [\text{satisfies ACI Sec. 11.7.1 as a deep beam}]$$

The idealized strut-and-tie model is shown in the illustration. The angle between the struts and the tie is

$$\theta = \tan^{-1}\left(\frac{21 \text{ in}}{42 \text{ in}}\right)$$
$$= 26.6°$$
$$> 25° \quad [\text{satisfies ACI Sec. A.2.5}]$$

The equivalent tie force is determined from the strut-and-tie model as

$$T = \frac{(50 \text{ kips})(42 \text{ in})}{21 \text{ in}}$$
$$= 100 \text{ kips}$$

The strength reduction factor is given by ACI Sec. 9.3.2.6 as

$$\phi = 0.75$$

The necessary reinforcement area is

$$A_{ts} = \frac{T}{\phi f_y}$$
$$= \frac{100 \text{ kips}}{(0.75)\left(60\ \frac{\text{kips}}{\text{in}^2}\right)}$$
$$= 2.22 \text{ in}^2$$

Use three no. 8 bars, which gives

$$A = 2.37 \text{ in}^2$$
$$> A_{ts} \quad [\text{satisfactory}]$$

As shown in the illustration, the dimensions of the nodal zone are w_t, l_b, and w_s. The equivalent tie width is

$$w_t = d_b + 2c$$
$$= 1 \text{ in} + (2)(2 \text{ in})$$
$$= 5 \text{ in}$$

The width of the equivalent support strut is

$$l_b = w_t \tan\theta$$
$$= (5 \text{ in}) \tan 26.6°$$
$$= 2.5 \text{ in}$$

The width of the equivalent concrete strut is

$$w_s = \frac{w_t}{\cos\theta}$$
$$= \frac{5 \text{ in}}{\cos 26.6°}$$
$$= 5.59 \text{ in}$$

The stress in the equivalent tie is

$$f_T = \frac{T}{bw_t}$$
$$= \frac{100 \text{ kips}}{(12 \text{ in})(5 \text{ in})}$$
$$= 1.67 \text{ kips/in}^2$$

For a hydrostatic nodal zone, the stress in the equivalent concrete strut is

$$f_C = f_T$$
$$= 1.67 \text{ kips/in}^2$$

The stress in the equivalent support strut is

$$f_R = f_T$$
$$= 1.67 \text{ kips/in}^2$$

For normal weight concrete and an unreinforced bottle-shaped strut, the design compressive strength of the concrete in the strut is given by ACI Eq. (A-3) as

$$\phi f_{ce} = (0.75)(0.85\beta_s f'_c)$$
$$= (0.75)(0.85)(0.6)(1.0)\left(5 \, \frac{\text{kips}}{\text{in}^2}\right)$$
$$= 1.91 \text{ kips/in}^2$$
$$> f_C \quad \text{[satisfactory]}$$

The design compressive strength of a nodal zone anchoring one layer of reinforcing bars without confining reinforcement is given by ACI Eq. (A-8) as

$$\phi f_{ce} = (0.75)(0.85\beta_n f'_c)$$
$$= (0.75)(0.85)(0.8)\left(5 \, \frac{\text{kips}}{\text{in}^2}\right)$$
$$= 2.55 \text{ kips/in}^2$$
$$> f_C \quad \text{[satisfactory]}$$

The anchorage length available for the tie reinforcement, using 2 in end cover, is

$$l_a = \frac{w_t}{2\tan\theta} + \frac{l_b}{2} + 6 \text{ in} - 2 \text{ in}$$
$$= \frac{5 \text{ in}}{2\tan 26.6°} + \frac{2.5 \text{ in}}{2} + 6 \text{ in} - 2 \text{ in}$$
$$= 10.2 \text{ in}$$

The development length for a grade 60, no. 8 bar with 2.5 in side cover and 2 in end cover, and with a standard 90° hook, is given by ACI Secs. 12.5.2 and 12.5.3 as

$$l_{dh} = \frac{(0.7)(1200)d_b}{\sqrt{f'_c}}$$
$$= \frac{(0.7)\left(1200 \, \frac{\text{lbf}}{\text{in}^2}\right)(1.0 \text{ in})}{\sqrt{5000 \, \frac{\text{lbf}}{\text{in}^2}}}$$
$$= 11.9 \text{ in}$$
$$> l_a \quad \text{[anchorage length is inadequate]}$$

Use an end plate to anchor the bars.

2. CORBELS

Nomenclature

a	shear span, distance between concentrated load and face of supports	ft
A_f	area of reinforcement in bracket or corbel resisting factored moment	in^2
A_h	area of closed stirrups parallel to primary tension reinforcement	in^2
A_n	area of reinforcement in bracket or corbel resisting tensile force N_{uc}	in^2
A_{sc}	area of primary tension reinforcement	in^2
A_{vf}	area of shear-friction reinforcement	in^2
b	width of compression face of member	in
h	overall thickness of member	in
f_y	specified yield strength of reinforcement	kips/in^2
M_u	factored moment at section	ft-kip
N_{uc}	factored tensile force applied at top of corbel	kips
V_c	nominal shear strength provided by concrete	kips
V_s	nominal shear strength provided by shear reinforcement	kips
V_u	factored shear force at section	kips

Symbols

λ	correction factor related to unit weight of concrete as given in ACI Sec. 8.6.1	–
μ	coefficient of friction	–
ρ	reinforcement ratio	–
ϕ	strength reduction factor, 0.75 for shear and flexure	–

As shown in Fig. 1.5, a corbel is a cantilever bracket supporting a load-bearing member. In accordance with ACI Sec. 11.8, the maximum allowable value of the ratio of the shear span to the effective depth is

$$\frac{a}{d} = 1$$

Figure 1.5 *Corbel Details*

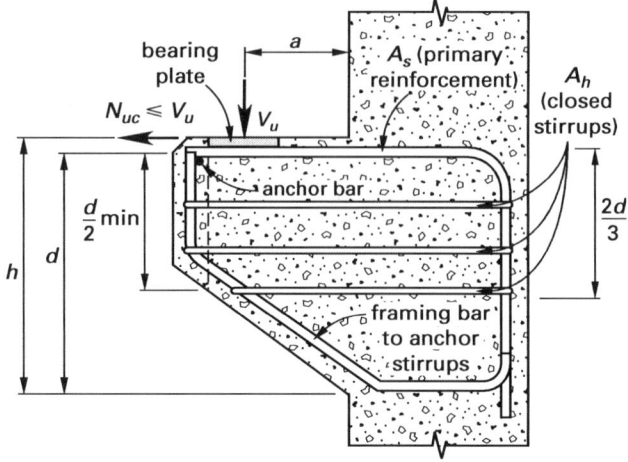

Similarly, the ratio of the horizontal tensile force to the vertical force is limited to a maximum value of

$$\frac{N_{uc}}{V_u} = 1$$

As specified in ACI Sec. 11.8.2, the depth of the corbel at the outside edge of bearing area shall not be less than half the effective depth.

The forces acting on the corbel at the face of the support are the shear force V_u, the tensile force N_{uc}, and the bending moment M_u. The applied factored shear force is

$$V_u \leq 0.2\phi f'_c b_w d \quad \text{[normal weight concrete]}$$
$$\leq 0.8\phi b_w d \quad \text{[normal weight concrete]}$$

From ACI Sec. 11.8.3.4, the horizontal tensile force caused by volume changes is

$$N_{uc} \geq 0.2 V_u$$

The bending moment caused by the applied shear force and the horizontal tensile force is

$$M_u = V_u a + N_{uc}(h - d)$$

The required area of shear friction reinforcement provided by the primary steel and horizontal closed stirrups is given by ACI Sec. R11.6.4.1 as

$$A_{vf} = \frac{V_u}{\phi f_y \mu}$$

ACI Sec. 11.6.4.3 gives the value of the coefficient of friction at the face of the support as $\mu = 1.4\lambda$ for concrete placed monolithically. ACI Sec. 11.6.4.3 defines the correction factor related to the unit weight of concrete as $\lambda = 1.0$ for normal weight concrete. ACI Sec. 11.8.3.1 gives the value of the strength reduction factor as $\phi = 0.75$ for all design calculations.

The required area of reinforcement to resist N_{uc} is given by ACI Sec. 11.8.3.4 as

$$A_n = \frac{N_{uc}}{\phi f_y}$$

The required area of flexural reinforcement, A_f, to resist the moment M_u is derived by the normal flexural theory given by ACI Sec. 10.3, using a strength reduction factor of $\phi = 0.75$ as specified by ACI Sec. 11.8.3.1.

The required total area of primary tension reinforcement is given by ACI Sec. 11.8.3.5 as

$$A_{sc} = A_f + A_n$$
$$\geq \frac{2 A_{vf}}{3} + A_n$$
$$\geq \frac{0.04 b_w d f'_c}{f_y}$$

The minimum required area of closed ties, distributed over a depth of two-thirds of the effective depth, is given by ACI Sec. 11.8.4 as

$$A_h = \frac{A_{sc} - A_n}{2}$$

Example 1.2

The reinforced concrete corbel shown in the illustration, with a width of 15 in, is reinforced with grade 60 bars. This corbel and has a concrete compressive strength of 3000 lbf/in². Determine whether the corbel is adequate for the applied factored loads indicated.

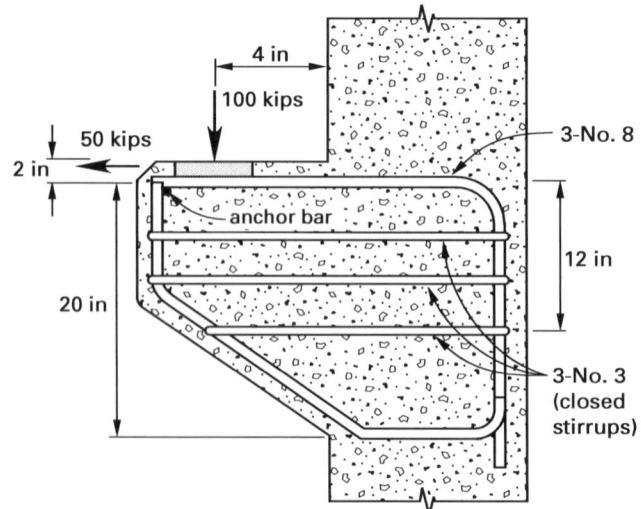

Solution

$$0.2\phi f'_c b_w d = (0.2)(0.75)\left(3 \ \frac{\text{kips}}{\text{in}^2}\right)(15 \text{ in})(20 \text{ in})$$
$$= 135 \text{ kips}$$
$$> V_u \quad \text{[satisfies ACI Sec. 11.8.3.2.1]}$$

$$0.8\phi b_w d = (0.8)(0.75)(15 \text{ in})(20 \text{ in})$$
$$= 180 \text{ kips}$$
$$> V_u \quad \text{[satisfies ACI Sec. 11.8.3.2.1]}$$

The shear friction reinforcement area is given by ACI Sec. R11.6.4.1 as

$$A_{vf} = \frac{V_u}{\phi f_y \mu} = \frac{100 \text{ kips}}{(0.75)\left(60 \ \frac{\text{kips}}{\text{in}^2}\right)(1.4)}$$
$$= 1.59 \text{ in}^2$$

The tension reinforcement area is given by ACI Sec. 11.8.3.4 as

$$A_n = \frac{N_{uc}}{\phi f_y}$$
$$= \frac{50 \text{ kips}}{(0.75)\left(60 \ \frac{\text{kips}}{\text{in}^2}\right)}$$
$$= 1.111 \text{ in}^2$$

The factored moment acting on the corbel is

$$M_u = V_u a + N_{uc}(h - d)$$
$$= (100 \text{ kips})(4 \text{ in}) + (50 \text{ kips})(22 \text{ in} - 20 \text{ in})$$
$$= 500 \text{ in-kips}$$

For a strength reduction factor of $\phi = 0.75$, the area of required flexural reinforcement is given by ACI Sec. 10.2 as

$$A_f = \frac{0.85 b_w d f'_c \left(1 - \sqrt{1 - \frac{M_u}{0.319 b d^2 f'_c}}\right)}{f_y}$$

$$= \frac{(0.85)(15 \text{ in})(20 \text{ in})\left(3 \ \frac{\text{kips}}{\text{in}^2}\right)}{60 \ \frac{\text{kips}}{\text{in}^2}}$$
$$\times \left(1 - \sqrt{1 - \frac{500 \text{ in-kips}}{(0.319)(15 \text{ in})(20 \text{ in})^2 \left(3 \ \frac{\text{kips}}{\text{in}^2}\right)}}\right)$$

$$= 0.568 \text{ in}^2$$

The primary reinforcement area required is given by ACI Sec. 11.8.3.5 as

$$A_{sc} = A_f + A_n$$
$$= 0.568 \text{ in}^2 + 1.111 \text{ in}^2$$
$$= 1.68 \text{ in}^2$$

Three no. 8 bars are provided, giving an area of

$$A_{s,\text{prov}} = 2.37 \text{ in}^2$$
$$> A_s \quad \text{[satisfactory]}$$

Also, from ACI Sec. 11.8.3.5, the area of primary reinforcement must not be less than

$$\frac{2A_{vf}}{3} + A_n = \frac{(2)(1.59 \text{ in}^2)}{3} + 1.111 \text{ in}^2$$
$$= 2.17 \text{ in}^2$$
$$< A_{s,\text{prov}} \quad \text{[satisfactory]}$$

The required area of closed stirrups is given by ACI Sec. 11.8.4 as

$$A_h = \frac{A_{sc} - A_n}{2}$$
$$= \frac{2.17 \text{ in}^2 - 1.111 \text{ in}^2}{2}$$
$$= 0.53 \text{ in}^2$$

Three no. 3 closed stirrups are provided, giving an area of

$$A_{h,\text{prov}} = 0.66 \text{ in}^2$$
$$> A_h \quad \text{[satisfactory]}$$

3. DESIGN FOR TORSION

Nomenclature

A_{cp}	area enclosed by outside perimeter of concrete cross section	in²
A_l	total area of longitudinal reinforcement to resist torsion	in²
A_o	gross area enclosed by shear flow	in²
A_{oh}	gross area enclosed by centerline of the outermost closed transverse torsional reinforcement	in²
A_t	area of one leg of a closed stirrup resisting torsion within a distance s	in²
A_v	area of shear reinforcement perpendicular to flexural tension reinforcement	in²
A_{v+t}	sum of areas of shear and torsion reinforcement	in²
b_w	web width	in
c	clear cover to tension reinforcement	in

d	distance from extreme compression fiber to centroid of tension reinforcement	in
d_b	diameter of reinforcement	in
f_y	yield strength of reinforcement	kips/in^2
f_{yt}	yield strength of transverse reinforcement	kips/in^2
h	overall thickness of member	in
$M_{\max}$	maximum factored moment at section caused by externally applied loads	in-kips
M_u	factored moment at section	in-kips
p_{cp}	outside perimeter of the concrete cross section	in
p_h	perimeter of centerline of outermost closed transverse torsional reinforcement	in
s	spacing of shear or torsion reinforcement	in
S_b	section modulus of the section referred to the bottom fiber	in^3
t	wall thickness before cracking	in
T_n	nominal torsional moment strength	in-kips
T_u	factored torsional moment at section	in-kips
V_c	nominal shear strength provided by concrete	kips
V_s	nominal shear strength provided by shear reinforcement	kips
V_u	factored shear force at section	kips

Symbols

τ	shear stress in walls	lbf/in^2
θ	angle of compression diagonals in truss analogy for torsion, 45° for a reinforced concrete section	deg
λ	correction factor related to unit weight of concrete	–
ϕ	strength reduction factor, 0.75 for shear and torsion	–

After torsional cracking occurs, the central core of a reinforced concrete member is largely ineffectual in resisting applied torsion and can be neglected. Hence, it is assumed in ACI Sec. R11.5 that a member behaves as a thin-walled tube when subjected to torsion. To maintain a consistent approach, before cracking, a member is also analyzed as a thin-walled tube. As shown in Fig. 1.6, the shear stress in the tube walls produces a uniform shear flow, acting at the midpoint of the walls, with a magnitude of

$$q = \tau t$$

The applied torsion is resisted by the moment of the shear flow in the walls about the centroid of the section and is

$$T = 2A_o q$$

The gross area enclosed by shear flow, A_o, is the area enclosed by the center line of the walls.

$$A_o = \frac{2A_{cp}}{3}$$

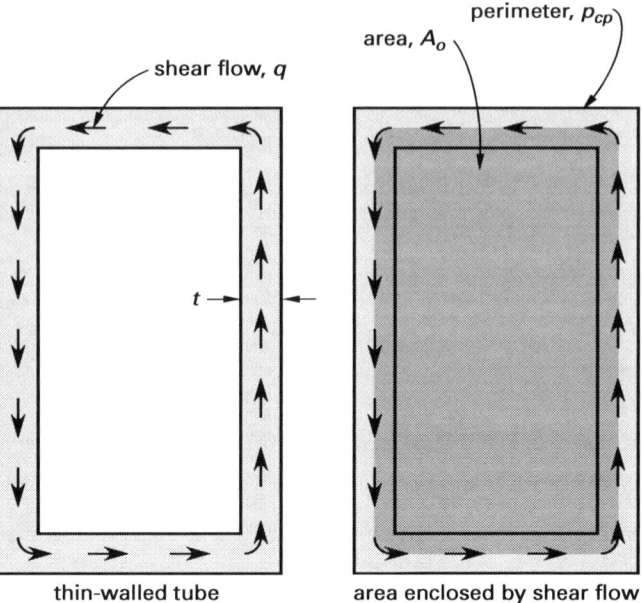

Figure 1.6 Torsion in a Rectangular Beam

The shear stress in the walls is

$$\tau = \frac{T}{2A_o t}$$

In accordance with ACI Sec. 11.5.2.4, the critical section for the calculation of torsion in a reinforced concrete beam is located at a distance from the support equal to the effective depth. When a concentrated torsion occurs within this distance, the critical section for design shall be at the face of the support.

In accordance with ACI Sec. R11.5.1, cracking is assumed to occur in a member when the principal tensile stress reaches a value of

$$p_t = 4\lambda\sqrt{f'_c} = \tau$$

The cracking torsion is

$$T_{cr} = 4\lambda\sqrt{f'_c}\left(\frac{A_{cp}^2}{p_{cp}}\right)$$

ACI Sec. 11.5.1 assumes that torsional effects may be neglected, and that closed stirrups and longitudinal torsional reinforcement are not required when the factored torque does not exceed the threshold torsion given by

$$T_u = \frac{\phi T_{cr}}{4}$$
$$= \phi\lambda\sqrt{f'_c}\left(\frac{A_{cp}^2}{p_{cp}}\right)$$

When the threshold value is exceeded, reinforcement must be provided to resist the full torsion, and the concrete is considered ineffective. When both shear and torsion reinforcements are required, the sum of the individual areas must be provided, as specified in ACI Sec. R11.5.3.8. Because the stirrup area A_v for shear is defined in terms of both legs of a stirrup, while the stirrup area A_t for torsion is defined in terms of one leg only, the summation of the areas is

$$\sum\left(\frac{A_{v+t}}{s}\right) = \frac{A_v}{s} + \frac{2A_t}{s}$$

The spacing of the reinforcement is limited by the minimum required spacing of either the shear or torsion reinforcement. As specified in ACI Sec. 11.4.5, the spacing of shear reinforcement must not exceed

$$s = \frac{d}{2} \quad [V_s \leq 4\sqrt{f'_c}b_w d]$$
$$s = \frac{d}{4} \quad [V_s > 4\sqrt{f'_c}b_w d]$$

As specified in ACI Sec. 11.5.6, the spacing of torsion reinforcement must not exceed

$$s = \frac{p_h}{8}$$
$$\leq 12 \text{ in}$$

After torsional cracking occurs, cracks are produced diagonally on all four faces of a member, forming a continuous spiral failure surface around the perimeter. For a reinforced concrete beam subjected to pure torsion, the principal tensile stresses and the cracks are produced at an angle θ. ACI Sec. 11.5.3.6 specifies that the angle θ may be assumed to be 45°. After cracking, the beam is idealized as a tubular space truss consisting of closed stirrups, longitudinal reinforcement, and concrete compression struts between the torsion cracks. To resist the applied torque, it is necessary to provide transverse and longitudinal reinforcement on each face. Closed stirrups are required to resist the vertical component, and longitudinal reinforcement is required to resist the horizontal component of the torsional stresses. After torsional cracking, the concrete outside the closed stirrups is ineffective in resisting the applied torsion, and the gross area enclosed by shear flow is redefined by ACI Sec. 11.5.3.6 as

$$A_o = 0.85 A_{oh}$$

In an indeterminate structure, redistribution of internal forces after cracking results in a reduction in torsional moment with a compensating redistribution of internal forces. This redistribution of forces is referred to as *compatibility torsion*. Non-prestressed members may be designed for a maximum factored torsional moment given by ACI Sec. 11.5.2.2 as

$$T_u = \phi 4\lambda\sqrt{f'_c}\left(\frac{A_{cp}^2}{p_{cp}}\right)$$

When the torsional moment cannot be reduced by redistribution of internal forces after cracking, the member must be designed for the full applied factored torque. This is referred to as *equilibrium torsion*.

When the threshold torsion value is exceeded, closed stirrups and longitudinal reinforcement must be provided to resist the appropriate value of the torque, and the concrete is considered ineffective.

ACI Sec. 11.5.3.6 specifies the required area of one leg of a closed stirrup as

$$\frac{A_t}{s} = \frac{T_u}{2\phi A_o f_{yt} \cot \theta} = \frac{T_u}{2\phi A_o f_{yt} \cot 45°}$$
$$= \frac{T_u}{1.7\phi A_{oh} f_{yt}}$$

The minimum combined area of stirrups for shear and torsion is given by ACI Sec. 11.5.5.2 as

$$\frac{A_v + 2A_t}{s} = 0.75\,\frac{\sqrt{f'_c}b_w}{f_{yt}} \quad \text{[ACI 11-23]}$$
$$\geq \frac{50 b_w}{f_{yt}}$$

The corresponding required area of longitudinal reinforcement is specified in ACI Secs. 11.5.3.7 and R11.5.3.10 as

$$A_l = \left(\frac{A_t}{s}\right)\left(\frac{p_h f_{yt}}{f_{yl}}\right)\cot^2 \theta \quad \text{[ACI 11-22]}$$
$$= \left(\frac{A_t}{s}\right)\left(\frac{p_h f_{yt}}{f_{yl}}\right)\cot^2 45°$$
$$= \left(\frac{A_t}{s}\right)\left(\frac{p_h f_{yt}}{f_{yl}}\right)$$

When the threshold torsional moment is exceeded, the minimum permissible area of longitudinal reinforcement is then given by ACI Sec. 11.5.5.3 as

$$A_{l,\min} = \frac{5 A_{cp}\sqrt{f'_c}}{f_y} - \left(\frac{A_t}{s}\right)p_h\left(\frac{f_{yt}}{f_y}\right) \quad \text{[ACI 11-24]}$$

In ACI Eq. (11-24), A_t/s must not be less than $25 b_w/f_{yt}$.

The bars are distributed around the inside perimeter of the closed stirrups at a maximum spacing of 12 in. A longitudinal bar is required in each corner of a closed stirrup.

The minimum diameter of longitudinal reinforcement is given by ACI Sec. 11.5.6.2 as

$$d_b = 0.042 s$$
$$\geq 3/8 \text{ in}$$

To prevent crushing of the concrete compression diagonals, the combined stress caused by the factored torsion and shear forces is limited by ACI Sec. 11.5.3.1. For solid sections, the dimensions of the section must be such that

$$\sqrt{\left(\frac{V_u}{b_w d}\right)^2 + \left(\frac{T_u p_h}{1.7 A_{oh}^2}\right)^2} \leq \phi\left(\frac{V_c}{b_w d} + 8\sqrt{f'_c}\right)$$

[ACI 11-18]

For hollow sections,

$$\frac{V_u}{b_w d} + \frac{T_u p_h}{1.7 A_{oh}^2} \leq \phi\frac{V_c}{b_w d} + 8\sqrt{f'_c} \quad \text{[ACI 11-19]}$$

Example 1.3

A simply supported reinforced concrete beam of normal weight concrete with an overall depth of $h = 23$ in, an effective depth of $d = 20$ in, and a width of $b_w = 14$ in, is reinforced with grade 60 bars, and has a concrete compressive strength of 3000 lbf/in². Determine the combined shear and torsion reinforcement required when the factored shear force is $V_u = 10$ kips and the factored torsion is $T_u = 200$ in-kips.

Solution

The design shear strength provided by the concrete is given by ACI Eq. (11-3) as

$$\phi V_c = 2\phi b_w d \lambda \sqrt{f'_c}$$

$$= \frac{(2)(0.75)(14\text{ in})(20\text{ in})(1.0)\sqrt{3000\ \frac{\text{lbf}}{\text{in}^2}}}{1000\ \frac{\text{lbf}}{\text{kip}}}$$

$$= 23\text{ kips}$$

$$> 2V_u$$

In accordance with ACI Sec. 11.4.6.1, shear reinforcement is not required.

The area enclosed by the outside perimeter of the beam is

$$A_{cp} = h b_w$$
$$= (23\text{ in})(14\text{ in})$$
$$= 322\text{ in}^2$$

The length of the outside perimeter of the beam is

$$p_{cp} = 2(h + b_w)$$
$$= (2)(23\text{ in} + 14\text{ in})$$
$$= 74\text{ in}$$

Torsional reinforcement is not required in accordance with ACI Sec. 11.5.1 when the factored torque does not exceed the threshold torsion given by

$$T_u = \phi\lambda\sqrt{f'_c}\left(\frac{A_{cp}^2}{p_{cp}}\right) = \frac{(0.75)(1.0)\sqrt{3000\ \frac{\text{lbf}}{\text{in}^2}}(322\text{ in}^2)^2}{(74\text{ in})\left(1000\ \frac{\text{lbf}}{\text{kip}}\right)}$$

$$= 58\text{ in-kips}$$

Because the applied torsion of 200 in-kips exceeds the threshold torsion, torsion reinforcement consisting of closed stirrups and longitudinal reinforcement is necessary. Using no. 3 stirrups with 1.5 in cover, the area enclosed by the center line of the stirrups is

$$A_{oh} = (23\text{ in} - 3\text{ in} - 0.375\text{ in})(14\text{ in} - 3\text{ in} - 0.375\text{ in})$$
$$= 208.52\text{ in}^2$$

From ACI Eq. (11-21), the required area of one arm of a closed stirrup is

$$\frac{A_t}{s} = \frac{T_u}{1.7\phi A_{oh} f_{yt}}$$

$$= \frac{200\text{ in-kips}}{(1.7)(0.75)(208.52\text{ in}^2)\left(60\ \frac{\text{kips}}{\text{in}^2}\right)}$$

$$= \left(0.0125\ \frac{\text{in}^2}{\text{in-arm}}\right)\left(12\ \frac{\text{in}}{\text{ft}}\right)$$

$$= 0.15\text{ in}^2/\text{ft-arm}$$

From ACI Eq. (11-23) the minimum area of closed stirrups is given by the lesser of

$$\frac{A_v + 2A_t}{s} = \frac{50 b_w}{f_{yt}}$$

$$= \frac{(50)(14\text{ in})\left(12\ \frac{\text{in}}{\text{ft}}\right)}{60{,}000\ \frac{\text{lbf}}{\text{in}^2}}$$

$$= 0.14\ \frac{\text{in}^2}{\text{ft}}$$

$$< 0.15\text{ in}^2/\text{ft-arm} \quad \text{[does not govern]}$$

$$\frac{A_v + 2A_t}{s} = \frac{0.75\sqrt{f'_c}\,b_w}{f_{yt}}$$

$$= \frac{0.75\sqrt{3000\ \frac{\text{lbf}}{\text{in}^2}}(14\text{ in})\left(12\ \frac{\text{in}}{\text{ft}}\right)}{60{,}000\ \frac{\text{lbf}}{\text{in}^2}}$$

$$= 0.12\text{ in}^2/\text{ft-arm}$$

$$< 0.15\text{ in}^2/\text{ft-arm} \quad \text{[does not govern]}$$

The perimeter of the center line of the closed stirrups is

$$p_h = (2)(23 \text{ in} + 14 \text{ in} - (2)(3.375 \text{ in}))$$
$$= 60.5 \text{ in}$$

The maximum permissible spacing of the closed stirrups is specified in ACI Sec. 11.5.6.1 as

$$s_{\max} = \frac{p_h}{8}$$
$$= \frac{60.5 \text{ in}}{8}$$
$$= 7.6 \text{ in}$$

Closed stirrups consisting of two arms of no. 3 bars at 6 in spacing provides an area of

$$\frac{A}{s} = 0.44 \; \frac{\text{in}^2}{\text{ft}}$$
$$> 0.15 \text{ in}^2/\text{ft} \quad \text{[satisfactory]}$$

The required area of the longitudinal reinforcement is given by ACI Eq. (11-22) as

$$A_l = \left(\frac{A_t}{s}\right) p_h \left(\frac{f_{yt}}{f_y}\right)$$
$$= \frac{\left(0.15 \; \frac{\text{in}^2}{\text{ft-arm}}\right)(60.5 \text{ in})}{12 \; \frac{\text{in}}{\text{ft}}}$$
$$= 0.76 \text{ in}^2/\text{arm}$$

Because the required value of $A_t/s = 0.0125 \text{ in}^2/\text{in-arm}$ is greater than $25 b_w/f_{yt} = 0.00583$, the minimum permissible area of longitudinal reinforcement is

$$A_{l,\min} = \frac{5 A_{cp} \sqrt{f'_c}}{f_y} - \left(\frac{A_t}{s}\right) p_h \left(\frac{f_{yt}}{f_y}\right)$$
$$= \frac{(5)(322 \text{ in}^2)\sqrt{3000 \; \frac{\text{lbf}}{\text{in}^2}}}{60{,}000 \; \frac{\text{lbf}}{\text{in}^2}}$$
$$- \left(0.0125 \; \frac{\text{in}^2}{\text{in-arm}}\right)(60.5 \text{ in})$$
$$= 0.714 \text{ in}^2 \quad \text{[does not govern]}$$
$$< 0.76 \text{ in}^2/\text{arm}$$

Using eight no. 3 bars distributed around the perimeter of the closed stirrups gives a longitudinal steel area of

$$A_l = 0.88 \text{ in}^2$$
$$> 0.76 \text{ in}^2 \quad \text{[satisfactory]}$$

PRACTICE PROBLEMS

For Probs. 1–3, refer to the illustration shown. The reinforced concrete beam has a width of 14 in and a concrete compressive strength of 5000 lbf/in². The factored applied force of 200 kips includes an allowance for the self-weight of the beam.

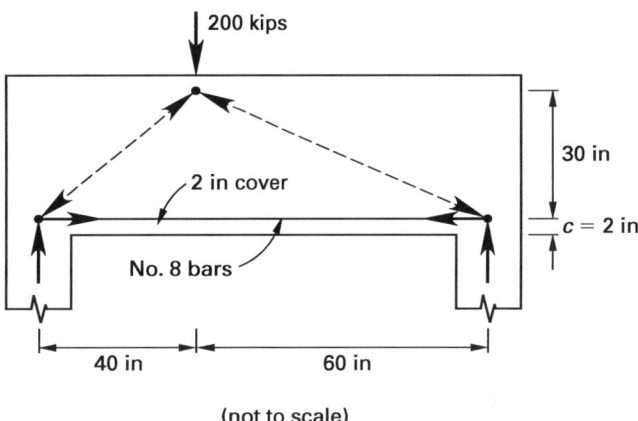

(not to scale)

1. Using the equivalent strut-and-tie model indicated, what is most nearly the area of grade 60 tension reinforcement required?

(A) 1.78 in²

(B) 2.67 in²

(C) 3.33 in²

(D) 3.56 in²

2. Is the equivalent, unreinforced bottle-shaped concrete strut at the right support compliant with the requirements of ACI App. A?

(A) No, the design compressive strength of the concrete is 1.91 kips/in².

(B) No, the design compressive strength of the concrete is 2.55 kips/in².

(C) Yes, the design compressive strength of the concrete is 2.29 kips/in².

(D) Yes, the design compressive strength of the concrete is 2.55 kips/in².

3. Is the equivalent nodal zone at the right support compliant with the requirements of ACI App. A?

(A) No, the design compressive strength is 1.90 kips/in^2

(B) No, the design compressive strength is 2.10 kips/in^2

(C) Yes, the design compressive strength is 2.55 kips/in^2

(D) Yes, the design compressive strength is 3.40 kips/in^2

Problems 4 and 5 refer to the reinforced concrete beam shown. The concrete strength is 3000 lbf/in^2, and all reinforcement is grade 60. The concrete section is adequate to support the applied shear force.

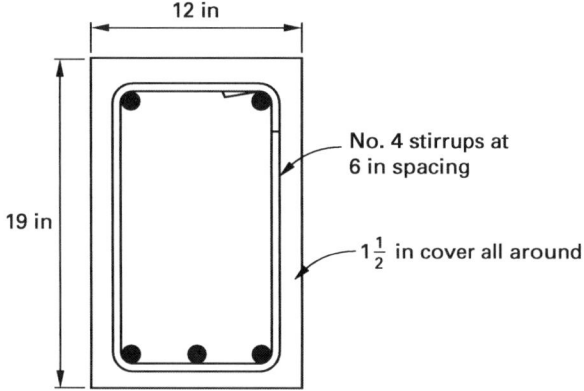

4. What is most nearly the maximum factored torsion that the beam can support?

(A) 340 in-kips

(B) 370 in-kips

(C) 420 in-kips

(D) 560 in-kips

5. What is most nearly the required area of longitudinal reinforcement for the beam?

(A) 1.40 in^2/arm

(B) 1.60 in^2/arm

(C) 1.90 in^2/arm

(D) 2.10 in^2/arm

SOLUTIONS

1. For the idealized strut-and-tie model shown in the illustration, the angle between the strut and tie at the right support is

$$\theta = \tan^{-1}\left(\frac{30 \text{ in}}{60 \text{ in}}\right)$$
$$= 26.6°$$
$$> 25° \quad \text{[satisfies ACI Sec. A.2.5]}$$

The reaction at the right support is

$$R = \frac{(200 \text{ kips})(40 \text{ in})}{100 \text{ in}} = 80 \text{ kips}$$

The equivalent tie force is determined from the strut-and-tie model as

$$T = \frac{(80 \text{ kips})(60 \text{ in})}{30 \text{ in}} = 160 \text{ kips}$$

The strength reduction factor is given by ACI Sec. 9.3.2.6 as

$$\phi = 0.75$$

The necessary reinforcement area is

$$A_{ts} = \frac{T}{\phi f_y} = \frac{160 \text{ kips}}{(0.75)\left(60 \dfrac{\text{kips}}{\text{in}^2}\right)}$$
$$= 3.56 \text{ in}^2$$

Use five no. 8 bars, which gives

$$A = 3.95 \text{ in}^2$$
$$> A_{ts} \quad \text{[satisfactory]}$$

The answer is (D).

2. The dimensions of the nodal zone are the equivalent tie width, w_t, and the width of the equivalent concrete strut, w_s.

$$w_t = d_b + 2c$$
$$= 1 \text{ in} + (2)(2 \text{ in})$$
$$= 5 \text{ in}$$
$$w_s = \frac{w_t}{\cos\theta}$$
$$= \frac{5 \text{ in}}{\cos 26.6°}$$
$$= 5.59 \text{ in}$$

The stress in the equivalent tie is

$$f_T = \frac{T}{bw_t} = \frac{160 \text{ kips}}{(14 \text{ in})(5 \text{ in})}$$
$$= 2.29 \text{ kips/in}^2$$

For a hydrostatic nodal zone, the stress in the equivalent concrete strut is

$$f_C = f_T = 2.29 \text{ kips/in}^2$$

For normal weight concrete and an unreinforced, bottle-shaped strut, the effective compressive strength of the concrete in the strut is given by ACI Eq. (A-3) as

$$f_{ce} = 0.85\beta_s f'_c = (0.85)(0.6)(1.0)\left(5 \frac{\text{kips}}{\text{in}^2}\right)$$
$$= 2.55 \text{ kips/in}^2$$

The design compressive stress is

$$\phi f_{ce} = (0.75)\left(2.55 \frac{\text{kips}}{\text{in}^2}\right)$$
$$= 1.91 \text{ kips/in}^2$$
$$< f_C \quad [\text{unsatisfactory}]$$

The answer is (A).

3. The nominal compressive strength of a nodal zone anchoring one layer of reinforcing bars without confining reinforcement is given by ACI Eq. (A-8) as

$$f_{ce} = 0.85\beta_n f'_c$$
$$= (0.85)(0.8)\left(5 \frac{\text{kips}}{\text{in}^2}\right)$$
$$= 3.40 \text{ kips/in}^2$$

The design compressive stress is

$$\phi f_{ce} = (0.75)\left(3.40 \frac{\text{kips}}{\text{in}^2}\right)$$
$$= 2.55 \text{ kips/in}^2$$
$$> f_C \quad [\text{satisfactory}]$$

The answer is (C).

4. The area enclosed by the center line of the stirrups is

$$A_{oh} = (19 \text{ in} - 3 \text{ in} - 0.5 \text{ in})(12 \text{ in} - 3 \text{ in} - 0.5 \text{ in})$$
$$= 131.75 \text{ in}^2$$

The area of one arm of the no. 4 closed stirrups provided at 6 in centers is

$$\frac{A_t}{s} = \frac{0.20 \text{ in}^2}{6 \text{ in-arm}}$$
$$= 0.033 \text{ in}^2/\text{in-arm}$$

From ACI Eq. (11-21) the maximum factored torsional moment that may be applied to the section is

$$T_u = \frac{1.7\phi A_{oh} f_{yt} A_t}{s}$$
$$= \frac{(1.7)(0.75)(131.75 \text{ in}^2)\left(60 \frac{\text{kips}}{\text{in}^2}\right)(0.20 \text{ in}^2)}{6 \text{ in}}$$
$$= 336 \text{ in-kips} \quad (340 \text{ in-kips})$$

The answer is (A).

5. The perimeter of the center line of the closed stirrups is

$$p_h = (2)(19 \text{ in} + 12 \text{ in} - (2)(3.5 \text{ in}))$$
$$= 48 \text{ in}$$

The required area of the longitudinal reinforcement is given by ACI Eq. (11-22) as

$$A_l = \left(\frac{A_t}{s}\right) p_h \left(\frac{f_{yt}}{f_y}\right)$$
$$= \left(0.033 \frac{\text{in}^2}{\text{in-arm}}\right)(48 \text{ in})$$
$$= 1.60 \text{ in}^2/\text{arm}$$

Using six no. 5 bars distributed around the perimeter of the closed stirrups gives a longitudinal steel area of

$$A_l = 1.86 \text{ in}^2$$
$$> 1.60 \quad [\text{satisfactory}]$$

The answer is (B).

Foundations

1. Eccentrically Loaded Column Bases 2-1
2. Combined Footings 2-10
3. Strap Footings 2-16
   Practice Problems 2-21
   Solutions 2-22

1. ECCENTRICALLY LOADED COLUMN BASES

Nomenclature

A_b	area of reinforcement in central band	in^2
A_s	total required reinforcement area	in^2
A_1	loaded area at base of column	ft^2
A_2	area of the base of the pyramid, with side slopes of 1:2, formed within the footing by the loaded area	ft^2
b_o	perimeter of critical section for punching shear	in
b_1	width of the critical perimeter measured in the direction of M_u	in
b_2	width of the critical perimeter measured perpendicular to b_1	in
B	length of short side of a rectangular footing	ft
c	length of side of column	in
c_1	length of short side of a rectangular column	in
c_2	length of long side of a rectangular column	in
d	average effective depth of reinforcement in the footing	in
d_b	bar diameter	in
e	eccentricity with respect to center of footing	in
e'	eccentricity with respect to edge of footing	in
f'_c	specified compressive strength of concrete	lbf/in^2
f_y	yield strength of reinforcement	lbf/in^2
h	depth of footing	in
K_u	design moment factor	lbf/in^2
L	length of long side of a rectangular footing, length of side of a square footing	ft
L	live load	kips
M	maximum unfactored moment caused by service loads	in-lbf
M_u	factored moment	ft-kips
P	column axial service load	kips
P_{bn}	nominal bearing strength	kips
P_T	total service load on the soil	kips
P_u	column axial factored load	kips
q	soil bearing pressure caused by service loads	lbf/ft^2
q_u	net factored pressure acting on footing	lbf/ft^2
S	section modulus of footing	in^3
V_c	nominal shear strength	kips
V_u	factored shear force at section	kips
W_b	weight of footing	kips
w_c	unit weight of concrete	lbf/ft^3
x	distance from edge of footing to critical section	ft
y	longer overall dimension of rectangular part of cross section	in

Symbols

α	reinforcement location factor	–
β	coating factor	–
β	ratio of the long side to the short side of footing	–
β_c	ratio of long side to short side of the column	–
β_1	compression zone factor given in ACI Sec. 10.2.7.3	–
λ	lightweight aggregate concrete factor	–
ρ	reinforcement ratio	–
ρ_t	reinforcement ratio for a tension-controlled section	–
ϕ	strength reduction factor	–

Soil Pressure Distribution

The eccentricity of a column on its footing determines the shape of the soil pressure distribution. As shown in Fig. 2.1, when the eccentricity of the column on the footing is less than $L/6$, the resulting soil pressure distribution is trapezoidal. The bending moment acting on the footing is

$$M = Pe$$

The maximum and minimum values of the soil pressure are

$$q_{\max} = \frac{P}{BL} + \frac{M}{S}$$

$$q_{\min} = \frac{P}{BL} - \frac{M}{S}$$

Figure 2.1 Eccentricity Less than L/6

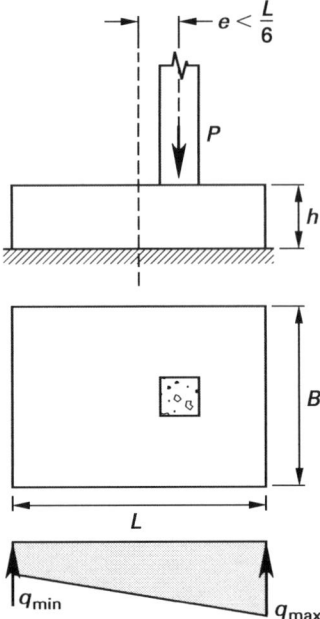

Figure 2.2 Eccentricity Equal to L/6

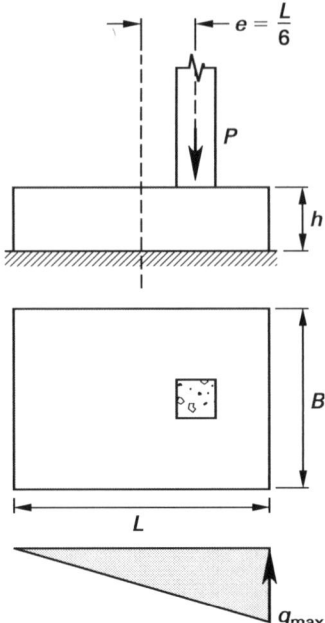

For a rectangular footing, the length of the short side is B and the length of the long side is L. The section modulus of the footing is

$$S = \frac{BL^2}{6}$$

The maximum and minimum values of the soil pressure are

$$q_{max} = \left(\frac{P}{BL}\right)\left(1 + \frac{6e}{L}\right)$$
$$q_{min} = \left(\frac{P}{BL}\right)\left(1 - \frac{6e}{L}\right)$$

When the eccentricity of the column on the footing equals one-sixth of the longest side, $L/6$, as shown in Fig. 2.2, the resulting soil pressure distribution is triangular. The bending moment acting on the footing is

$$M = \frac{PL}{6}$$

The maximum and minimum values of the soil pressure are

$$q_{max} = \frac{P}{BL} + \left(\frac{PL}{6}\right)\left(\frac{6}{BL^2}\right)$$
$$= \frac{2P}{BL}$$
$$q_{min} = \frac{P}{BL} - \left(\frac{PL}{6}\right)\left(\frac{6}{BL^2}\right)$$
$$= 0$$

It is not possible to develop tensile stress at the interface of the soil and the soffit of the footing. Hence, when the eccentricity of the column on the footing exceeds one-sixth of the longest side, $L/6$, as shown in Fig. 2.3, the triangular soil pressure distribution extends over only a portion of the base. For equilibrium, the centroid of the pressure distribution must coincide with the line of action of the column axial service load, P. The length of the interface in compression is

$$x = 3e'$$
$$e' = \frac{L}{2} - e$$

Figure 2.3 Eccentricity Greater than L/6

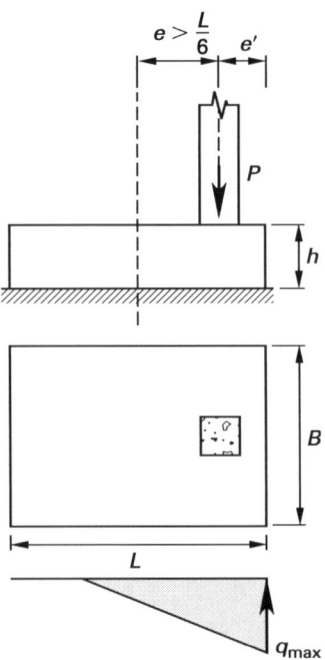

Equating vertical forces gives

$$P = \frac{Bxq_{\max}}{2}$$

The maximum value of the soil pressure is

$$q_{\max} = \frac{2P}{Bx}$$
$$= \frac{2P}{3Be'}$$

Example 2.1

The 9 ft × 5 ft reinforced normal weight concrete footing shown in the illustration supports a column with an axial load of $P = 60$ kips and a bending moment of $M = 60$ ft-kips. Determine the maximum pressure in the soil.

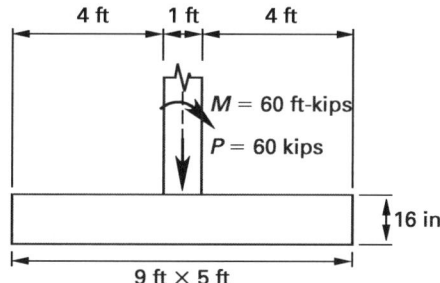

Solution

The weight of the footing is

$$W_b = w_c LBh$$
$$= \frac{\left(150 \,\frac{\text{lbf}}{\text{ft}^3}\right)(9 \text{ ft})(5 \text{ ft})(16 \text{ in})}{\left(1000 \,\frac{\text{lbf}}{\text{kip}}\right)\left(12 \,\frac{\text{in}}{\text{ft}}\right)}$$
$$= 9 \text{ kips}$$

The total load on the soil is

$$P_T = P + W_b$$
$$= 60 \text{ kips} + 9 \text{ kips}$$
$$= 69 \text{ kips}$$

The equivalent eccentricity of the total load is

$$e = \frac{M}{P_T}$$
$$= \frac{60 \text{ ft-kips}}{69 \text{ kips}}$$
$$= 0.87 \text{ ft}$$
$$< L/6 \quad \text{[within middle third of the base]}$$

The maximum soil pressure is

$$q_{\max} = \left(\frac{P_T}{BL}\right)\left(1 + \frac{6e}{L}\right)$$
$$= \left(\frac{69 \text{ kips}}{(5 \text{ ft})(9 \text{ ft})}\right)\left(1 + \frac{(6)(0.87 \text{ ft})}{9 \text{ ft}}\right)$$
$$= 2.4 \text{ kips/ft}^2$$

Factored Soil Pressure

A reinforced concrete footing must be designed for punching shear, flexural shear, and flexure. The critical section for each of these effects is located at a different position in the footing, and each must be designed for the applied factored loads. Hence, the soil pressure distribution caused by factored loads must be determined. Because the self-weight of the footing produces an equal and opposite pressure in the soil, the footing is designed for the net pressure from the column load only, and the weight of the footing is not included.

Example 2.2

The 9 ft × 5 ft reinforced concrete footing described in Ex. 2.1 supports a column with a factored axial load of $P_u = 100$ kips. It has a factored bending moment of $M_u = 100$ ft-kips. Determine the factored pressure distribution in the soil. The effective depth is $d = 12$ in.

Solution

The equivalent eccentricity of the factored column loads is

$$e = \frac{M_u}{P_u}$$
$$= \frac{100 \text{ ft-kips}}{100 \text{ kips}}$$
$$= 1 \text{ ft}$$
$$< L/6 \quad \text{[within middle third of the base]}$$

The maximum factored soil pressure is

$$q_{u,\max} = \left(\frac{P_u}{BL}\right)\left(1 + \frac{6e}{L}\right)$$
$$= \left(\frac{100 \text{ kips}}{(5 \text{ ft})(9 \text{ ft})}\right)\left(1 + \frac{(6)(1.0 \text{ ft})}{9 \text{ ft}}\right)$$
$$= 3.70 \text{ kips/ft}^2$$

The minimum factored soil pressure is

$$q_{u,\min} = \left(\frac{P_u}{BL}\right)\left(1 - \frac{6e}{L}\right)$$
$$= \left(\frac{100 \text{ kips}}{(5 \text{ ft})(9 \text{ ft})}\right)\left(1 - \frac{(6)(1.0 \text{ ft})}{9 \text{ ft}}\right)$$
$$= 0.74 \text{ kips/ft}^2$$

The relevant values at critical sections are shown in the illustration.

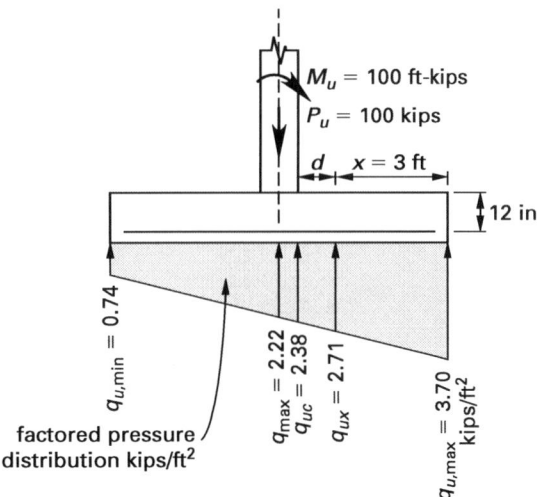

Design for Punching Shear

The depth of footing is usually governed by the punching shear capacity. The critical perimeter for punching shear is specified in ACI Secs. 15.5.2 and 11.11.1.2 and illustrated in Fig. 2.4. For a concrete column, the critical perimeter is a distance from the face of the column equal to half the effective depth. The length of the critical perimeter is

$$b_o = 2(b_1 + b_2)$$
$$= 2(c_1 + c_2) + 4d$$
$$= 4(c + d) \quad [c_1 = c_2 = c]$$

Figure 2.4 Critical Perimeter for Punching Shear

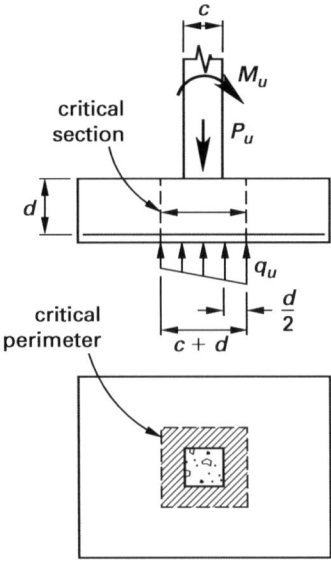

The width of the critical perimeter measured in the direction of M_u is

$$b_1 = c_1 + d$$

The width of the critical perimeter measured perpendicular to b_1 is

$$b_2 = c_2 + d$$

The value of c_1 is the width of the column measured in the direction of M_u, and c_2 is the width of the column measured perpendicular to c_1.

When the column supports only an axial load, P_u, shear stress at the critical perimeter is uniformly distributed around the critical perimeter, and is

$$v_u = \frac{V_u}{db_o}$$

The factored shear force acting on the critical perimeter is

$$V_u = P_u\left(1 - \frac{b_1 b_2}{BL}\right)$$

L is the length of the footing measured in the direction of M_u. B is the length of the footing measured perpendicular to L.

When, in addition to the axial load, a bending moment, M_u, is applied to the column, an eccentric shear stress is also introduced into the critical section with the maximum value occurring on the face nearest the largest bearing pressure. The maximum value of this shear stress is given by ACI Sec. R11.11.7.2 as

$$v_u = \frac{\gamma_v M_u y}{J_c}$$

The distance from the centroid of the critical perimeter to edge of the critical perimeter is

$$y = \frac{b_1}{2} \quad \text{[footing with central column]}$$

The fraction of the applied moment transferred by shear, as specified by ACI Secs. 11.11.7.1 and 13.5.3.2, is

$$\gamma_v = 1 - \frac{1}{1 + \left(\frac{2}{3}\right)\sqrt{\frac{b_1}{b_2}}}$$

J_c is the polar moment of inertia of the critical perimeter, and

$$\frac{J_c}{y} = \frac{b_1 d(b_1 + 3b_2) + d^3}{3}$$

$$\begin{bmatrix} \text{footing with central column} \\ \text{as specified by ACI Sec. R11.11.7.2} \end{bmatrix}$$

When both axial load and bending moment occur, the shear stresses caused by both conditions are combined, as specified in ACI Sec. R11.11.7.2, to give a maximum value of

$$v_u = \frac{V_u}{db_o} + \frac{\gamma_v M_u y}{J_c}$$

The design punching shear strength of the footing is determined by ACI Sec 11.11.2.1 as

$$\phi V_c = 4\phi db_o \lambda \sqrt{f'_c} \quad [\text{ACI 11-33}]$$

When $\beta_c > 2$, the design punching shear strength is

$$\phi V_c = \phi db_o \lambda \left(2 + \frac{4}{\beta_c}\right) \sqrt{f'_c} \quad [\text{ACI 11-31}]$$
$$\phi = 0.75$$

Example 2.3

The 9 ft × 5 ft reinforced concrete footing described in Ex. 2.1 supports a column with a factored axial load of $P_u = 100$ kips and a bending moment of $M_u = 100$ ft-kips. For a concrete strength of 3000 lbf/in², determine whether the punching shear capacity of the footing is satisfactory. The effective depth is $d = 12$ in.

Solution

The length of the critical perimeter is

$$b_o = 4(c + d)$$
$$= (4)(12 \text{ in} + 12 \text{ in})$$
$$= \frac{96 \text{ in}}{12 \frac{\text{in}}{\text{ft}}}$$
$$= 8 \text{ ft}$$

The average factored soil pressure acting on the area bounded by the critical perimeter is obtained from Ex. 2.2 as

$$q_{u,\text{ave}} = 2.22 \text{ kips/ft}^2$$

Shear at the critical perimeter caused by the axial load is

$$V_u = P_u - (q_{u,\text{ave}})(c + d)^2$$
$$= 100 \text{ kips} - \left(2.22 \frac{\text{kips}}{\text{ft}^2}\right)$$
$$\quad \times \left((12 \text{ in} + 12 \text{ in})\left(\frac{1 \text{ ft}}{12 \text{ in}}\right)\right)^2$$
$$= 91 \text{ kips}$$

The polar moment of inertia of the critical perimeter is

$$\frac{J_c}{y} = \frac{b_1 d(b_1 + 3b_2) + d^3}{3} \quad \begin{bmatrix} \text{footing with a} \\ \text{central column} \end{bmatrix}$$
$$= \frac{(24 \text{ in})(12 \text{ in})\left(24 \text{ in} + (3)(24 \text{ in})\right) + (12 \text{ in})^3}{3}$$
$$= 9792 \text{ in}^3$$

The fraction of the column moment transferred by shear is

$$\gamma_v = 1 - \frac{1}{1 + \left(\frac{2}{3}\right)\sqrt{\frac{b_1}{b_2}}}$$
$$= 1 - \frac{1}{1 + \left(\frac{2}{3}\right)\sqrt{\frac{12 \text{ in}}{12 \text{ in}}}}$$
$$= 0.40$$

The combined shear stress from the applied axial load and the column moment is

$$v_u = \frac{V_u}{db_o} + \frac{\gamma_v M_u y}{J_c}$$
$$= \frac{(91 \text{ kips})\left(1000 \frac{\text{lbf}}{\text{kip}}\right)}{(12 \text{ in})(96 \text{ in})}$$
$$\quad + \frac{(0.4)(1200 \text{ in-kips})\left(1000 \frac{\text{lbf}}{\text{kip}}\right)}{9792 \text{ in}^3}$$
$$= 79 \frac{\text{lbf}}{\text{in}^2} + 49 \frac{\text{lbf}}{\text{in}^2}$$
$$= 128 \text{ lbf/in}^2$$

The ratio of the long side to the short side of the column is

$$\beta_c = \frac{c_2}{c_1}$$
$$= \frac{12 \text{ in}}{12 \text{ in}}$$
$$= 1.00$$
$$< 2$$

The allowable shear stress for two-way action is given by ACI Eq. (11-33) as

$$\phi v_c = 4\phi \lambda \sqrt{f'_c}$$
$$= (4)(0.75)(1.0)\sqrt{3000 \frac{\text{lbf}}{\text{in}^2}}$$
$$= 164 \text{ lbf/in}^2$$
$$> v_u \quad [\text{satisfactory}]$$

Design for Flexural Shear

For concrete columns, the location of the critical section for flexural shear is defined in ACI Secs. 15.5.2 and 11.1.3.1 as being located a distance, d, from the face of the concrete column, as shown in Fig. 2.5. The design flexural shear strength of the footing is given by ACI Sec. 11.2.1.1 as

$$\phi V_c = 2\phi bd\lambda\sqrt{f'_c} \quad \text{[ACI 11-3]}$$

Figure 2.5 Critical Section for Flexural Shear

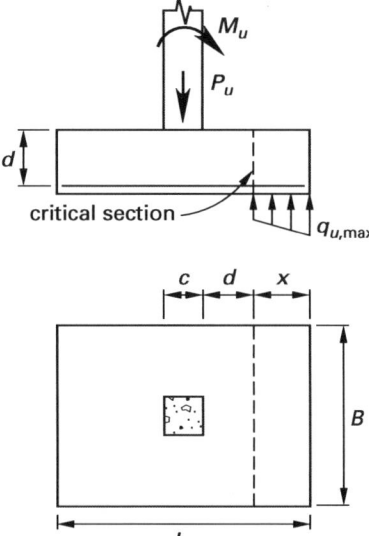

Example 2.4

The 9 ft × 5 ft reinforced concrete footing described in Ex. 2.1 supports a column with a factored axial load of $P_u = 100$ kips and a bending moment of $M_u = 100$ ft-kips. For a concrete strength is 3000 lbf/in^2, determine whether the flexural shear capacity of the footing is satisfactory. The effective depth is $d = 12$ in.

Solution

The distance of the critical section for flexural shear from the edge of the footing is

$$x = \frac{L}{2} - \frac{c}{2} - d$$
$$= \frac{9 \text{ ft}}{2} - \frac{1 \text{ ft}}{2} - 1 \text{ ft}$$
$$= 3 \text{ ft}$$

The net factored pressure on the footing at this section is obtained in Ex. 2.2 as

$$q_{ux} = 2.71 \text{ kips/ft}^2$$

The factored shear force at the critical section is

$$V_u = \frac{Bx(q_{u,\max} + q_{ux})}{2}$$
$$= (5 \text{ ft})(3 \text{ ft})\left(\frac{3.70 \frac{\text{kips}}{\text{ft}^2} + 2.71 \frac{\text{kips}}{\text{ft}^2}}{2}\right)$$
$$= 48 \text{ kips}$$

The design flexural shear capacity of the footing is given by ACI Eq. (11-3) as

$$\phi V_c = 2\phi Bd\lambda\sqrt{f'_c}$$
$$= \frac{(2)(0.75)(60 \text{ in})(12 \text{ in})(1.0)\sqrt{3000 \frac{\text{lbf}}{\text{in}^2}}}{1000 \frac{\text{lbf}}{\text{kip}}}$$
$$= 59 \text{ kips}$$
$$> V_u \quad \text{[satisfactory]}$$

Design for Flexure

ACI Sec. 15.4.2 defines the critical section for flexure to be located at the face of a concrete column, as shown in Fig. 2.6. The required reinforcement area is determined in accordance with ACI Sec. 10.2. The minimum ratio, $\rho_{\min}$, of reinforcement area to gross concrete area is specified in ACI Sec. 7.12.2, for both main reinforcement and distribution reinforcement, as 0.0018 for grade 60 bars. As specified in ACI Sec. 10.5.4, the maximum spacing of the main reinforcement shall not exceed 18 in, or three times the footing depth. The diameter of bar provided must be such that the development length does not

Figure 2.6 Critical Section for Flexure

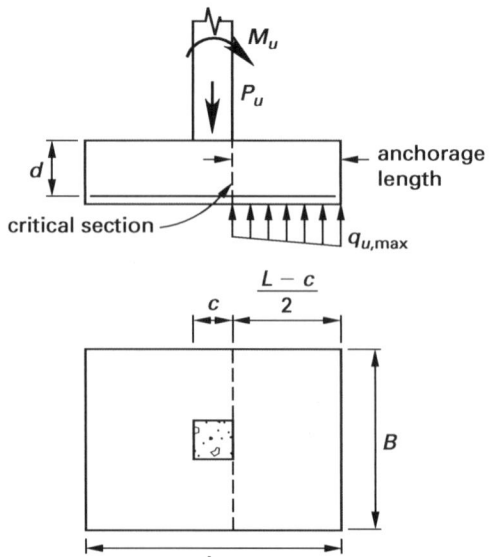

exceed the available anchorage length. Distribution reinforcement may be spaced at a maximum of 18 in, or five times the footing depth.

Example 2.5

The 9 ft × 5 ft reinforced concrete footing described in Ex. 2.1 supports a column with a factored axial load of $P_u = 100$ kips and a bending moment of $M_u = 100$ ft-kips. For a concrete strength of 3000 lbf/in^2, determine the area of grade 60 reinforcement required in the longitudinal direction. The effective depth is $d = 12$ in.

Solution

The net factored pressure on the footing at the face of the column is obtained in Ex. 2.2 as

$$q_{uc} = 2.38 \text{ kips/ft}^2$$

The factored moment at the face of the column is

$$M_u = \frac{B\left(\frac{L}{2} - \frac{c}{2}\right)^2 (2q_{u,\max} + q_{uc})}{6}$$

$$= \frac{(5 \text{ ft})\left(\frac{9 \text{ ft}}{2} - \frac{1 \text{ ft}}{2}\right)^2 \times \left((2)\left(3.70 \frac{\text{kips}}{\text{ft}^2}\right) + 2.38 \frac{\text{kips}}{\text{ft}^2}\right)}{6}$$

$$= 130.40 \text{ ft-kips}$$

The design moment factor is

$$K_u = \frac{M_u}{Bd^2}$$

$$= \frac{(130.40 \text{ ft-kips})\left(12 \frac{\text{in}}{\text{ft}}\right)\left(1000 \frac{\text{lbf}}{\text{kip}}\right)}{(60 \text{ in})(12 \text{ in})^2}$$

$$= 181.11 \text{ lbf/in}^2$$

For a tension-controlled section the required reinforcement ratio is

$$\rho = 0.85 f'_c \frac{1 - \sqrt{1 - \frac{K_u}{0.383 f'_c}}}{f_y}$$

$$= (0.85)\left(3 \frac{\text{kips}}{\text{in}^2}\right)$$

$$\times \left(\frac{1 - \sqrt{1 - \frac{181.11 \frac{\text{lbf}}{\text{in}^2}}{(0.383)\left(3000 \frac{\text{lbf}}{\text{in}^2}\right)}}}{60 \frac{\text{kips}}{\text{in}^2}}\right)$$

$$= 0.00349$$

The compression zone factor is given by ACI Sec. 10.2.7.3 as

$$\beta_1 = 0.85$$

The maximum allowable reinforcement ratio for a tension-controlled section is obtained from ACI Sec. 10.3.4 as

$$\rho_t = \frac{0.319 \beta_1 f'_c}{f_y}$$

$$= \frac{(0.319)(0.85)\left(3 \frac{\text{kips}}{\text{in}^2}\right)}{60 \frac{\text{kips}}{\text{in}^2}}$$

$$= 0.014$$

$$> \rho \quad \text{[satisfactory, the section is tension controlled]}$$

The required area of reinforcement is

$$A_s = \rho B d$$

$$= (0.00349)(60 \text{ in})(12 \text{ in})$$

$$= 2.51 \text{ in}^2$$

The minimum allowable reinforcement area for a footing is given by ACI Sec. 7.12.2 as

$$A_{s,\min} = 0.0018 B h$$

$$= (0.0018)(60 \text{ in})(16 \text{ in})$$

$$= 1.73 \text{ in}^2 \quad \text{[does not govern]}$$

Providing six no. 6 bars gives an area of

$$A_{s,\text{prov}} = 2.64 \text{ in}^2$$

$$> A_s \quad \text{[satisfactory]}$$

The anchorage length provided to the reinforcement is

$$l_a = \frac{L}{2} - \frac{c}{2} - \text{end cover}$$

$$= \left(\frac{9 \text{ ft}}{2} - \frac{1 \text{ ft}}{2}\right)\left(12 \frac{\text{in}}{\text{ft}}\right) - 3 \text{ in}$$

$$= 45 \text{ in}$$

The clear spacing of the bars exceeds twice the bar diameter, the clear cover exceeds the bar diameter, and the development length of the no. 6 bars is given by ACI Secs. 12.2.2 and 12.2.4 as

$$l_d = \frac{\psi_t \psi_e d_b f_y}{25 \lambda \sqrt{f'_c}}$$

$$\psi_t = 1.0$$

$$\psi_e = 1.0$$

$$\lambda = 1.0$$

The development length of the no. 6 bars is

$$l_d = \frac{d_b f_y}{25\sqrt{f'_c}}$$

$$= \frac{(0.75)\left(60{,}000\,\frac{\text{lbf}}{\text{in}^2}\right)}{25\sqrt{3000\,\frac{\text{lbf}}{\text{in}^2}}}$$

$$= 33 \text{ in } \quad [\text{for no. 6 bar}]$$

$$< 45 \text{ in anchorage length provided} \quad [\text{satisfactory}]$$

Transverse Reinforcement Band Width

Bending moments are calculated at the critical sections in both the longitudinal and transverse directions. The reinforcement required in the longitudinal direction is distributed uniformly across the width of the footing. Part of the reinforcement required in the transverse direction is concentrated in a central band width equal to the length of the short side of the footing, as shown in Fig. 2.7.

Figure 2.7 Transverse Reinforcement Band Width and Areas

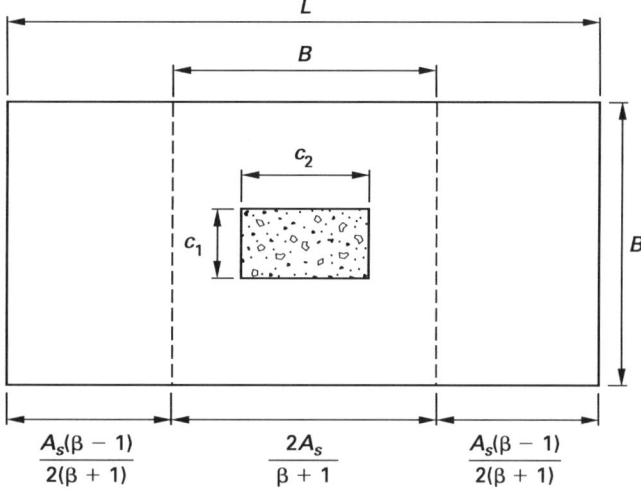

The area of reinforcement required in the central band is given by ACI Sec. 15.4.4.2 as

$$A_b = \frac{2A_s}{\beta + 1}$$

The remainder of the reinforcement required in the transverse direction is

$$A_r = \frac{A_s(\beta - 1)}{\beta + 1}$$

This is distributed uniformly on each side of the center band.

Example 2.6

The 9 ft × 5 ft reinforced concrete footing described in Ex. 2.1 supports a column with a factored axial load of $P_u = 100$ kips and a bending moment of $M_u = 100$ ft-kips. For a concrete strength of 3000 lbf/in², determine the area of grade 60 reinforcement required in the transverse direction. The effective depth is $d = 12$ in.

Solution

The factored moment in the transverse direction at the critical section, which is at the face of the column, is

$$M_u = \frac{L q_{u,\text{ave}} \left(\frac{B}{2} - \frac{c_1}{2}\right)^2}{2}$$

$$= \frac{(9 \text{ ft})\left(2.22\,\frac{\text{kips}}{\text{ft}^2}\right)\left(\frac{5 \text{ ft}}{2} - \frac{1 \text{ ft}}{2}\right)^2}{2}$$

$$= 39.96 \text{ ft-kips}$$

The design moment factor is

$$K_u = \frac{M_u}{Ld^2}$$

$$= \frac{(39.96 \text{ ft-kips})\left(12\,\frac{\text{in}}{\text{ft}}\right)\left(1000\,\frac{\text{lbf}}{\text{kip}}\right)}{(108 \text{ in})(12 \text{ in})^2}$$

$$= 30.83 \text{ lbf/in}^2$$

For a tension-controlled section the required reinforcement ratio is

$$\rho = 0.85 f'_c \frac{1 - \sqrt{1 - \dfrac{K_u}{0.383 f'_c}}}{f_y}$$

$$= (0.85)\left(3\,\frac{\text{kips}}{\text{in}^2}\right) \frac{1 - \sqrt{1 - \dfrac{30.83\,\frac{\text{lbf}}{\text{in}^2}}{(0.383)\left(3000\,\frac{\text{lbf}}{\text{in}^2}\right)}}}{60\,\frac{\text{kips}}{\text{in}^2}}$$

$$= 0.00057$$

The required area of reinforcement is

$$A_s = \rho L d$$

$$= (0.00057)(108 \text{ in})(12 \text{ in})$$

$$= 0.74 \text{ in}^2$$

The reinforcement required in the central 5 ft band width is

$$A_b = \frac{2A_s}{\beta + 1}$$
$$= \frac{(2)(0.74 \text{ in}^2)}{\dfrac{9 \text{ ft}}{5 \text{ ft}} + 1}$$
$$= 0.53 \text{ in}^2$$

The minimum allowable reinforcement area is given by ACI Sec. 7.12.2 as

$$A_{s,\text{min}} = 0.0018Bh$$
$$= (0.0018)(60 \text{ in})(16 \text{ in})$$
$$= 1.73 \text{ in}^2$$
$$> A_b$$

Minimum reinforcement governs, and will be equally spaced along the length of the footing. The total required reinforcement is

$$A_{s,\text{min}} = 0.0018Lh$$
$$= (0.0018)(108 \text{ in})(16 \text{ in})$$
$$= 3.11 \text{ in}^2$$

Providing 29 no. 3 bars gives an area of

$$A_{s,\text{prov}} = 3.19 \text{ in}^2$$
$$> A_{s,\text{min}} \quad [\text{satisfactory}]$$

The anchorage length provided to the reinforcement is

$$l_a = \frac{B}{2} - \frac{c}{2} - \text{end cover}$$
$$= 30 \text{ in} - 6 \text{ in} - 3 \text{ in}$$
$$= 21 \text{ in}$$

The clear spacing of the bars exceeds twice the bar diameter, the clear cover exceeds the bar diameter, and the development length of the no. 3 bars is obtained from Ex. 2.5 as

$$l_d = 43.8d_b = (43.8)(0.375 \text{ in})$$
$$= 16.4 \text{ in} \quad [\text{no. 3 bar}]$$
$$< 21 \text{ in anchorage length provided} \quad [\text{satisfactory}]$$

Transfer of Force at Base of Column

Load transfer between a reinforced concrete column and a footing may be effected by bearing on concrete and by reinforcement. The design bearing capacity of the column concrete at the interface is given by ACI Sec. 10.14.1 as

$$\phi P_{bn} = 0.85\phi f'_c A_1$$
$$= 0.553 f'_c A_1 \quad [\text{for } \phi = 0.65]$$

As shown in Fig. 2.8, the bearing capacity of the footing concrete at the interface is given by ACI Sec. 10.14.1 as

$$\phi P_{bn} = 0.85\phi f'_c A_1 \sqrt{\frac{A_2}{A_1}}$$
$$\leq (2)(0.85\phi f'_c A_1)$$

Figure 2.8 Transfer of Load to Footing

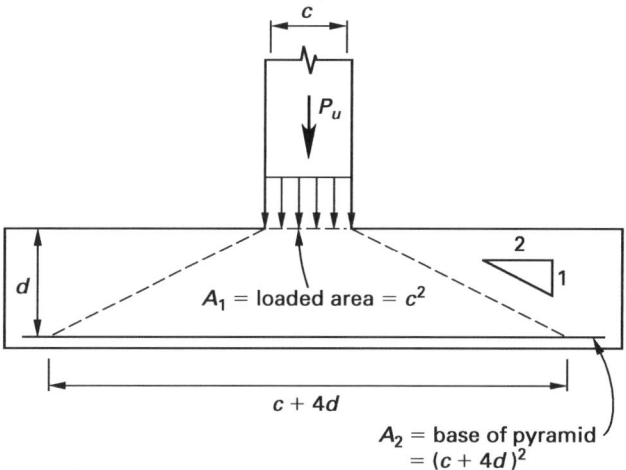

In accordance with ACI Sec. R15.8.1.1, when the bearing strength at the base of the column or at the top of the footing is exceeded, reinforcement must be provided to carry the excess load. This reinforcement may be provided by dowels or extended longitudinal bars, and the capacity of this reinforcement is

$$\phi P_s = \phi A_s f_y$$

A minimum area of reinforcement is required across the interface, and this is given by ACI Sec. 15.8.2.1 as

$$A_{s,\text{min}} = 0.005 A_1$$

Example 2.7

The 9 ft × 5 ft reinforced concrete footing described in Ex. 2.1 supports a column with a factored axial load only of $P_u = 100$ kips. The concrete strength in both the column and the footing is $f'_c = 3000$ lbf/in^2. Determine whether the bearing capacity at the interface is adequate.

Solution

The bearing capacity of the column concrete is given by ACI Sec. 10.14.1 as

$$\phi P_{bn} = 0.553 f'_c A_1$$
$$= (0.553)\left(3 \ \frac{\text{kips}}{\text{in}^2}\right)(144 \ \text{in}^2)$$
$$= 239 \ \text{kips}$$
$$> P_u \quad \text{[satisfactory]}$$

The minimum dowel area required at the interface is given by ACI Sec. 15.8.2.1 as

$$A_{s,\text{min}} = 0.005 A_1$$
$$= (0.005)(144 \ \text{in}^2)$$
$$= 0.72 \ \text{in}^2$$

Provide four no. 4 bars to give an area of

$$A_s = 0.80 \ \text{in}^2$$
$$> A_{s,\text{min}} \quad \text{[satisfactory]}$$

The excess reinforcement factor for the dowel bars is

$$\epsilon = \frac{A_{s,\text{min}}}{A_s}$$
$$= \frac{0.72 \ \text{in}^2}{0.80 \ \text{in}^2}$$
$$= 0.90$$

The development length of the dowels in the column and in the footing is given by ACI Sec. 12.3.2 as the larger of

$$l_{dc} = \epsilon 0.0003 d_b f_y$$
$$= (0.90)\left(0.0003 \ \frac{\text{in}^2}{\text{lbf}}\right)(0.50 \ \text{in})\left(60{,}000 \ \frac{\text{lbf}}{\text{in}^2}\right)$$
$$= 8.1 \ \text{in}$$

$$l_{dc} = \frac{\epsilon 0.02 d_b f_y}{\lambda \sqrt{f'_c}}$$
$$= \frac{(0.90)(0.02)(0.50 \ \text{in})\left(60{,}000 \ \frac{\text{lbf}}{\text{in}^2}\right)}{(1.0)\sqrt{3000 \ \frac{\text{lbf}}{\text{in}^2}}}$$
$$= 9.9 \ \text{in}$$

Use a length of 10 in.

The length of the base of the pyramid with side slopes of 1:2, formed within the footing by the loaded area, is

$$L_p = c + 4d = 12 \ \text{in} + (4)(12 \ \text{in})$$
$$= 60 \ \text{in}$$
$$= B \quad \text{[satisfactory]}$$

The area of the base of the pyramid is

$$A_2 = L_p^2 = (60 \ \text{in})^2$$
$$= 3600 \ \text{in}^2$$
$$\sqrt{\frac{A_2}{A_1}} = \sqrt{\frac{3600 \ \text{in}^2}{144 \ \text{in}^2}}$$
$$= 5$$

Use a maximum value of 2, as specified in ACI Sec. 10.14.1.

The bearing capacity of the footing concrete is given by ACI Sec. 10.14.1 as

$$\phi P_{bn} = (2)(0.553 f'_c A_1)$$
$$= (2)(0.553)\left(3 \ \frac{\text{kips}}{\text{in}^2}\right)(144 \ \text{in}^2)$$
$$= 480 \ \text{kips}$$
$$> P_u \quad \text{[satisfactory]}$$

2. COMBINED FOOTINGS

Nomenclature

A_b	area of reinforcement in central band	in^2
A_s	total required reinforcement area	in^2
b_o	perimeter of critical section for punching shear	in
B	length of short side of a footing	ft
c	length of side of column	in
d	average effective depth of reinforcement in the footing	in
d_b	bar diameter	in
f'_c	specified compressive strength of concrete	lbf/in^2
h	depth of footing	in
K_u	design moment factor	lbf/in^2
l	distance between column centers	ft
L	length of long side of a footing	ft
M_u	factored moment	ft-kips
P	column axial service load	kips
P_u	column axial factored load	kips
q	soil bearing pressure caused by service loads	lbf/ft^2
q_b	soil bearing pressure caused by weight of footing	lbf/ft^2
q_e	equivalent bearing pressure	lbf/ft^2
q_u	net factored pressure acting on footing	lbf/ft^2
V_c	nominal shear strength	kips

w_c	unit weight of concrete	lbf/ft^3
x	distance from edge of footing or center of column to critical section	ft
x_o	distance from edge of property line to centroid of service loads	ft

Symbols

β	ratio of the long side to the short side of footing	–
β_c	ratio of the long side to the short side of the column	–
β_1	compression zone factor given in ACI Sec. 10.2.7.3	–
λ	lightweight aggregate concrete factor	–
ρ	reinforcement ratio	–
ρ_t	reinforcement ratio for a tension-controlled section	–
ϕ	strength reduction factor	–

Soil Pressure Distribution

A combined footing is used, as shown in Fig. 2.9, when a column must be located adjacent to the property line. A second column is placed on the combined footing, and the length of the footing is adjusted until its centroid coincides with the centroid of the service loads on the two columns. Hence, a uniformly distributed soil pressure is produced under the combined footing. The footing width is adjusted to ensure that the soil bearing pressure does not exceed the allowable pressure.

Example 2.8

Determine the plan dimensions required to provide a uniform soil bearing pressure of $q = 3000$ lbf/ft^2 under the service loads indicated for the combined footing shown in the illustration.

Solution

The weight of the footing produces a uniformly distributed pressure on the soil of

$$q_b = w_c h = \frac{\left(150 \frac{\text{lbf}}{\text{ft}^3}\right)(2 \text{ ft})}{1000 \frac{\text{lbf}}{\text{kip}}}$$

$$= 0.3 \text{ kips/ft}^2$$

The maximum allowable equivalent soil bearing pressure is

$$q_e = q - q_b = 3 \frac{\text{kips}}{\text{ft}^2} - 0.3 \frac{\text{kips}}{\text{ft}^2}$$

$$= 2.7 \text{ kips/ft}^2$$

The centroid of the column service loads is located a distance x_o from the property line. The distance is obtained by taking moments about the property line, and given by

$$x_o = \frac{0.5 P_1 + 12.5 P_2}{P_1 + P_2}$$

$$= \frac{(0.5 \text{ ft})(100 \text{ kips}) + (12.5 \text{ ft})(200 \text{ kips})}{100 \text{ kips} + 200 \text{ kips}}$$

$$= 8.5 \text{ ft}$$

Hence, the length of footing required to produce a uniform bearing pressure on the soil is

$$L = 2x_o = (2)(8.5 \text{ ft})$$

$$= 17 \text{ ft}$$

Figure 2.9 Soil Pressure for Service Loads

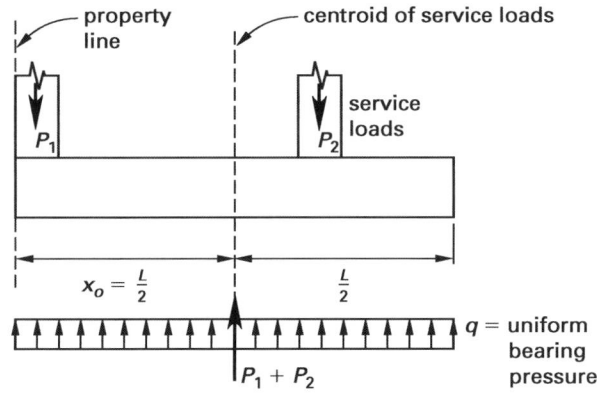

Illustration for Example 2.8

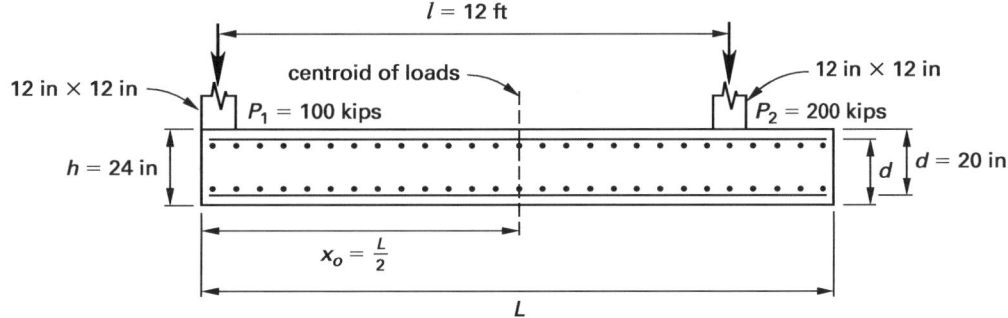

The width of footing required to produce a uniform pressure on the soil of 3000 lbf/ft² is

$$B = \frac{P_1 + P_2}{q_e L} = \frac{100 \text{ kips} + 200 \text{ kips}}{\left(2.7 \, \dfrac{\text{kips}}{\text{ft}^2}\right)(17 \text{ ft})}$$

$$= 6.5 \text{ ft}$$

Factored Soil Pressure

The reinforced concrete footing must be designed for punching shear, flexural shear, and flexure. The critical section for each of these effects is located at a different position in the footing, and each must be designed for the applied factored loads. The soil pressure distribution due to factored loads must be determined as shown in Fig. 2.10. The soil pressure will not necessarily be uniformly distributed unless the ratios of the factored loads to service loads on both columns are identical. Because the self-weight of the footing produces an equal and opposite pressure in the soil, the footing is designed for the net pressure from the column loads only, and the weight of the footing is not included.

Figure 2.10 Soil Pressure for Factored Loads

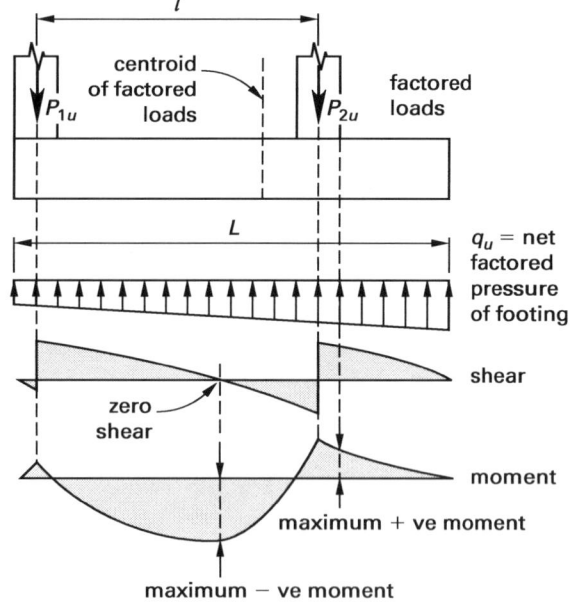

Example 2.9

The normal weight concrete combined footing described in Ex. 2.8 has a factored load on each column that is 1.5 times the service load. Determine the factored pressure distribution in the soil.

Solution

The ratios of the factored loads to the service loads on both columns are identical. The net factored pressure on the footing is uniform, and has a value of

$$q_u = 1.5 q_e$$

$$= (1.5)\left(2.7 \, \dfrac{\text{kips}}{\text{ft}^2}\right)$$

$$= 4.05 \text{ kips/ft}^2$$

For column no. 1, the factored load is

$$P_{1u} = 1.5 P_1$$

$$= (1.5)(100 \text{ kips})$$

$$= 150 \text{ kips}$$

For column no. 2, the factored load is

$$P_{2u} = 1.5 P_2$$

$$= (1.5)(200 \text{ kips})$$

$$= 300 \text{ kips}$$

The shear force and bending moment diagrams for the combined footing are shown in the illustration.

Design for Punching Shear

The depth of footing may be governed by the punching shear capacity. The critical perimeter for punching shear is specified in ACI Secs. 15.5.2 and 11.11.1.2 and illustrated in Fig. 2.11. For a concrete column, the critical perimeter is a distance from the face of the column equal to half the effective depth. For the interior column, the length of the critical perimeter is

$$b_o = 4(c + d)$$

Figure 2.11 Critical Perimeter for Punching Shear

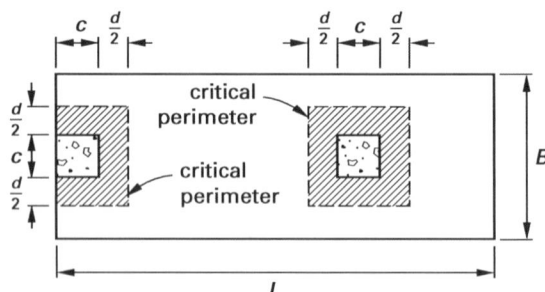

For the end column, the length of the critical perimeter is

$$b_o = (c + d) + 2\left(c + \dfrac{d}{2}\right)$$

Illustration for Example 2.9

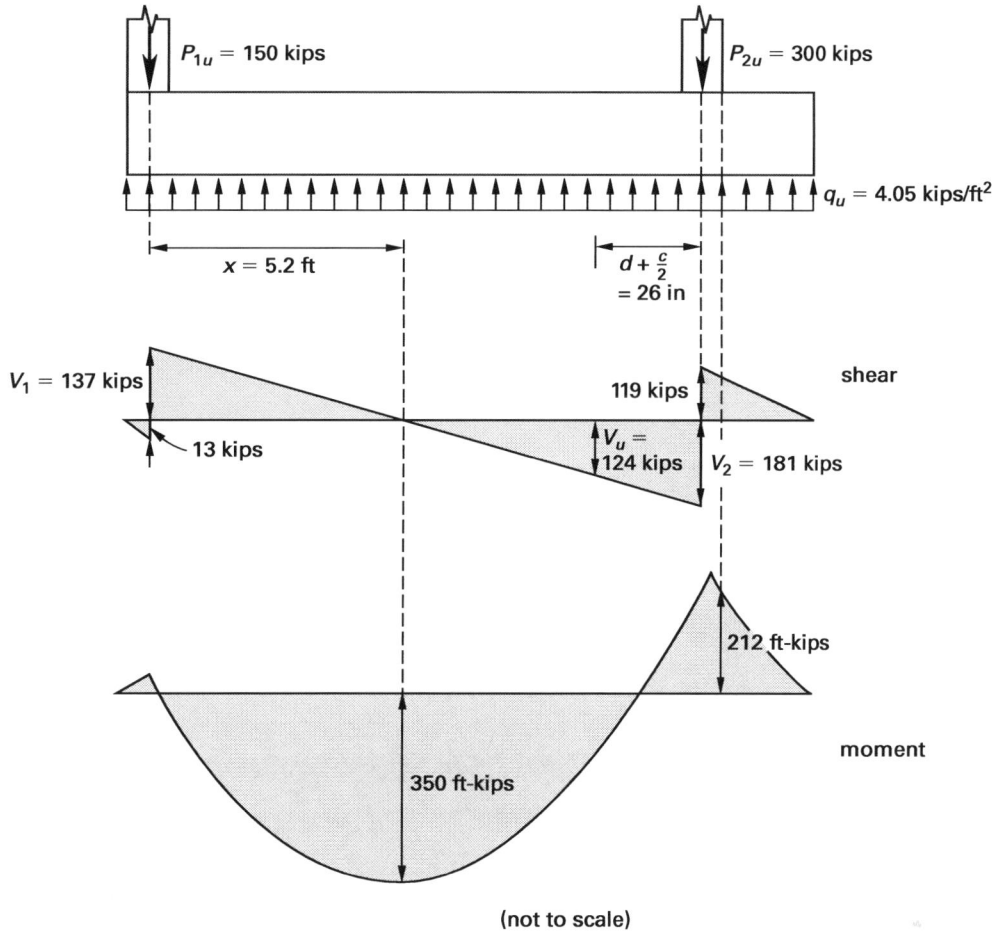

(not to scale)

The design punching shear strength of the footing is determined by ACI Sec. 11.11.2.1 as

$$\phi V_c = 4\phi d b_o \lambda \sqrt{f'_c} \quad \text{[ACI 11-33]}$$

When $\beta_c > 2$, the design punching shear strength is

$$\phi V_c = \phi d b_o \lambda \left(2 + \frac{4}{\beta_c}\right) \sqrt{f'_c} \quad \text{[ACI 11-31]}$$

$$\phi = 0.75$$

Example 2.10

For the combined footing described in Ex. 2.9, determine whether the punching shear capacity is adequate. The concrete strength is 3000 lbf/in^2, and the effective depth is $d = 20$ in.

Solution

For column no. 1, the length of the critical perimeter is

$$\begin{aligned} b_o &= (c + d) + 2\left(c + \frac{d}{2}\right) \\ &= (12 \text{ in} + 20 \text{ in}) + (2)\left(12 \text{ in} + \frac{20 \text{ in}}{2}\right) \\ &= 76 \text{ in} \end{aligned}$$

The punching shear force at the critical perimeter is

$$\begin{aligned} V_u &= P_{1u} - q_u(c+d)\left(c + \frac{d}{2}\right) \\ &= 150 \text{ kips} \\ &\quad - \frac{\left(4.05 \dfrac{\text{kips}}{\text{ft}^2}\right)(12 \text{ in} + 20 \text{ in})\left(12 \text{ in} + \dfrac{20 \text{ in}}{2}\right)}{\left(12 \dfrac{\text{in}}{\text{ft}}\right)^2} \\ &= 130 \text{ kips} \end{aligned}$$

The design punching shear capacity is given by ACI Eq. (11-33) as

$$\phi V_c = 4\phi d b_o \lambda \sqrt{f'_c}$$

$$= \frac{(4)(0.75)(20 \text{ in})(76 \text{ in})(1.0)\sqrt{3000 \frac{\text{lbf}}{\text{in}^2}}}{1000 \frac{\text{lbf}}{\text{kip}}}$$

$$= 250 \text{ kips}$$

$$> V_u \quad [\text{satisfactory}]$$

For column no. 2 the length of the critical perimeter is

$$b_o = 4(c + d)$$
$$= (4)(12 \text{ in} + 20 \text{ in})$$
$$= 128 \text{ in}$$

The punching shear force at the critical perimeter is

$$V_u = P_{2u} - q_u(c+d)^2$$

$$= 300 \text{ kips} - \frac{\left(4.05 \frac{\text{kips}}{\text{ft}^2}\right)(12 \text{ in} + 20 \text{ in})^2}{\left(12 \frac{\text{in}}{\text{ft}}\right)^2}$$

$$= 271 \text{ kips}$$

The design punching shear capacity is given by ACI Eq. (11-33) as

$$\phi V_c = 4\phi d b_o \lambda \sqrt{f'_c}$$

$$= \frac{(4)(0.75)(20 \text{ in})(128 \text{ in})(1.0)\sqrt{3000 \frac{\text{lbf}}{\text{in}^2}}}{1000 \frac{\text{lbf}}{\text{kip}}}$$

$$= 421 \text{ kips}$$

$$> V_u \quad [\text{satisfactory}]$$

Design for Flexural Shear

For combined footings, the location of the critical section for flexural shear is defined in ACI Secs. 15.5.2 and 11.1.3.1 as being located a distance, d, from the face of the concrete column, as shown in Fig. 2.5. The shear force at the critical section is obtained from the shear force diagram for the footing. The design flexural shear strength of the footing is given by ACI Sec. 11.2.1.1 as

$$\phi V_c = 2\phi b d \lambda \sqrt{f'_c} \quad [\text{ACI 11-3}]$$

Example 2.11

For the combined footing described in Ex. 2.9, determine whether the flexural shear capacity is adequate. The concrete strength is 3000 lbf/in^2, and the effective depth is $d = 20$ in.

Solution

At the center of column no. 1 the shear force is

$$V_1 = P_{1u} - \frac{q_u B c}{2}$$

$$= 150 \text{ kips} - \frac{\left(4.05 \frac{\text{kips}}{\text{ft}^2}\right)(6.54 \text{ ft})(1.0 \text{ ft})}{2}$$

$$= 137 \text{ kips}$$

At the center of column no. 2 the shear force is

$$V_2 = V_1 - q_u B l$$

$$= 137 \text{ kips} - \left(4.05 \frac{\text{kips}}{\text{ft}^2}\right)(6.54 \text{ ft})(12 \text{ ft})$$

$$= -181 \text{ kips}$$

The shear force diagram is shown in Ex. 2.9. The critical flexural shear is at a distance from the center of column no. 2 given by

$$d + \frac{c}{2} = \frac{20 \text{ in} + \frac{12 \text{ in}}{2}}{12 \frac{\text{in}}{\text{ft}}} = 2.17 \text{ ft}$$

The critical flexural shear at this section is

$$V_u = V_2 - q_u B\left(d + \frac{c}{2}\right)$$

$$= -181 \text{ kips} + \left(4.05 \frac{\text{kips}}{\text{ft}^2}\right)(6.54 \text{ ft})(2.17 \text{ ft})$$

$$= -124 \text{ kips}$$

The flexural shear capacity of the footing is given by ACI Eq. (11-3) as

$$\phi V_c = 2\phi b d \lambda \sqrt{f'_c}$$

$$= (2)(0.75)(6.54 \text{ ft})\left(12 \frac{\text{in}}{\text{ft}}\right)(20 \text{ in})$$

$$\times (1.0)\left(\frac{\sqrt{3000 \frac{\text{lbf}}{\text{in}^2}}}{1000 \frac{\text{lbf}}{\text{kip}}}\right)$$

$$= 129 \text{ kips}$$

$$> V_u \quad [\text{satisfactory}]$$

Design for Flexure

The footing is designed in the longitudinal direction as a beam continuous over two supports. As shown in Fig. 2.10, the maximum negative moment occurs at the section of zero shear. The maximum positive moment occurs at the outside face of column no. 2. In the transverse direction, it is assumed that the footing cantilevers about the face of both columns. The required reinforcement is concentrated under each column in a band width equal to the length of the shorter side. The area of reinforcement required in the band width under column no. 1 is given by ACI Sec. 15.4.4.2 as

$$A_{1b} = \frac{2A_s P_{1u}}{(\beta+1)(P_{1u}+P_{2u})}$$

Example 2.12

Determine the longitudinal grade 60 reinforcement required for the combined footing of Ex. 2.9. The effective depth is $d=20$ in, and the concrete strength is 3000 lbf/in^2.

Solution

From Ex. 2.9, the point of zero shear is a distance from the center of column no. 1 given by

$$x = \frac{V_1}{q_u B} = \frac{137 \text{ kips}}{\left(4.05 \dfrac{\text{kips}}{\text{ft}^2}\right)(6.54 \text{ ft})} = 5.2 \text{ ft}$$

The maximum negative moment at this point is

$$M_u = P_{1u}x - \frac{q_u B\left(x+\frac{c}{2}\right)^2}{2}$$
$$= (150 \text{ kips})(5.2 \text{ ft})$$
$$\quad - \frac{\left(4.05 \dfrac{\text{kips}}{\text{ft}^2}\right)(6.54 \text{ ft})\left(5.2 \text{ ft} + \dfrac{1 \text{ ft}}{2}\right)^2}{2}$$
$$= 350 \text{ ft-kips}$$

The design moment factor is

$$K_u = \frac{M_u}{Bd^2}$$
$$= \left(\frac{350 \text{ ft-kips}}{(78.5 \text{ in})(20 \text{ in})^2}\right)\left(12 \dfrac{\text{in}}{\text{ft}}\right)\left(1000 \dfrac{\text{lbf}}{\text{kip}}\right)$$
$$= 134 \text{ lbf/in}^2$$

For a tension-controlled section, the required reinforcement ratio is

$$\rho = 0.85 f'_c \left(\frac{1-\sqrt{1-\dfrac{K_u}{0.383 f'_c}}}{f_y}\right)$$

$$= (0.85)\left(3 \dfrac{\text{kips}}{\text{in}^2}\right) \left(\frac{1-\sqrt{1-\dfrac{134 \dfrac{\text{lbf}}{\text{in}^2}}{(0.383)\left(3000 \dfrac{\text{lbf}}{\text{in}^2}\right)}}}{60 \dfrac{\text{kips}}{\text{in}^2}}\right)$$

$$= 0.00256$$

The required area of reinforcement is

$$A_s = \rho B d$$
$$= (0.00256)(78.5 \text{ in})(20 \text{ in})$$
$$= 4.0 \text{ in}^2$$

The compression zone factor is given by ACI Sec. 10.2.7.3 as

$$\beta_1 = 0.85$$

The maximum allowable reinforcement ratio for a tension-controlled section is obtained from ACI Sec. 10.3.4 as

$$\rho_t = \frac{0.319 \beta_1 f'_c}{f_y}$$
$$= \frac{(0.319)(0.85)\left(3 \dfrac{\text{kips}}{\text{in}^2}\right)}{60 \dfrac{\text{kips}}{\text{in}^2}}$$
$$= 0.014$$
$$< \rho \quad \text{[satisfactory, the section is tension-controlled]}$$

The minimum allowable reinforcement area for a footing is given by ACI Sec. 7.12.2 as

$$A_{s,\min} = 0.0018 Bh$$
$$= (0.0018)(78.5 \text{ in})(24 \text{ in})$$
$$= 3.39 \text{ in}^2 \quad \text{[does not govern]}$$
$$< A_s$$

Providing 10 no. 6 bars gives an area of

$$A_{s,\text{prov}} = 4.4 \text{ in}^2$$
$$> A_s \quad \text{[satisfactory]}$$

The maximum positive moment at the outside face of column no. 2 is

$$M_u = \frac{q_u B(L-l-c)^2}{2}$$
$$= \frac{\left(4.05 \dfrac{\text{kips}}{\text{ft}^2}\right)(6.54 \text{ ft})(17 \text{ ft} - 12 \text{ ft} - 1 \text{ ft})^2}{2}$$
$$= 212 \text{ ft-kips}$$

The design moment factor is

$$K_u = \frac{M_u}{Bd^2}$$
$$= \left(\frac{212 \text{ ft-kips}}{(78.5 \text{ in})(20 \text{ in})^2}\right)\left(12 \dfrac{\text{in}}{\text{ft}}\right)\left(1000 \dfrac{\text{lbf}}{\text{kip}}\right)$$
$$= 81 \text{ lbf/in}^2$$

For a tension-controlled section the required reinforcement ratio is

$$\rho = 0.85 f'_c \left(\frac{1 - \sqrt{1 - \frac{K_u}{0.383 f'_c}}}{f_y} \right)$$

$$= (0.85)\left(3 \frac{\text{kips}}{\text{in}^2}\right) \left(\frac{1 - \sqrt{1 - \frac{81 \frac{\text{lbf}}{\text{in}^2}}{(0.383)\left(3000 \frac{\text{lbf}}{\text{in}^2}\right)}}}{60 \frac{\text{kips}}{\text{in}^2}} \right)$$

$$= 0.00153$$

Hence, the required area of reinforcement is

$$A_s = \rho B d = (0.00153)(78.5 \text{ in})(20 \text{ in})$$
$$= 2.4 \text{ in}^2$$

Providing 12 no. 4 bars gives an area of

$$A_{s,\text{prov}} = 2.4 \text{ in}^2$$
$$= A_s \quad [\text{satisfactory}]$$

3. STRAP FOOTINGS

Nomenclature

A_1	base area of pad footing no. 1	in^2
A_2	base area of pad footing no. 2	in^2
B_S	length of short side of strap	ft
B_1	length of short side of pad footing no. 1	ft
B_2	length of short side of pad footing no. 2	ft
h_S	depth of strap	in
h_1	depth of pad footing no. 1	in
h_2	depth of pad footing no. 2	in
l	distance between column centers	ft
l_R	distance between soil reactions	ft
L_S	length of long side of strap	ft
L_1	length of long side of pad footing no. 1	ft
L_2	length of long side of pad footing no. 2	ft
q	soil pressure	lbf/ft^2
R_1	soil reaction under pad footing no. 1	kips
R_2	soil reaction under pad footing no. 2	kips
V_c	nominal shear strength provided by concrete	lbf
w_c	unit weight of concrete	lbf/ft^3
W_S	weight of strap beam	kips
W_1	weight of pad footing no. 1	kips
W_2	weight of pad footing no. 2	kips

Soil Pressure Distribution

As shown in Fig. 2.12, a strap footing consists of two pad footings connected by a strap beam. The strap beam is underlaid by a layer of Styrofoam™ so that the soil pressure under the strap may be considered negligible. It is further assumed that the strap and pad footings act as a rigid body producing uniform soil pressure under the pad footings. The base areas of the two pad footings are adjusted to produce an identical soil pressure under both footings.

The total acting service load is obtained from Fig. 2.12 as

$$\sum P = P_1 + P_2 + W_1 + W_2 + W_S$$

The soil pressure under the pad footings is

$$q = \frac{\sum P}{A_1 + A_2}$$

The soil reactions act at the center of the pad footings, and are

$$R_1 = qA_1$$
$$R_2 = qA_2$$

Pad footing no. 2 is located symmetrically with respect to column no. 2 so that the lines of action of P_2 and R_2 are coincident. The distance between soil reactions is

$$l_R = l + \frac{c_1}{2} - \frac{B_1}{2}$$

The total length of the strap is

$$L_S = l_R - \frac{B_1 + B_2}{2}$$

Equating vertical forces gives

$$R_2 = \sum P - R_1 \quad [\text{equilibrium equation no. 1}]$$

Taking moments about the center of pad footing no. 2 gives

$$R_1 = \frac{P_1 l + W_1 l_R + \frac{W_S(L_S + B_2)}{2}}{l_R}$$

[equilibrium equation no. 2]

To determine dimensions that will give a soil bearing pressure equal to the allowable pressure, q, suitable values are selected for h_1, h_2, h_S, B_1, B_2, and B_S. The dimensions l_R and L_S are determined, and

$$W_S = w_c L_S B_S h_S$$

An initial estimate is made of R_1, and

$$A_1 = \frac{R_1}{q}$$
$$W_1 = w_c A_1 h_1$$

Figure 2.12 Strap Footing With Applied Service Loads

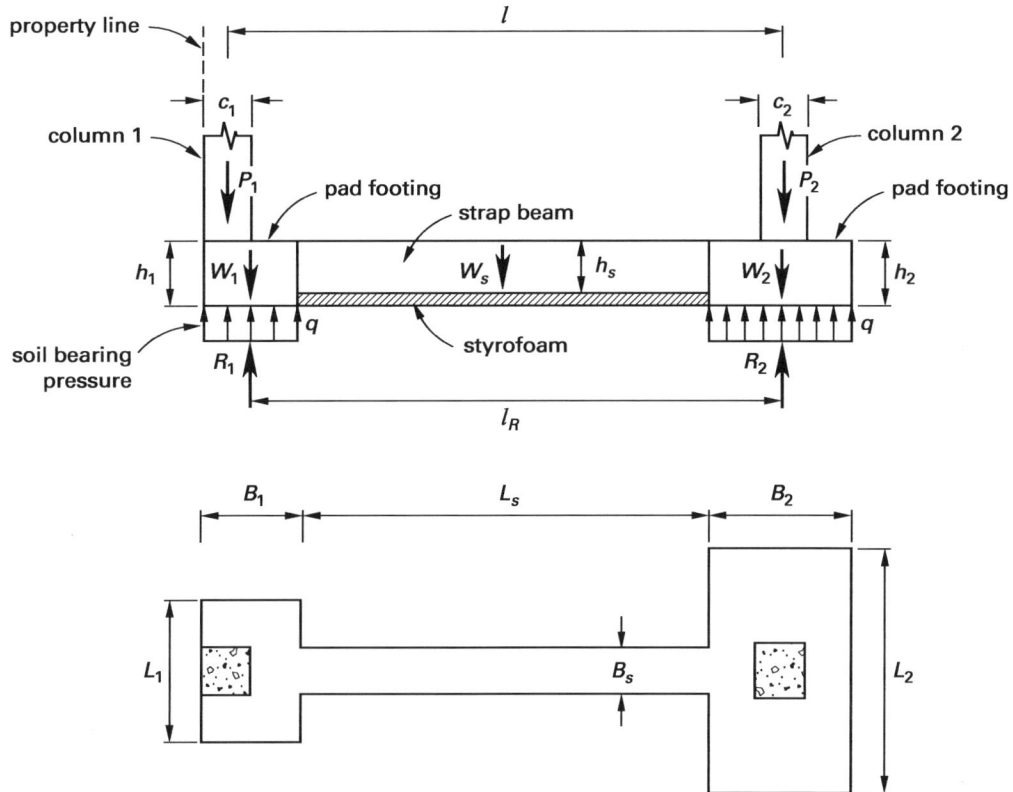

An initial estimate is made of R_2, and

$$A_2 = \frac{R_2}{q}$$
$$W_2 = w_c A_2 h_2$$
$$\sum P = P_1 + P_2 + W_1 + W_2 + W_S$$

Substituting in the two equilibrium equations provides revised estimates of R_1 and R_2, and the process is repeated until convergence is reached.

Example 2.13

Determine the dimensions required for the strap footing shown in the illustration to provide a uniform bearing pressure of 4000 lbf/ft² under both pad footings for the service loads indicated.

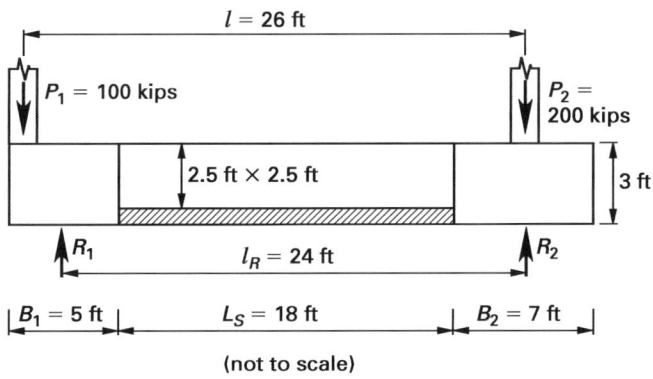

(not to scale)

Solution

From the dimensions indicated in the illustration, the weight of the strap is

$$W_S = w_c L_S B_S h_S$$
$$= \left(0.15 \ \frac{\text{kips}}{\text{ft}^3}\right)(18 \ \text{ft})(2.5 \ \text{ft})(2.5 \ \text{ft})$$
$$= 16.9 \ \text{kips}$$

As an initial estimate, assume that soil reaction under pad footing no. 1 is

$$R_1 = 131 \ \text{kips}$$

Then, the area and weight of pad footing no. 1 are

$$A_1 = \frac{R_1}{q} = \frac{131 \ \text{kips}}{4 \ \frac{\text{kips}}{\text{ft}^2}}$$
$$= 32.8 \ \text{ft}^2$$
$$W_1 = w_c A_1 h_1 = \left(0.15 \ \frac{\text{kips}}{\text{ft}^3}\right)(32.8 \ \text{ft}^2)(3 \ \text{ft})$$
$$= 14.8 \ \text{kips}$$

As an initial estimate, assume that soil reaction under pad footing no. 2 is

$$R_2 = 225 \ \text{kips}$$

Then, the area and weight of pad footing no. 2 are

$$A_2 = \frac{R_2}{q} = \frac{225 \text{ kips}}{4 \frac{\text{kips}}{\text{ft}^2}}$$

$$= 56.3 \text{ ft}^2$$

$$W_2 = w_c A_2 h_2 = \left(0.15 \frac{\text{kips}}{\text{ft}^3}\right)(56.3 \text{ ft}^2)(3 \text{ ft})$$

$$= 25.3 \text{ kips}$$

The total applied load is

$$\sum P = P_1 + P_2 + W_1 + W_2 + W_S$$
$$= 100 \text{ kips} + 200 \text{ kips} + 14.7 \text{ kips}$$
$$+ 25.3 \text{ kips} + 16.9 \text{ kips}$$
$$= 356.9 \text{ kips}$$

Taking moments about the center of pad footing no. 2 gives

$$R_1 = \frac{P_1 l + W_1 l_R + \frac{W_S(L_S + B_2)}{2}}{l_R}$$

$$= \frac{(100 \text{ kips})(26 \text{ ft}) + (14.7 \text{ kips})(24 \text{ ft}) + \frac{(16.9 \text{ ft})(18 \text{ ft} + 7 \text{ ft})}{2}}{24 \text{ ft}}$$

$$= 132 \text{ kips}$$

$$\approx 131 \text{ kips} \quad [\text{satisfactory}]$$

Equating vertical forces gives

$$R_2 = \sum P - R_1 = 356.9 \text{ kips} - 132 \text{ kips}$$
$$= 224.9 \text{ kips}$$
$$\approx 225 \text{ kips} \quad [\text{satisfactory}]$$

The initial estimates are sufficiently accurate, and the required pad footing areas are

$$A_1 = 32.8 \text{ ft}^2$$
$$A_2 = 56.3 \text{ ft}^2$$

The required widths of the pad footings are

$$L_1 = \frac{A_1}{B_1} = \frac{32.8 \text{ ft}^2}{5 \text{ ft}}$$
$$= 6.56 \text{ ft}$$
$$L_2 = \frac{A_2}{B_2} = \frac{56.3 \text{ ft}^2}{7 \text{ ft}}$$
$$= 8.04 \text{ ft}$$

Factored Soil Pressure

The reinforced concrete strap must be designed for flexural shear and flexure. The critical section for each of these effects is located at a different position in the strap, and each must be designed for the applied factored loads. The factored loads must be determined as shown in Fig. 2.13. The soil pressure will not necessarily be identical under both pad footings.

Figure 2.13 Soil Pressure for Factored Loads

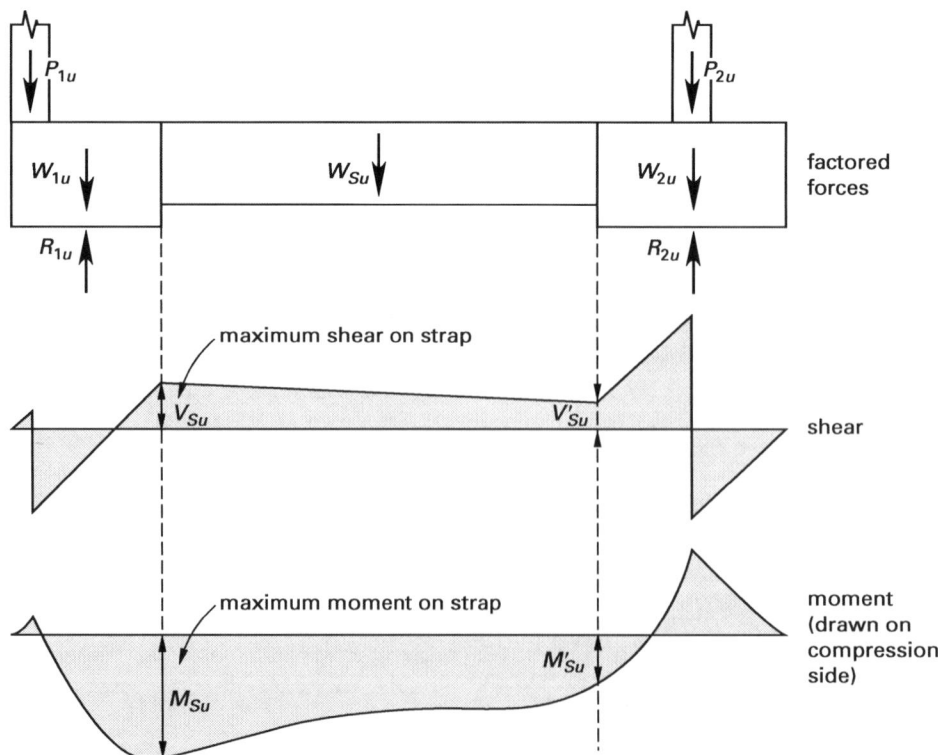

Example 2.14

The strap footing described in Ex. 2.13 has a factored load on each column that is 1.5 times the service load. Determine the factored soil reaction under each pad footing.

Solution

For column no. 1, the factored load is

$$P_{1u} = 1.5 P_1 = (1.5)(100 \text{ kips})$$
$$= 150 \text{ kips}$$

For column no. 2, the factored load is

$$P_{2u} = 1.5 P_2 = (1.5)(200 \text{ kips})$$
$$= 300 \text{ kips}$$

The factored weights of the component parts are

$$W_{1u} = 1.2 W_1 = (1.2)(14.7 \text{ kips})$$
$$= 17.6 \text{ kips}$$

$$W_{2u} = 1.2 W_2 = (1.2)(25.3 \text{ kips})$$
$$= 30.4 \text{ kips}$$

$$W_{Su} = 1.2 W_S$$
$$= (1.2)(16.9 \text{ kips})$$
$$= 20.3 \text{ kips}$$

The total factored load on the footing is

$$\sum P_u = P_{1u} + P_{2u} + W_{1u} + W_{2u} + W_{Su}$$
$$= 150 \text{ kips} + 300 \text{ kips} + 17.6 \text{ kips}$$
$$\quad + 30.4 \text{ kips} + 20.3 \text{ kips}$$
$$= 518 \text{ kips}$$

Taking moments about the center of pad footing no. 2 gives

$$R_{1u} = \frac{\sum P_{1u} l + W_{1u} l_R + \dfrac{W_{Su}(L_S + B_2)}{2}}{l_R}$$

$$= \frac{\begin{aligned}&(150 \text{ kips})(26 \text{ ft}) + (17.6 \text{ kips})(24 \text{ ft}) \\ &\quad + \frac{(20.3 \text{ kips})(18 \text{ ft} + 7 \text{ ft})}{2}\end{aligned}}{24 \text{ ft}}$$

$$= 191 \text{ kips}$$

Equating vertical forces gives

$$R_{2u} = \sum P_u - R_{1u}$$
$$= 518 \text{ kips} - 191 \text{ kips}$$
$$= 327 \text{ kips}$$

Design of Strap Beam for Shear

The factored forces acting on the footing are shown in Fig. 2.13. The total factored load on the footing is

$$\sum P_u = P_{1u} + P_{2u} + W_{1u} + W_{2u} + W_{Su}$$

Taking moments about the center of pad footing no. 2 gives

$$R_{1u} = \frac{P_{1u} l + w_{1u} l_R + \dfrac{W_{su}(L_S + B_2)}{2}}{l_R}$$

Equating vertical forces gives

$$R_{2u} = \sum P_u - R_{1u}$$

The shear at the left end of the strap is

$$V_{Su} = R_{1u} - P_{1u} - W_{1u}$$

The shear at the right end of the strap is

$$V'_{Su} = P_{2u} + W_{2u} - R_{2u}$$

The design flexural shear strength of the strap beam is

$$\phi V_c = 2\phi b d \lambda \sqrt{f'_c} \quad \text{[ACI 11-3]}$$

Example 2.15

The normal weight concrete strap footing for Ex. 2.13 has a concrete strength of 3000 lbf/in^2 and a factored load on each column that is 1.5 times the service load. The strap beam has an effective depth of 27 in. Determine whether the shear capacity of the strap beam is adequate.

Solution

The factored soil reactions under the pad footings are obtained from Ex. 2.14 as

$$R_{1u} = 191 \text{ kips}$$
$$R_{2u} = 327 \text{ kips}$$

The factored shear force at the right end of the strap is

$$V'_{Su} = P_{2u} + W_{2u} - R_{2u}$$
$$= 300 \text{ kips} + 30.4 \text{ kips} - 327 \text{ kips}$$
$$= 3.4 \text{ kips}$$

The factored shear force at the left end of the strap is

$$V_{Su} = R_{1u} - P_{1u} - W_{1u}$$
$$= 191 \text{ kips} - 150 \text{ kips} - 17.6 \text{ kips}$$
$$= 23.4 \text{ kips} \quad \text{[governs]}$$

The design shear capacity of the strap beam is given by ACI Eq. (11-3) as

$$\phi V_c = 2\phi b d \lambda \sqrt{f'_c}$$

$$= \frac{(2)(0.75)(30 \text{ in})(27 \text{ in})(1.0)\sqrt{3000 \frac{\text{lbf}}{\text{in}^2}}}{1000 \frac{\text{lbf}}{\text{kip}}}$$

$$= 67 \text{ kips}$$

$$> 2V_{Su} \quad \text{[no shear reinforcement is required]}$$

Design of Strap Beam for Flexure

From Fig. 2.13, the factored moment at the left end of the strap is

$$M_{Su} = P_{1u}\left(B_1 - \frac{c_1}{2}\right) - \frac{(R_{1u} - W_{1u})B_1}{2}$$

The factored moment at the right end of the strap is

$$M'_{Su} = \frac{(R_{2u} - W_{2u} - P_{2u})B_2}{2}$$

Example 2.16

Determine the required grade 60 flexural reinforcement for the strap beam of Ex. 2.13. The strap beam has an effective depth of 27 in.

Solution

The factored moment at the right end of the strap is

$$M'_{Su} = \frac{(R_{2u} - W_{2u} - P_{2u})B_2}{2}$$

$$= \frac{(327 \text{ kips} - 30.4 \text{ kips} - 300 \text{ kips})(7 \text{ ft})}{2}$$

$$= -11.9 \text{ ft-kips}$$

The factored moment at the left end of the strap is

$$M_{Su} = P_{1u}\left(B_1 - \frac{c_1}{2}\right) - \frac{(R_{1u} - W_{1u})B_1}{2}$$

$$= (150 \text{ kips})\left(5 \text{ ft} - \frac{1 \text{ ft}}{2}\right)$$

$$\quad - \frac{(191 \text{ kips} - 17.6 \text{ kips})(5 \text{ ft})}{2}$$

$$= 242 \text{ ft-kips} \quad \text{[governs]}$$

The design moment factor is

$$K_u = \frac{M_{Su}}{bd^2}$$

$$= \frac{(242 \text{ ft-kips})\left(12 \frac{\text{in}}{\text{ft}}\right)\left(1000 \frac{\text{lbf}}{\text{kip}}\right)}{(30 \text{ in})(27 \text{ in})^2}$$

$$= 133 \text{ lbf/in}^2$$

For a tension-controlled section, the required reinforcement ratio is

$$\rho = \frac{0.85 f'_c\left(1 - \sqrt{1 - \frac{K_u}{0.383 f'_c}}\right)}{f_y}$$

$$= \frac{(0.85)\left(3 \frac{\text{kips}}{\text{in}^2}\right)\left(1 - \sqrt{1 - \frac{133 \frac{\text{lbf}}{\text{in}^2}}{(0.383)\left(3000 \frac{\text{lbf}}{\text{in}^2}\right)}}\right)}{60 \frac{\text{kips}}{\text{in}^2}}$$

$$= 0.00254$$

The controlling minimum reinforcement ratio is given by ACI Sec. 10.5.1 as the greater of

$$\rho_{\min} = \frac{3\sqrt{f'_c}}{f_y} = \frac{3\sqrt{3000 \frac{\text{lbf}}{\text{in}^2}}}{60{,}000 \frac{\text{lbf}}{\text{in}^2}}$$

$$= 0.00274$$

$$\rho_{\min} = \frac{200}{f_y} = \frac{200 \frac{\text{lbf}}{\text{in}^2}}{60{,}000 \frac{\text{lbf}}{\text{in}^2}}$$

$$= 0.0033 \quad \text{[governs]}$$

In accordance with ACI Sec. 10.5.3, the reinforcement ratio need not exceed

$$\rho_s = \frac{4\rho}{3} = \frac{(4)(0.00254)}{3}$$

$$= 0.0034 \quad \text{[does not control]}$$

The required area of reinforcement in the top of the strap beam is

$$A_s = \rho_{\min} bd = (0.0033)(30 \text{ in})(27 \text{ in})$$

$$= 2.67 \text{ in}^2$$

Providing six no. 6 bars gives an area of 2.64 in^2, which is satisfactory.

PRACTICE PROBLEMS

Problems 1–8 refer to the combined footing shown. The factored load on each column is 1.4 times the service load, the effective depth is $d = 20$ in, and the concrete strength is 3000 lbf/in^2.

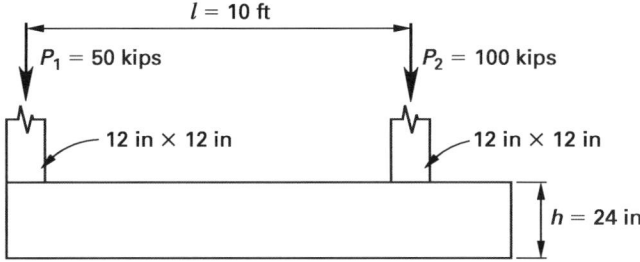

1. What are most nearly the plan dimensions required to provide a uniform soil bearing pressure of $q = 2000$ lbf/ft^2 under the service loads indicated?

(A) 14.3 ft × 5.25 ft

(B) 14.3 ft × 6.17 ft

(C) 21.5 ft × 3.49 ft

(D) 21.5 ft × 4.10 ft

2. What is most nearly the factored pressure distribution in the soil?

(A) 2.38 kips/ft^2

(B) 2.80 kips/ft^2

(C) 3.13 kips/ft^2

(D) 4.37 kips/ft^2

3. What is most nearly the punching shear force for each column?

(A) column no. 1: 48.6 kips, column no. 2: 109 kips

(B) column no. 1: 56.3 kips, column no. 2: 120 kips

(C) column no. 1: 54.7 kips, column no. 2: 118 kips

(D) column no. 1: 58.4 kips, column no. 2: 123 kips

4. What is most nearly the punching shear capacity for each column?

(A) column no. 1: 200 kips, column no. 2: 400 kips

(B) column no. 1: 230 kips, column no. 2: 410 kips

(C) column no. 1: 250 kips, column no. 2: 421 kips

(D) column no. 1: 260 kips, column no. 2: 433 kips

5. What is most nearly the shear force at the center of each column?

(A) column no. 1: 60 kips, column no. 2: −80 kips

(B) column no. 1: 63 kips, column no. 2: −84 kips

(C) column no. 1: 66 kips, column no. 2: −88 kips

(D) column no. 1: 70 kips, column no. 2: −95 kips

6. What is most nearly the critical flexural shear?

(A) −52 kips

(B) −54 kips

(C) −58 kips

(D) −62 kips

7. What is most nearly the design flexural shear capacity?

(A) 120 kips

(B) 122 kips

(C) 124 kips

(D) 126 kips

8. What is most nearly the longitudinal grade 60 reinforcement required in the top of the combined footing?

(A) 1.5 in^2

(B) 2.0 in^2

(C) 2.6 in^2

(D) 3.2 in^2

SOLUTIONS

1. The weight of the footing produces a uniformly distributed pressure on the soil of

$$q_b = w_c h$$
$$= \frac{\left(150 \ \frac{\text{lbf}}{\text{ft}^3}\right)(2 \ \text{ft})}{1000 \ \frac{\text{lbf}}{\text{kip}}}$$
$$= 0.3 \ \text{kips/ft}^2$$

The maximum allowable equivalent soil bearing pressure is

$$q_e = q - q_b$$
$$= 2 \ \frac{\text{kips}}{\text{ft}^2} - 0.3 \ \frac{\text{kips}}{\text{ft}^2}$$
$$= 1.7 \ \text{kips/ft}^2$$

The centroid of the column service loads is located a distance x_o from the property line. The distance x_o is obtained by taking moments about the property line, and is

$$x_o = \frac{0.5 P_1 + 10.5 P_2}{P_1 + P_2}$$
$$= \frac{(0.5 \ \text{ft})(50 \ \text{kips}) + (10.5 \ \text{ft})(100 \ \text{kips})}{50 \ \text{kips} + 100 \ \text{kips}}$$
$$= 7.17 \ \text{ft}$$

The length of footing required to produce a uniform bearing pressure on the soil is

$$L = 2x_o$$
$$= (2)(7.17 \ \text{ft})$$
$$= 14.3 \ \text{ft}$$

The width of footing required to produce a uniform pressure on the soil of 2000 lbf/ft² is

$$B = \frac{P_1 + P_2}{q_e L}$$
$$= \frac{50 \ \text{kips} + 100 \ \text{kips}}{\left(1.7 \ \frac{\text{kips}}{\text{ft}^2}\right)(14.3 \ \text{ft})}$$
$$= 6.17 \ \text{ft}$$

The answer is (B).

2. Because the ratio of the factored loads to service loads on both columns is identical, the net factored pressure on the footing is uniform and has a value of

$$q_u = 1.4 q_e$$
$$= (1.4)\left(1.7 \ \frac{\text{kips}}{\text{ft}^2}\right)$$
$$= 2.38 \ \text{kips/ft}^2$$

The answer is (A).

3. For column no. 1, the factored load is

$$P_{1u} = 1.4 P_1$$
$$= (1.4)(50 \ \text{kips})$$
$$= 70 \ \text{kips}$$

For column no. 2, the factored load is

$$P_{2u} = 1.4 P_2$$
$$= (1.4)(100 \ \text{kips})$$
$$= 140 \ \text{kips}$$

For column no. 1, the length of the critical perimeter is

$$b_o = (c + d) + 2\left(c + \frac{d}{2}\right)$$
$$= (12 \ \text{in} + 20 \ \text{in}) + (2)\left(12 \ \text{in} + \frac{20 \ \text{in}}{2}\right)$$
$$= 76 \ \text{in}$$

The punching shear force at the critical perimeter is

$$V_u = P_{1u} - q_u(c + d)\left(c + \frac{d}{2}\right)$$
$$= 70 \ \text{kips}$$
$$- \frac{\left(2.38 \ \frac{\text{kips}}{\text{ft}^2}\right)(12 \ \text{in} + 20 \ \text{in})\left(12 \ \text{in} + \frac{20 \ \text{in}}{2}\right)}{\left(12 \ \frac{\text{in}}{\text{ft}}\right)^2}$$
$$= 58.4 \ \text{kips}$$

For column no. 2, the length of the critical perimeter is

$$b_o = 4(c + d)$$
$$= (4)(12 \ \text{in} + 20 \ \text{in})$$
$$= 128 \ \text{in}$$

The punching shear force at the critical perimeter is

$$V_u = P_{2u} - qu(c+d)^2$$

$$= 140 \text{ kips} - \frac{\left(2.38 \dfrac{\text{kips}}{\text{ft}^2}\right)(12 \text{ in} + 20 \text{ in})^2}{\left(12 \dfrac{\text{in}}{\text{ft}}\right)^2}$$

$$= 123 \text{ kips}$$

The answer is (D).

4. For column no. 1, the design punching shear capacity is given by ACI Eq. (11-33) as

$$\phi V_c = 4\phi d b_o \lambda \sqrt{f'_c}$$

$$= \frac{(4)(0.75)(20 \text{ in})(76 \text{ in})(1.0)\sqrt{3000 \dfrac{\text{lbf}}{\text{in}^2}}}{1000 \dfrac{\text{lbf}}{\text{kip}}}$$

$$= 250 \text{ kips}$$

$$> V_u \quad \text{[satisfactory]}$$

For column no. 2, the design punching shear capacity is given by ACI Eq. (11-33) as

$$\phi V_c = 4\phi d b_o \lambda \sqrt{f'_c}$$

$$= \frac{(4)(0.75)(20 \text{ in})(128 \text{ in})(1.0)\sqrt{3000 \dfrac{\text{lbf}}{\text{in}^2}}}{1000 \dfrac{\text{lbf}}{\text{kip}}}$$

$$= 421 \text{ kips}$$

$$> V_u \quad \text{[satisfactory]}$$

The answer is (C).

5. At the center of column no. 1, the shear force is

$$V_1 = P_{1u} - \frac{q_u B c}{2}$$

$$= 70 \text{ kips} - \frac{\left(2.38 \dfrac{\text{kips}}{\text{ft}^2}\right)(6.17 \text{ ft})(1 \text{ ft})}{2}$$

$$= 62.7 \text{ kips} \quad (63 \text{ kips})$$

At the center of column no. 2 the shear force is

$$V_2 = V_1 - q_u B l$$

$$= 62.7 \text{ kips} - \left(2.38 \dfrac{\text{kips}}{\text{ft}^2}\right)(6.17 \text{ ft})(10 \text{ ft})$$

$$= -84.1 \text{ kips} \quad (-84 \text{ kips}) \quad \text{[governs]}$$

The answer is (B).

6. The critical flexural shear is at a distance from the center of column no. 2 given by

$$\left(d + \frac{c}{2}\right) = 20 \text{ in} + \frac{12 \text{ in}}{2}$$

$$= \frac{26 \text{ in}}{12 \dfrac{\text{in}}{\text{ft}}}$$

$$= 2.17 \text{ ft}$$

The critical flexural shear at this section is

$$V_u = V_2 - q_u B \left(d + \frac{c}{2}\right)$$

$$= -84.1 \text{ kips} + \left(2.38 \dfrac{\text{kips}}{\text{ft}^2}\right)(6.17 \text{ ft})(2.17 \text{ ft})$$

$$= -52 \text{ kips}$$

The answer is (A).

7. The design flexural shear capacity of the footing is given by ACI Eq. (11-3) as

$$\phi V_c = 2\phi b d \lambda \sqrt{f'_c}$$

$$= \frac{(2)(0.75)(6.17 \text{ ft})(20 \text{ in})(1.0)\sqrt{3000 \dfrac{\text{lbf}}{\text{in}^2}}\left(12 \dfrac{\text{in}}{\text{ft}}\right)}{1000 \dfrac{\text{lbf}}{\text{kip}}}$$

$$= 122 \text{ kips}$$

$$> V_u \quad \text{[satisfactory]}$$

The answer is (B).

8. From Prob. 4, the point of zero shear is a distance x from the center of column no. 1, given by

$$x = \frac{V_1}{q_u B}$$

$$= \frac{62.7 \text{ kips}}{\left(2.38 \, \frac{\text{kips}}{\text{ft}^2}\right)(6.17 \text{ ft})}$$

$$= 4.3 \text{ ft}$$

The maximum negative moment at this point is

$$M_u = P_{1u} x - \frac{q_u B \left(x + \frac{c}{2}\right)^2}{2}$$

$$= (70 \text{ kips})(4.3 \text{ ft})$$

$$- \frac{\left(2.38 \, \frac{\text{kips}}{\text{ft}^2}\right)(6.17 \text{ ft})\left(4.3 \text{ ft} + \frac{1 \text{ ft}}{2}\right)^2}{2}$$

$$= 132 \text{ ft-kips}$$

The design moment factor is

$$K_u = \frac{M_u}{Bd^2}$$

$$= \frac{(132 \text{ ft-kips})\left(12 \, \frac{\text{in}}{\text{ft}}\right)\left(1000 \, \frac{\text{lbf}}{\text{kip}}\right)}{(74.0 \text{ in})(20 \text{ in})^2}$$

$$= 53.5 \text{ lbf/in}^2$$

For a tension-controlled section the required reinforcement ratio is

$$\rho = \frac{0.85 f'_c \left(1 - \sqrt{1 - \frac{K_u}{0.383 f'_c}}\right)}{f_y}$$

$$= \frac{(0.85)\left(3 \, \frac{\text{kips}}{\text{in}^2}\right)\left(1 - \sqrt{1 - \frac{53.5 \, \frac{\text{lbf}}{\text{in}^2}}{(0.383)\left(3000 \, \frac{\text{lbf}}{\text{in}^2}\right)}}\right)}{60 \, \frac{\text{kips}}{\text{in}^2}}$$

$$= 0.00100$$

The required area of reinforcement is

$$A_s = \rho B d$$

$$= (0.00100)(74.0 \text{ in})(20 \text{ in})$$

$$= 1.48 \text{ in}^2$$

The minimum allowable reinforcement area is given by ACI Sec. 7.12.2 as

$$A_{s,\text{min}} = 0.0018 B h$$

$$= (0.0018)(74.0 \text{ in})(24 \text{ in})$$

$$= 3.2 \text{ in}^2 \quad [\text{governs}]$$

$$> A_s$$

Providing eight no. 6 bars gives an area of

$$A_{s,\text{prov}} = 3.52 \text{ in}^2$$

$$> A_{s,\text{min}} \quad [\text{satisfactory}]$$

The answer is (D).

3 Prestressed Concrete Design

1. Strength Design of Flexural Members 3-1
2. Design for Shear and Torsion 3-7
3. Prestress Losses 3-15
4. Composite Construction 3-19
5. Load Balancing Procedure 3-24
6. Concordant Cable Profile 3-26
   Practice Problems 3-28
   Solutions 3-29

1. STRENGTH DESIGN OF FLEXURAL MEMBERS

Nomenclature

a	depth of equivalent rectangular stress block	in
A_{ct}	area of concrete section between the centroid and extreme tension fiber	in^2
A_g	area of concrete section	in^2
A_{ps}	area of prestressed reinforcement in tension zone	in^2
A_s	area of nonprestressed tension reinforcement	in^2
A'_s	area of compression reinforcement	in^2
b	width of compression face of member	in
c	distance from extreme compression fiber to neutral axis	in
C_u	total compression force in equivalent rectangular stress block, $0.85abf'_c$	lbf
d	distance from extreme compression fiber to centroid of nonprestressed reinforcement	in
d'	distance from extreme compression fiber to centroid of compression reinforcement	in
d_p	distance from extreme compression fiber to centroid of prestressed reinforcement as defined in Fig. 3.1	in
e	eccentricity of prestressing force	in
E_c	modulus of elasticity of concrete, $57\sqrt{f'_c}$	kips/in^2
E_p	modulus of elasticity of prestressing tendon	kips/in^2
f'_c	specified compressive strength of concrete	lbf/in^2
f_{ps}	stress in prestressed reinforcement at nominal strength	kips/in^2
f_{pu}	specified tensile strength of prestressing tendons	kips/in^2
f_{py}	specified yield strength of prestressing tendons	kips/in^2
f_r	modulus of rupture of concrete, $7.5\sqrt{f'_c}$	lbf/in^2
f_{se}	effective stress in prestressed reinforcement after allowance for all losses	kips/in^2
f_y	specified yield strength of nonprestressed reinforcement	kips/in^2
h	height of section	in
I_g	moment of inertia of gross concrete section	in^4
l	span length	ft
M_{cr}	cracking moment strength	in-kips
M_n	nominal flexural strength	in-kips
M_u	factored moment	in-kips
N_c	tensile force in concrete caused by unfactored dead load plus live load	lbf
P_e	force in the prestressing tendons at service loads after allowance for all losses	kips
S_b	section modulus of the concrete section referred to the bottom fiber	in^3
T_u	total tensile force in the prestressing tendons	kips

Symbols

β_1	compression zone factor	–
γ_p	factor for type of prestressing tendon	–
ϵ_c	strain at extreme compression fiber at nominal strength, 0.003	–
ϵ_t	strain produced in prestressed reinforcement by the ultimate loading	–
ρ	ratio of nonprestressed tension reinforcement	–
ρ'	ratio of compression reinforcement	–
ρ_p	ratio of prestressed reinforcement	–
λ	correction factor related to unit weight of concrete as given in ACI Sec. 8.6.1	–
ϕ	strength reduction factor	–
ω	reinforcement index of nonprestressed tension reinforcement	–
ω'	reinforcement index of compression reinforcement	–
ω_p	reinforcement index of prestressed reinforcement	–

Basic Principles

ACI Sec. 18.7.1 specifies that the design moment strength of a prestressed beam shall be computed by the strength design method. As shown in Fig. 3.1, ACI Sec. 10.2 assumes that the strain distribution over the depth of a prestressed beam is linear, and that a rectangular stress block is formed in the concrete at the ultimate load. The maximum useable compressive strain in the concrete is specified as

$$\epsilon_c = 0.003$$

Figure 3.1 Prestressed Beam at Ultimate Load

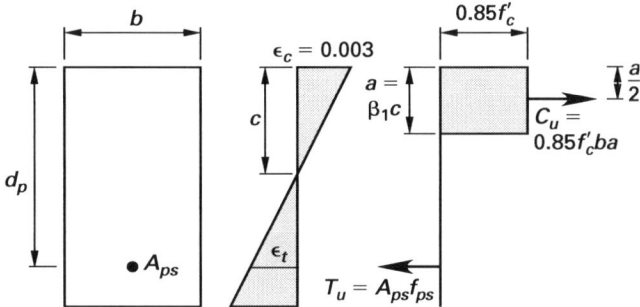

The stress in the stress block is $0.85f'_c$ as defined by ACI Sec. 10.2.7.1. The depth of the stress block is

$$a = \beta_1 c$$

The compression zone factor is defined in ACI Sec. 10.2.7.3 as

$$\beta_1 = 0.85 \quad [f'_c \leq 4000 \text{ lbf/in}^2]$$
$$= 0.85 - \frac{f'_c - 4000}{20{,}000} \quad \begin{bmatrix} 4000 \text{ lbf/in}^2 < f'_c \\ \leq 8000 \text{ lbf/in}^2 \end{bmatrix}$$
$$= 0.65 \text{ minimum} \quad [f'_c > 8000 \text{ lbf/in}^2]$$

Equating compressive and tensile forces on the section gives

$$0.85f'_c ab = A_{ps}f_{ps}$$

The depth of the stress block is derived as

$$a = \frac{A_{ps}f_{ps}}{0.85f'_c b}$$

The nominal flexural strength of the member is

$$M_n = A_{ps}f_{ps}\left(d_p - \frac{a}{2}\right)$$
$$= A_{ps}f_{ps}\left(\frac{d_p - 0.59A_{ps}f_{ps}}{bf'_c}\right)$$

Nonprestressed auxiliary reinforcement, conforming to ASTM A615, A706, and A996, may be added in the tensile zone to increase the nominal strength. The nominal strength becomes

$$M_n = A_{ps}f_{ps}\left(d_p - \frac{a}{2}\right) + A_s f_y\left(d - \frac{a}{2}\right)$$
$$a = \frac{A_{ps}f_{ps} + A_s f_y}{0.85f'_c b}$$

Compression reinforcement conforming to ASTM A615, A706, and A996, may also be added in the compression zone to increase the nominal strength. The nominal strength then becomes

$$M_n = A_{ps}f_{ps}\left(d_p - \frac{a}{2}\right) + A_s f_y\left(d - \frac{a}{2}\right)$$
$$+ A'_s f_y\left(\frac{a}{2} - d'\right)$$
$$a = \frac{A_{ps}f_{ps} + A_s f_y - A'_s f_y}{0.85f'_c b}$$

In ACI Sec. 10.3.4, a section is defined as tension controlled if the net tensile strain in the extreme tension steel, ϵ_t, is not less than 0.005 when the concrete reaches its maximum useable compressive strain of 0.003. The strength reduction factor is then given by ACI Sec. 9.3.2 as

$$\phi = 0.9$$

The following relationships may be derived from Fig. 3.1.

$$\epsilon_t = 0.005$$
$$\frac{c}{d_p} = 0.375$$
$$a = 0.375\beta_1 d_p$$

In ACI Sec. 10.3.3, a section is defined as compression-controlled if the net tensile strain in the extreme tension steel is not more than 0.002 when the concrete reaches its maximum useable compressive strain of 0.003. The strength reduction factor is then given by ACI Sec. 9.3.2 as

$$\phi = 0.65$$

The following relationships may then be derived from Fig. 3.1.

$$\epsilon_t = 0.002$$
$$\frac{c}{d_p} = 0.600$$
$$a = 0.600\beta_1 d_p$$

Members with a value of tensile strain between 0.002 and 0.005 are in the transition zone: The value of the strength reduction factor may be interpolated as shown in Fig. 3.2.

Figure 3.2 Variation of ϕ with ϵ_t

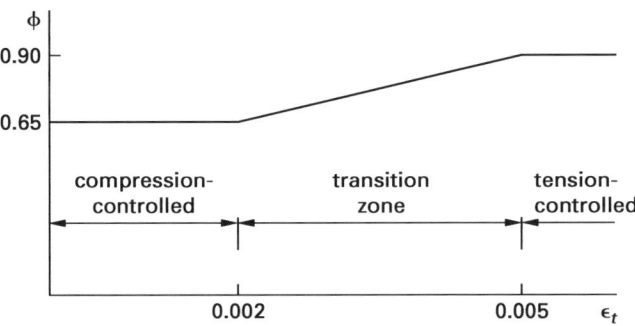

To prevent sudden failure when the modulus of rupture of the concrete is exceeded and the first flexural crack forms, ACI Sec. 18.8.2 requires the provision of sufficient prestressed and nonprestressed reinforcement to give a design flexural strength of

$$\phi M_n \geq 1.2 M_{cr}$$

The cracking moment of the member after all prestress losses have occurred is

$$M_{cr} = S_b(P_e R_b + f_r)$$
$$R_b = \frac{1}{A_g} + \frac{e}{S_b}$$
$$f_r = 7.5\lambda\sqrt{f'_c} \quad \text{[from ACI Sec. 9.5.2.3, Eq. (9-10)]}$$

This provision may be waived for

- two-way, unbonded post-tensioned slabs
- members with shear and flexural design strength at least twice the required strength

Example 3.1

The normal weight concrete prestressed beam shown in the illustration is simply supported over a span of 30 ft. It has a 28 day concrete strength of 6000 lbf/in². The area of the low relaxation prestressing tendons provided is $A_{ps}=0.765$ in² with a specified tensile strength of $f_{pu}=270$ kips/in², a yield strength of $f_{py}=243$ kips/in², and an effective stress of $f_{se}=150$ kips/in² after all losses. Determine the cracking moment of the beam.

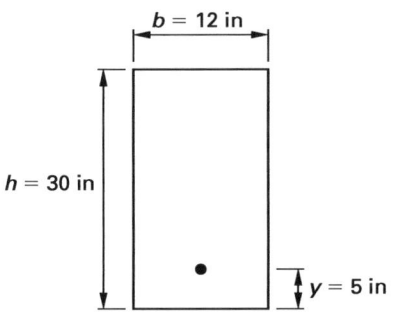

Solution

The relevant properties of the beam are

$$A_g = hb$$
$$= (30 \text{ in})(12 \text{ in})$$
$$= 360 \text{ in}^2$$

$$S_b = \frac{A_g h}{6}$$
$$= \frac{(360 \text{ in}^2)(30 \text{ in})}{6}$$
$$= 1800 \text{ in}^3$$

$$e = \frac{h}{2} - y$$
$$= \frac{30 \text{ in}}{2} - 5 \text{ in}$$
$$= 10 \text{ in}$$

$$R_b = \frac{1}{A_g} + \frac{e}{S_b}$$
$$= \frac{1}{360 \text{ in}^2} + \frac{10 \text{ in}}{1800 \text{ in}^3}$$
$$= 0.00833 \text{ in}^{-2}$$

The force in the prestressing tendons at service loads after allowance for all losses is

$$P_e = A_{ps} f_{se}$$
$$= (0.765 \text{ in}^2)\left(150 \frac{\text{kips}}{\text{in}^2}\right)$$
$$= 114.75 \text{ kips}$$

The modulus of rupture of the concrete is given by ACI Eq. (9-10) as

$$f_r = 7.5\lambda\sqrt{f'_c}$$
$$= (7.5)(1.0)\sqrt{6000 \frac{\text{lbf}}{\text{in}^2}}$$
$$= 581 \text{ lbf/in}^2$$

The cracking moment is

$$M_{cr} = S_b(P_e R_b + f_r)$$
$$= (1800 \text{ in}^3)$$
$$\times \left((114.75 \text{ kips})(0.00833 \text{ in}^{-2}) + 0.581 \frac{\text{kips}}{\text{in}^2}\right)$$
$$= 2766 \text{ in-kips}$$

Flexural Strength of Members with Bonded Tendons

When the effective stress after all losses, f_{se}, is not less than half the tensile strength of the tendons, f_{pu}, ACI Sec. 18.7.2 gives the stress in the tendons at nominal strength as

$$f_{ps} = f_{pu}\left(1 - \frac{\gamma_p}{\beta_1}\left(\rho_p \frac{f_{pu}}{f'_c} + \frac{d}{d_p}(\omega - \omega')\right)\right)$$

[ACI 18-3]

$\gamma_p = 0.55$ [deformed bars with $f_{py}/f_{pu} \geq 0.80$]

$\quad\;\; = 0.40$ $\begin{bmatrix}\text{stress-relieved wire and strands,}\\ \text{and plain bars with } f_{py}/f_{pu} \geq 0.85\end{bmatrix}$

$\quad\;\; = 0.28$ $\begin{bmatrix}\text{low-relaxation wire and strands}\\ \text{with } f_{py}/f_{pu} \geq 0.90\end{bmatrix}$

The reinforcement index of nonprestressed tension reinforcement is

$$\omega = \frac{\rho f_y}{f'_c}$$

The reinforcement index of compression reinforcement is

$$\omega' = \frac{\rho' f_y}{f'_c}$$

The ratio of nonprestressed tension reinforcement is

$$\rho = \frac{A_s}{bd}$$

The ratio of compression reinforcement is

$$\rho' = \frac{A'_s}{bd}$$

The ratio of prestressed reinforcement is

$$\rho_p = \frac{A_{ps}}{bd_p}$$

When compression reinforcement is provided in the section

$$0.17 \geq \rho_p\left(\frac{f_{pu}}{f'_c}\right) + \frac{d(\omega - \omega')}{d_p}$$

$$d' \leq 0.15 d_p$$

When the section contains no auxiliary reinforcement, the value for the stress in prestressed reinforcement at nominal strength reduces to

$$f_{ps} = f_{pu}\left(1 - \frac{\gamma_p \rho_p f_{pu}}{\beta_1 f'_c}\right)$$

Example 3.2

The fully bonded pretensioned beam described in Ex. 3.1 is simply supported over a span of 30 ft, and has a 28 day concrete strength of 6000 lbf/in^2. The area of the low relaxation prestressing tendons provided is 0.765 in^2, with a specified tensile strength of 270 kips/in^2, a yield strength of 243 kips/in^2, and an effective stress of 150 kips/in^2 after all losses. Calculate the design flexural strength of the beam and determine whether the beam complies with ACI Sec. 18.8.2.

Solution

The compression zone factor for a specified concrete strength of $f'_c = 6000$ lbf/in^2, given by ACI Sec. 10.2.7.3, is

$$\beta_1 = 0.75$$

For a low-relaxation strand with a ratio of specified yield strength to specified tensile strength $f_{py}/f_{pu} \geq 0.90$, the tendon factor is given by ACI Sec. 18.7.2 as

$$\gamma_p = 0.28$$

The ratio of prestressed reinforcement is

$$\rho_p = \frac{A_{ps}}{bd_p}$$

$$= \frac{0.765 \text{ in}^2}{(12 \text{ in})(25 \text{ in})}$$

$$= 0.00255$$

$$\frac{\gamma_p \rho_p f_{pu}}{\beta_1 f'_c} = \frac{(0.28)(0.00255)\left(270 \frac{\text{kips}}{\text{in}^2}\right)}{(0.75)\left(6 \frac{\text{kips}}{\text{in}^2}\right)}$$

$$= 0.043$$

The ratio of effective stress to tensile strength of the tendons is

$$\frac{f_{se}}{f_{pu}} = \frac{150 \frac{\text{kips}}{\text{in}^2}}{270 \frac{\text{kips}}{\text{in}^2}}$$

$$= 0.56$$

$$> 0.5$$

The stress in the tendons at nominal strength, for a section without auxiliary reinforcement, may be derived from ACI Eq. (18-3) as

$$f_{ps} = f_{pu}\left(1 - \frac{\gamma_p \rho_p f_{pu}}{\beta_1 f'_c}\right)$$

$$= \left(270 \ \frac{\text{kips}}{\text{in}^2}\right)\left(1 - \frac{(0.28)(0.00255)\left(270 \ \frac{\text{kips}}{\text{in}^2}\right)}{(0.75)\left(6 \ \frac{\text{kips}}{\text{in}^2}\right)}\right)$$

$$= 258 \ \text{kips/in}^2$$

The depth of the stress block is

$$a = \frac{A_{ps} f_{ps}}{0.85 f'_c b} = \frac{(0.765 \ \text{in}^2)\left(258 \ \frac{\text{kips}}{\text{in}^2}\right)}{(0.85)\left(6 \ \frac{\text{kips}}{\text{in}^2}\right)(12 \ \text{in})}$$

$$= 3.23 \ \text{in}$$

The maximum depth of the stress block for a tension-controlled section is given by ACI Sec. 10.3.4 as

$$a_t = 0.375 \beta_1 d_p$$
$$= (0.375)(0.75)(25 \ \text{in})$$
$$= 7.03 \ \text{in}$$
$$> a$$

The section is tension-controlled and

$$\phi = 0.9$$

The nominal flexural strength of the section is

$$M_n = A_{ps} f_{ps}\left(d_p - \frac{a}{2}\right)$$
$$= (0.765 \ \text{in}^2)\left(258 \ \frac{\text{kips}}{\text{in}^2}\right)\left(25 \ \text{in} - \frac{3.23 \ \text{in}}{2}\right)$$
$$= 4615 \ \text{in-kips}$$

The design flexural strength of the beam is

$$\phi M_n = (0.9)(4615 \ \text{in-kips})$$
$$= 4154 \ \text{in-kips}$$

The cracking moment of the beam is determined in Ex. 3.1 as

$$M_{cr} = 2766 \ \text{in-kips}$$
$$\frac{\phi M_n}{M_{cr}} = \frac{4154 \ \text{in-kips}}{2766 \ \text{in-kips}}$$
$$= 1.5$$
$$> 1.2$$

The beam complies with ACI Sec. 18.8.2.

Flexural Strength of Members with Unbonded Tendons

When the effective stress after all losses, f_{se}, is not less than half the tensile strength of the tendons, f_{pu}, ACI Sec. 18.7.2 permits the stress in the tendons at nominal strength to be calculated by the following methods.

For unbonded tendons and a span-to-depth ratio of 35 or less, ACI Sec. 18.7.2 gives the stress in the tendons at nominal strength as

$$f_{ps} = f_{se} + 10{,}000 + \frac{f'_c}{100 \rho_p} \quad [\text{ACI 18-4}]$$
$$\leq f_{py}$$
$$\leq f_{se} + 60{,}000$$

f_{se} and f'_c are in lbf/in^2.

For unbonded tendons and a span-to-depth ratio greater than 35, ACI Sec. 18.7.2 gives the stress in the tendons at nominal strength as

$$f_{ps} = f_{se} + 10{,}000 + \frac{f'_c}{300 \rho_p} \quad [\text{ACI 18-5}]$$
$$\leq f_{py}$$
$$\leq f_{se} + 30{,}000$$

f_{se} and f'_c are in lbf/in^2.

In accordance with ACI Sec. 18.9, auxiliary bonded reinforcement is required near the extreme tension fiber in all flexural members with unbonded tendons. The minimum required area is independent of the grade of steel, and is given by ACI Sec. 18.9.2 as

$$A_s = 0.004 A_{ct} \quad [\text{ACI 18-6}]$$

A_{ct} is the area of the concrete section between the centroid of the section and the extreme tension fiber, as shown in Fig. 3.3.

Figure 3.3 Bonded Reinforcement Area

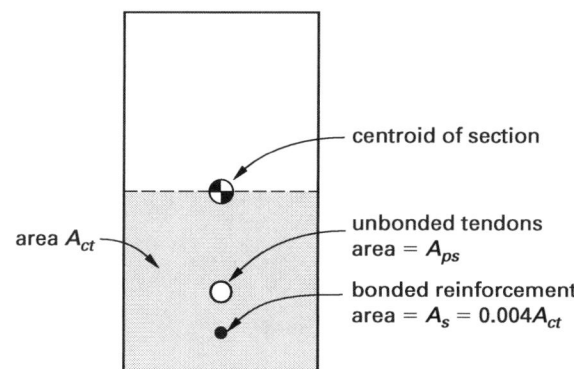

In two-way slabs, when the tensile stress caused by dead load + live load exceeds $2\sqrt{f'_c}$, ACI Sec. 18.9.3.2 requires the provision of auxiliary reinforcement with a minimum area of

$$A_s = \frac{N_c}{0.5 f_y} \quad \text{[ACI 18-7]}$$

Example 3.3

The post-tensioned unbonded beam described in Ex. 3.1 is simply supported over a span of 30 ft, and has a 28 day concrete strength of 6000 lbf/in². The area of the low-relaxation prestressing tendons provided is 0.765 in², with a specified tensile strength of 270 kips/in², a yield strength of 243 kips/in², and an effective stress of 150 kips/in² after all losses. An area of 0.8 in² of grade 60 auxiliary reinforcement is provided at a height of 3 in above the beam soffit. Calculate the design flexural strength of the beam, and determine whether the beam complies with ACI Sec. 18.8.2.

Solution

Because $f_{se}/f_{pu} > 0.5$, the method of ACI Sec. 18.7.2 may be used. The ratio of prestressed reinforcement is

$$\rho_p = \frac{A_{ps}}{b d_p}$$
$$= \frac{0.765 \text{ in}^2}{(12 \text{ in})(25 \text{ in})}$$
$$= 0.00255$$

For unbonded tendons and a span-to-depth ratio of 35 or less, the stress in the unbonded tendons at nominal strength is given by ACI Sec. 18.7.2 as

$$f_{ps} = f_{se} + 10 + \frac{f'_c}{100 \rho_p} \quad \text{[ACI 18-4]}$$
$$= 150 \, \frac{\text{kips}}{\text{in}^2} + 10 \, \frac{\text{kips}}{\text{in}^2} + \frac{6 \, \frac{\text{kips}}{\text{in}^2}}{(100)(0.00255)}$$
$$= 184 \text{ kips/in}^2$$
$$< f_{py} \quad \text{[satisfactory]}$$
$$< f_{se} + 60 \text{ kips/in}^2 \quad \text{[satisfactory]}$$

The required minimum area of auxiliary reinforcement is specified by ACI Sec. 18.9.2 as

$$A_s = 0.004 A_{ct} \quad \text{[ACI 18-6]}$$
$$= (0.004)(12 \text{ in})(15 \text{ in})$$
$$= 0.72 \text{ in}^2$$
$$< 0.80 \text{ in}^2 \text{ provided} \quad \text{[satisfactory]}$$

Assuming full utilization of the auxiliary reinforcement, the depth of the stress block is

$$a = \frac{A_{ps} f_{ps} + A_s f_y}{0.85 f'_c b}$$
$$= \frac{(0.765 \text{ in}^2)\left(184 \, \frac{\text{kips}}{\text{in}^2}\right) + (0.8 \text{ in}^2)\left(60 \, \frac{\text{kips}}{\text{in}^2}\right)}{(0.85)\left(6 \, \frac{\text{kips}}{\text{in}^2}\right)(12 \text{ in})}$$
$$= 3.08 \text{ in}$$

The maximum depth of the stress block for a tension-controlled section is given by ACI Sec. 10.3.4 as

$$a_t = 0.375 \beta_1 d_p$$
$$= (0.375)(0.75)(25 \text{ in})$$
$$= 7.03 \text{ in}$$
$$> a$$

The section is tension-controlled and

$$\phi = 0.9$$

The nominal flexural strength of the section is

$$M_n = A_{ps} f_{ps}\left(d_p - \frac{a}{2}\right) + A_s f_y\left(d - \frac{a}{2}\right)$$
$$= (0.765 \text{ in}^2)\left(184 \, \frac{\text{kips}}{\text{in}^2}\right)\left(25 \text{ in} - \frac{3.08 \text{ in}}{2}\right)$$
$$\quad + (0.80 \text{ in}^2)\left(60 \, \frac{\text{kips}}{\text{in}^2}\right)\left(27 \text{ in} - \frac{3.08 \text{ in}}{2}\right)$$
$$= 4524 \text{ in-kips}$$

The design flexural strength of the beam is

$$\phi M_n = (0.9)(4524 \text{ in-kips})$$
$$= 4072 \text{ in-kips}$$

The cracking moment of the beam is determined in Ex. 3.1 as

$$M_{cr} = 2766 \text{ in-kips}$$
$$\frac{\phi M_n}{M_{cr}} = \frac{4072 \text{ in-kips}}{2766 \text{ in-kips}}$$
$$= 1.47$$
$$> 1.2$$

The beam complies with ACI Sec. 18.8.2.

2. DESIGN FOR SHEAR AND TORSION

Nomenclature

A_{cp}	area enclosed by outside perimeter of concrete cross section	in^2
A_l	total area of longitudinal reinforcement to resist torsion	in^2
A_o	gross area enclosed by shear flow	in^2
A_{oh}	gross area enclosed by the center line of the outermost closed transverse torsional reinforcement	in^2
A_{ps}	area of prestressed reinforcement in tension zone	in^2
A_t	area of one leg of a closed stirrup resisting torsion within a spacing s	in^2
A_v	area of shear reinforcement within a spacing s	in^2
A_{v+t}	sum of areas of shear and torsion reinforcement	in^2
b_w	web width	in
d	distance from extreme compression fiber to centroid of prestressed and nonprestressed tension reinforcement, as defined in ACI Sec. 11.4.1 (need not be less than $0.8h$ for prestressed members)	in
d_{bl}	diameter of longitudinal torsional reinforcement	in
d_p	actual distance from extreme compression fiber to centroid of prestressing tendons	in
f_{pc}	compressive stress in the concrete, caused by the final prestressing force at the centroid of the section	kips/in^2
f_{pu}	specified strength of prestressing tendons	kips/in^2
f_{se}	effective stress in prestressing reinforcement after allowance for all losses	kips/in^2
f_y	specified yield strength of reinforcement	kips/in^2
f_y	yield strength of longitudinal torsional reinforcement	kips/in^2
f_{yt}	yield strength of transverse reinforcement	kips/in^2
h	overall thickness of member	in
l	span length	in
$M_{\max}$	maximum factored moment at section caused by externally applied loads	in-kips
M_u	factored moment at section	in-kips
p_{cp}	outside perimeter of the concrete cross section	in
p_h	perimeter of centerline of outermost closed transverse torsional reinforcement	in
s	spacing of shear or torsion reinforcement	in
S_b	section modulus of the section referred to the bottom fiber	in^3
t	wall thickness	in
T	applied torsion	in-kips
T_n	nominal torsional moment strength	in-kips
T_u	factored torsional moment at section	in-kips
V_c	nominal shear strength provided by concrete	kips
V_n	nominal shear strength	kips
V_s	nominal shear strength provided by shear reinforcement	kips
V_u	factored shear force at section	kips
w_u	factored load per unit length of beam or one way slab	kips/ft
x	distance from the support	in
x_o	distance between the center lines of vertical legs of closed stirrups	in
y_o	distance between the center lines of horizontal legs of closed stirrups	in

Symbols

ϕ	strength reduction factor, 0.75 for shear and torsion	–
τ	shear stress in the walls	lbf/in^2

Design for Shear

Critical Section for Shear

As shown in Fig. 3.4, the critical section for the calculation of shear in a prestressed beam is located at a distance from the support equal to half the overall depth. The maximum design factored shear force is

$$V_u = R_u - w_u\left(\frac{h}{2}\right)$$

Figure 3.4 Critical Section for Shear

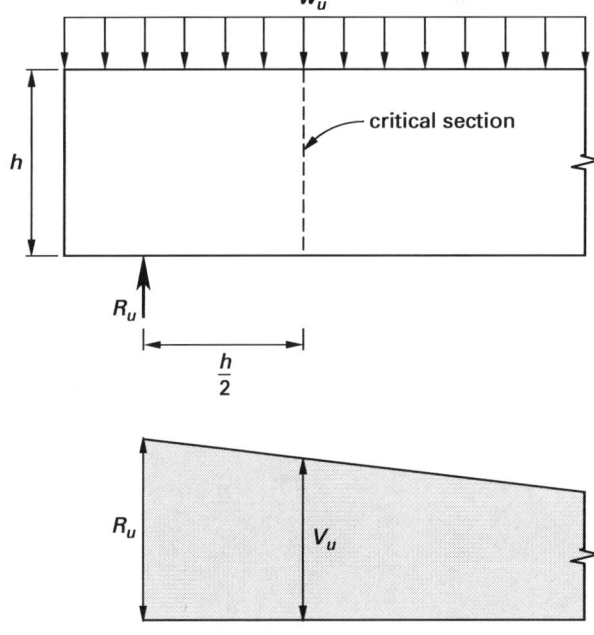

The beam may be designed for this shear force at a distance from the face of the support equal to half the overall depth. As specified in ACI Sec. 11.1.3, this is permitted provided that

- the support reaction produces a compressive stress in the end of the beam
- loads are applied at or near the top of the beam
- concentrated loads are not located closer to the support than half the overall depth

Shear Capacity of the Concrete Section

The nominal shear capacity provided by the concrete section may be obtained from ACI Sec. 11.3.2, provided that the effective prestressing force after all losses is not less than 40% of the tensile strength of the flexural reinforcement.

$$f_{se} \geq 0.4 f_{pu}$$

This provides a simplified method for determining a nominal shear strength that is conservative and usually sufficiently accurate. This method gives a value for the nominal shear capacity of the concrete section of

$$V_c = \left(0.6\lambda\sqrt{f'_c} + 700\frac{V_u d_p}{M_u}\right) b_w d \quad \text{[ACI 11-9]}$$
$$\leq 5 b_w d \lambda \sqrt{f'_c}$$
$$\geq 2 b_w d \lambda \sqrt{f'_c}$$
$$\frac{V_u d_p}{M_u} \leq 1.0$$
$$\sqrt{f'_c} \leq 100 \; \frac{\text{lbf}}{\text{in}^2} \quad \text{[from ACI Sec. 11.1.2]}$$

The factored shear force at the section is V_u, and the factored moment occurring simultaneously with V_u at the section is M_u. The distance from the extreme compression fiber to the centroid of the prestressed and nonprestressed tension reinforcement is defined as d in ACI Sec. 11.3.1. However, it need not be less than $0.8h$ for prestressed members. The actual distance from the extreme compression fiber to the centroid of the prestressing tendons is d_p.

For a simply supported beam with a uniformly distributed applied load, ACI Sec. R11.3.2 provides the expression

$$\frac{V_u d_p}{M_u} = \frac{d_p(l - 2x)}{x(l - x)}$$

As specified in ACI Sec. 11.4.6.1, a minimum area of shear reinforcement must be provided in all prestressed beams where the factored shear force exceeds one-half the design shear strength provided by the concrete section. This minimum area of shear reinforcement is specified by ACI Secs. 11.4.6.3 and 11.4.6.4 as the least value given by

$$A_{v,\min} = \frac{50 b_w s}{f_{yt}}$$
$$\geq \frac{A_{ps} f_{pu} s \sqrt{\dfrac{d}{b_w}}}{80 f_{yt} d} \quad \text{[ACI 11-14 for } f_{se} > 0.4 f_{pu}\text{]}$$
$$\geq \frac{0.75\sqrt{f'_c} b_w s}{f_{yt}} \quad \text{[ACI 11-13]}$$

f_{yt}, f_{pu}, and f'_c are in lbf/in². As stated in ACI Sec. 11.4.3, d need not be taken less than $0.8h$.

ACI Sec. 11.4.5 limits the spacing of the stirrups to a maximum value of 0.75 times the overall thickness of the member, h, but not less than 24 in. When the value of the nominal shear strength, V_s, exceeds $4 b_w d\sqrt{f'_c}$, the spacing is reduced to a maximum value of $0.375h$, or 12 in.

Example 3.4

A normal weight concrete prestressed beam is simply supported over a span of 30 ft, and has a 28 day concrete strength of 6000 lbf/in². The area of the low-relaxation prestressing tendons provided is $A_{ps} = 0.765$ in², with a specified tensile strength of $f_{pu} = 270$ kips/in², a yield strength of $f_{py} = 243$ kips/in², and an effective stress of $f_{se} = 150$ kips/in² after all losses. As shown in the illustration, the centroid of the prestressing tendons is 10 in above the soffit of the beam at a distance of 15 in from the support. The beam supports a uniformly distributed factored load, including its self-weight, of 5 kips/ft. Determine whether the shear capacity of the beam is adequate.

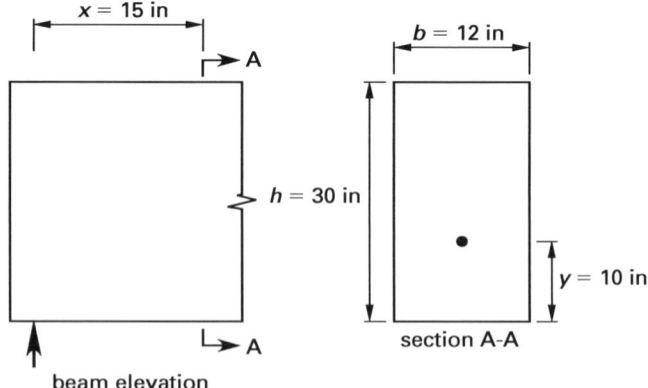

Solution

The critical section for shear occurs at a distance from the support given by

$$x = \frac{h}{2}$$
$$= 15 \text{ in}$$

The actual depth of the cable at this location is

$$d_p = h - y$$
$$= 30 \text{ in} - 10 \text{ in}$$
$$= 20 \text{ in}$$

The effective depth of the section is

$$d = 0.8h$$
$$= (0.8)(30 \text{ in})$$
$$= 24 \text{ in} \quad \text{[governs]}$$

The support reaction is

$$R_u = \frac{w_u l}{2}$$
$$= \frac{\left(5 \frac{\text{kips}}{\text{ft}}\right)(30 \text{ ft})}{2}$$
$$= 75 \text{ kips}$$

The maximum design factored shear force at the critical section is

$$V_u = R_u - w_u \left(\frac{h}{2}\right)$$
$$= 75 \text{ kips} - \frac{\left(5 \frac{\text{kips}}{\text{ft}}\right)\left(\frac{30 \text{ in}}{2}\right)}{12 \frac{\text{in}}{\text{ft}}}$$
$$= 68.75 \text{ kips}$$

From ACI Sec. R11.3.2,

$$\frac{V_u d_p}{M_u} = \frac{d_p(l - 2x)}{x(l - x)}$$
$$= \frac{(20 \text{ in})(360 \text{ in} - (2)(15 \text{ in}))}{(15 \text{ in})(360 \text{ in} - 15 \text{ in})}$$
$$= 1.28$$

From ACI Sec. 11.3.2, use a maximum value of

$$\frac{V_u d_p}{M_u} = 1.0$$

The ratio of the effective stress to the tensile strength of the tendons is

$$\frac{f_{se}}{f_{pu}} = \frac{150 \frac{\text{kips}}{\text{in}^2}}{270 \frac{\text{kips}}{\text{in}^2}}$$
$$= 0.56$$
$$> 0.4$$

ACI Sec. 11.3.2 is applicable and the nominal shear capacity provided by the concrete is

$$V_c = \left(0.6\lambda\sqrt{f'_c} + 700 \frac{V_u d_p}{M_u}\right) b_w d \quad [d = 0.8h = 24 \text{ in}]$$
$$= \left((0.6)(1.0)\sqrt{6000 \frac{\text{lbf}}{\text{in}^2}} + (700)(1.0)\right)$$
$$\times \left(\frac{(12 \text{ in})(24 \text{ in})}{1000 \frac{\text{lbf}}{\text{kip}}}\right)$$
$$= 215 \text{ kips}$$

From ACI Sec. 11.3.2, the maximum permitted value of the nominal shear capacity is

$$V_c = 5 b_w d \lambda \sqrt{f'_c}$$
$$= \frac{(5)(12 \text{ in})(24 \text{ in})(1.0)\sqrt{6000 \frac{\text{lbf}}{\text{in}^2}}}{1000 \frac{\text{lbf}}{\text{kip}}}$$
$$= 111.54 \text{ kips} \quad \text{[governs]}$$

From ACI Sec. 9.3.2.3, the design shear capacity is

$$\phi V_c = 0.75 V_c$$
$$= (0.75)(111.54 \text{ kips})$$
$$= 83.66 \text{ kips}$$
$$> V_u \quad \text{[satisfactory]}$$

Because the factored shear force exceeds one-half the design shear strength provided by the concrete section, a minimum area of shear reinforcement is required by ACI Sec. 11.4.6.1. The minimum area of shear reinforcement is given by the least value of

$$\frac{A_{v,\min}}{s} = \frac{50 b_w}{f_{yt}}$$
$$= \frac{(50)(12 \text{ in})\left(12 \frac{\text{in}}{\text{ft}}\right)}{60{,}000 \frac{\text{lbf}}{\text{in}^2}}$$
$$= 0.12 \text{ in}^2/\text{ft}$$

$$\frac{A_{v,\min}}{s} = \frac{A_{ps} f_{pu} \sqrt{\frac{d}{b_w}}}{80 f_{yt} d} \quad \text{[ACI 11-14]}$$
$$= \frac{(0.765 \text{ in}^2)\left(270 \frac{\text{kips}}{\text{in}^2}\right)\sqrt{\frac{24 \text{ in}}{12 \text{ in}}}\left(12 \frac{\text{in}}{\text{ft}}\right)}{(80)\left(60 \frac{\text{kips}}{\text{in}^2}\right)(24 \text{ in})}$$
$$= 0.030 \text{ in}^2/\text{ft} \quad \text{[governs]}$$

$$\frac{A_{v,\min}}{s} = 0.75\sqrt{f'_c}\left(\frac{b_w}{f_{yt}}\right)$$

$$= (0.75)\sqrt{6000\ \frac{\text{lbf}}{\text{in}^2}}\left(\frac{(12\text{ in})\left(12\ \frac{\text{in}}{\text{ft}}\right)}{60{,}000\ \frac{\text{lbf}}{\text{in}^2}}\right)$$

$$= 0.14\text{ in}^2/\text{ft}$$

Provide shear reinforcement consisting of two vertical legs of no. 3 stirrups at 18 in spacing, which gives

$$\frac{A_v}{s} = \frac{(0.22\text{ in}^2)\left(12\ \frac{\text{in}}{\text{ft}}\right)}{18\text{ in}}$$

$$= 0.15\text{ in}^2/\text{ft}$$

$$> 0.030\text{ in}^2/\text{ft}\quad [\text{satisfactory}]$$

The maximum spacing permitted by ACI Sec. 11.4.5.1 is the lesser of

$$s = 24\text{ in}$$

$$s = 0.75h$$

$$= (0.75)(30\text{ in})$$

$$= 22.5\text{ in}$$

$$> 18\text{ in}\quad [\text{satisfactory}]$$

Shear Capacity of Shear Reinforcement

The nominal shear capacity of shear reinforcement perpendicular to the member is given by ACI Sec. 11.4.7.2 as

$$V_s = \frac{A_v f_{yt} d}{s}\quad [\text{ACI 11-15}]$$

$$\leq 8 b_w d \sqrt{f'_c}$$

ACI Sec. 11.4.5.1 limits the spacing of the stirrups to a maximum value of three-quarters of h or 24 in. When the value of the nominal shear strength provided by the shear reinforcement, V_s, exceeds $4 b_w d \sqrt{f'_c}$, the spacing is reduced to a maximum value of $0.375d$, or 12 in.

In accordance with ACI Sec. 11.1.1, the nominal shear strength of the concrete section and of the shear reinforcement are additive, and the combined nominal shear capacity of the concrete and the stirrups is

$$V_n = V_c + V_s$$

$$\phi V_n = \phi V_c + \phi V_s$$

When the applied factored shear force V_u is less than $\phi V_c/2$, the concrete section is adequate to carry the shear without any shear reinforcement. Within the range $\phi V_c/2 \leq V_u \leq \phi V_c$, the minimum area of shear reinforcement is specified by ACI Secs. 11.4.6.3 and 11.4.6.4.

Example 3.5

The prestressed beam described in Ex. 3.4 is simply supported over a span of 30 ft, and has a 28 day concrete strength of 6000 lbf/in^2. The area of the low-relaxation prestressing tendons provided is $A_{ps} = 0.765$ in^2, with a specified tensile strength of $f_{pu} = 270$ kips/in^2, a yield strength of $f_{py} = 243$ kips/in^2, and an effective stress of $f_{se} = 150$ kips/in^2 after all losses. The centroid of the prestressing tendons is 10 in above the soffit of the beam, at a distance of 15 in from the support. The beam supports a uniformly distributed factored load, including its self-weight, of 8 kips/ft. Determine the shear reinforcement required.

Solution

From Ex. 3.4, the factored shear force at the critical section is

$$V_u = \frac{(68.75\text{ kips})\left(8\ \frac{\text{kips}}{\text{ft}}\right)}{5\ \frac{\text{kips}}{\text{ft}}}$$

$$= 110\text{ kips}$$

The design shear strength of the concrete section is derived in Ex. 3.4 as

$$\phi V_c = 83.66\text{ kips}$$

The design shear strength required from stirrups is

$$\phi V_s = V_u - \phi V_c$$

$$= 110\text{ kips} - 83.66\text{ kips}$$

$$= 26.34\text{ kips}$$

The design shear strength of required shear reinforcement is limited by ACI Sec. 11.4.7.9 to a maximum value of

$$\phi V_s = \phi 8 b_w d \sqrt{f'_c} = \frac{(0.75)(8)(12\text{ in})(24\text{ in})\sqrt{6000\ \frac{\text{lbf}}{\text{in}^2}}}{1000\ \frac{\text{lbf}}{\text{kips}}}$$

$$= 134\text{ kips}$$

$$> 26.35\text{ kips}\quad [\text{satisfactory}]$$

The area of shear reinforcement is given by ACI Sec. 11.4.7.2 as

$$\frac{A_v}{s} = \frac{\phi V_s}{\phi f_{yt} d} = \frac{(26.34\text{ kips})\left(12\ \frac{\text{in}}{\text{ft}}\right)}{(0.75)\left(60\ \frac{\text{kips}}{\text{in}^2}\right)(24\text{ in})}$$

$$= 0.293\text{ in}^2/\text{ft}$$

Provide shear reinforcement consisting of two vertical legs of no. 3 stirrups at 9 in spacing, which gives

$$\frac{A_v}{s} = \frac{(0.22 \text{ in}^2)\left(12 \frac{\text{in}}{\text{ft}}\right)}{9 \text{ in}}$$
$$= 0.293 \text{ in}^2/\text{ft} \quad [\text{satisfactory}]$$

Design for Torsion

General Principles

After torsional cracking occurs, the central core of a prestressed concrete member is largely ineffectual in resisting applied torsion and can be neglected. Hence, it is assumed in ACI Sec. R11.5 that a member behaves as a thin-walled tube when subjected to torsion. To maintain a consistent approach, a member is also analyzed as a thin-walled tube before cracking. As shown in Fig. 3.5, the shear stress in the tube walls produces a uniform shear flow, acting at the midpoint of the walls, with a magnitude of

$$q = \tau t$$

The applied torsion is resisted by the moment of the shear flow in the walls about the centroid of the section, and is

$$T = 2A_o q$$

The gross area enclosed by shear flow is the area A_o enclosed by the center line of the walls, given by

$$A_o = \frac{2A_{cp}}{3}$$

Figure 3.5 Thin-Walled Tube Analogy

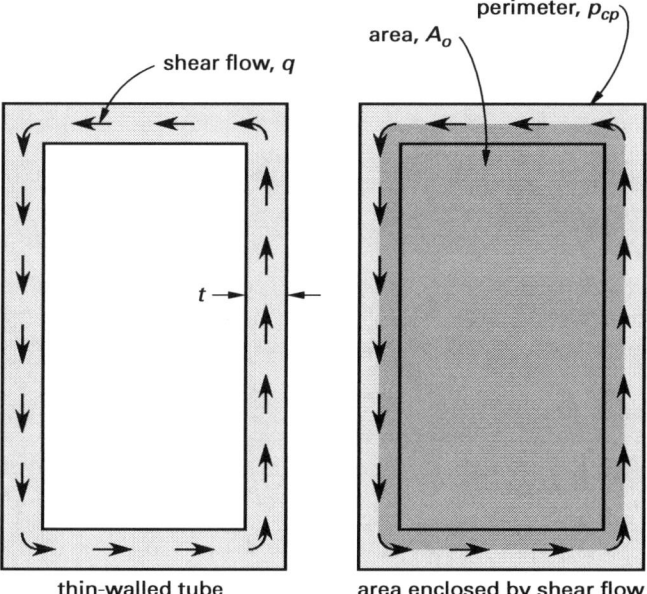

The shear stress in the walls is

$$\tau = \frac{T}{2A_o t}$$

For flanged sections, the value of the overhanging flange width (used in calculating the value of the area A_{cp} enclosed by the outside perimeter of the concrete cross section) and the value of p_{cp} (the outside perimeter of the concrete cross section) are determined as specified in ACI Secs. 11.5.1.1 and 13.2.4. The overhanging flange width is shown in Fig. 3.6.

Figure 3.6 Overhanging Flange Width

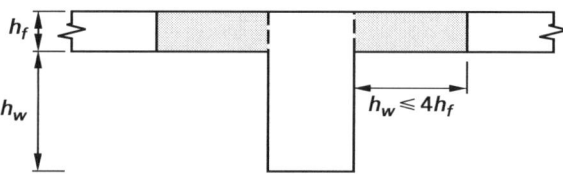

In accordance with ACI Sec. 11.5.2.5, the critical section for the calculation of torsion in a prestressed beam is located at a distance from the support equal to half the overall depth. When a concentrated torsion occurs within this distance, the critical section for design shall be at the face of the support.

Cracking is assumed to occur in a member when the principal tensile stress reaches a value of

$$p_t = 4\sqrt{f'_c}$$
$$= \tau$$

The cracking torsion is

$$T_{cr} = 4\lambda\sqrt{f'_c}\left(\frac{A_{cp}^2}{p_{cp}}\right)\sqrt{1 + \frac{f_{pc}}{4\lambda\sqrt{f'_c}}}$$

It is assumed in ACI Sec. 11.5.1 that torsional effects may be neglected, and closed stirrups and longitudinal torsional reinforcement are not required when the factored torque does not exceed the threshold torsion given by

$$T_u = \frac{\phi T_{cr}}{4}$$
$$= \phi\lambda\sqrt{f'_c}\left(\frac{A_{cp}^2}{p_{cp}}\right)\sqrt{1 + \frac{f_{pc}}{4\lambda\sqrt{f'_c}}}$$

When the threshold value is exceeded, reinforcement must be provided to resist the full torsion, and the concrete is considered ineffective. When both shear and torsion reinforcements are required, the sum of the individual areas must be provided as specified in ACI Sec. 11.5.3.8. Because the stirrup area A_v for shear is defined in terms of both legs of a stirrup, while the stirrup area A_t for torsion is defined in terms of one leg only, the summation of the areas is

$$\sum \frac{A_{v+t}}{s} = \frac{A_v}{s} + \frac{2A_t}{s}$$

The spacing of the reinforcement is limited by the minimum required spacing of either the shear or torsion reinforcement.

Example 3.6

The prestressed beam described in Ex. 3.4 is simply supported over a span of 30 ft, and has a 28 day concrete strength of 6000 lbf/in². The area of the low-relaxation prestressing tendons provided is $A_{ps} = 0.765$ in², with a specified tensile strength of $f_{pu} = 270$ kips/in², a yield strength of $f_{py} = 243$ kips/in², and an effective stress of $f_{se} = 150$ kips/in² after all losses. The centroid of the prestressing tendons is 10 in above the soffit of the beam at a distance of 15 in from the support. The beam supports a uniformly distributed factored load, including its self-weight, of 8 kips/ft. Determine the threshold torsion of the section.

Solution

The compressive stress in the concrete, caused by the final prestressing force at the centroid of the section, is

$$f_{pc} = \frac{P_e}{A_c}$$
$$= \frac{A_{ps}f_{se}}{A_c}$$
$$= \frac{(0.765 \text{ in}^2)\left(150 \frac{\text{kips}}{\text{in}^2}\right)}{360 \text{ in}^2}$$
$$= 0.319 \text{ kips/in}^2$$

The area enclosed by the outside perimeter of the beam is

$$A_{cp} = hb$$
$$= (30 \text{ in})(12 \text{ in})$$
$$= 360 \text{ in}^2$$

The length of the outside perimeter of the beam is

$$p_{cp} = 2(h+b)$$
$$= (2)(30 \text{ in} + 12 \text{ in})$$
$$= 84 \text{ in}$$
$$\frac{A_{cp}^2}{p_{cp}} = \frac{(360 \text{ in}^2)^2}{84 \text{ in}}$$
$$= 1543 \text{ in}^3$$

The threshold torsion is given by ACI Sec. 11.5.1 as

$$T_u = \phi\lambda\sqrt{f'_c}\left(\frac{A_{cp}^2}{p_{cp}}\right)\sqrt{1 + \frac{f_{pc}}{4\lambda\sqrt{f'_c}}}$$

$$= \frac{(0.75)(1.0)\sqrt{6000 \frac{\text{lbf}}{\text{in}^2}}(1543 \text{ in}^3)}{1000 \frac{\text{lbf}}{\text{kip}}} \times \sqrt{1 + \frac{319 \frac{\text{lbf}}{\text{in}}}{(4)(1.0)\sqrt{6000 \frac{\text{lbf}}{\text{in}^2}}}}$$

$$= 128 \text{ in-kips}$$

Design for Torsion Reinforcement

After torsional cracking occurs, cracks are produced diagonally on all four faces of a member, forming a continuous spiral failure surface around the perimeter. For a prestressed concrete beam subjected to pure torsion, the principal tensile stresses and the cracks are produced at an angle θ. ACI Sec. 11.5.3.6 specifies that the angle θ may be assumed to be 37.5°. After cracking, the beam is idealized as a tubular space truss consisting of closed stirrups, longitudinal reinforcement, and concrete compression struts between the torsion cracks. To resist the applied torque, it is necessary to provide transverse and longitudinal reinforcement on each face. Closed stirrups are required to resist the vertical component, and longitudinal reinforcement is required to resist the horizontal component of the torsional stresses. After torsional cracking, the concrete outside the closed stirrups is ineffectual in resisting the applied torsion, and the gross area enclosed by shear flow is redefined by ACI Sec. 11.5.3.6 as

$$A_o = 0.85 A_{oh}$$

In an indeterminate structure, redistribution of internal forces after cracking results in a reduction in torsional moment with a compensating redistribution of internal forces. This is referred to as *compatibility torsion*, and the member may be designed for a maximum factored torsional moment of

$$T_u = 4\phi\lambda\sqrt{f'_c}\left(\frac{A_{cp}^2}{p_{cp}}\right)\sqrt{1 + \frac{f_{pc}}{4\lambda\sqrt{f'_c}}}$$

When the torsional moment cannot be reduced by redistribution of internal forces after cracking, the member must be designed for the full applied factored torque. This is referred to as *equilibrium torsion*.

When the threshold torsion value is exceeded, closed stirrups and longitudinal reinforcement must be provided to resist the appropriate value of the torque, and the concrete is considered ineffective.

ACI Sec. 11.5.3.6 specifies the required area of one leg of a closed stirrup as

$$\frac{A_t}{s} = \frac{T_u}{2\phi A_o f_{yt} \cot \theta}$$

$$= \frac{T_u}{2\phi A_o f_{yt} \cot 37.5°}$$

$$= \frac{T_u}{1.7\phi A_{oh} f_{yt} \cot 37.5°}$$

The minimum combined area of stirrups for shear and torsion is given by ACI Sec. 11.5.5.2 as

$$\frac{A_v + 2A_t}{s} = 0.75\sqrt{f'_c} \frac{b_w}{f_{yt}}$$

$$\geq \frac{50 b_w}{f_{yt}}$$

The maximum spacing of closed stirrups is given by ACI Sec. 11.5.6.1 as

$$s = \frac{p_h}{8}$$

$$\leq 12 \text{ in}$$

The corresponding required area of longitudinal reinforcement is specified in ACI Secs. 11.5.3.7 and R11.5.3.10 as

$$A_l = \left(\frac{A_t}{s}\right)\left(\frac{p_h f_{yt}}{f_y}\right) \cot^2 \theta$$

$$= \left(\frac{A_t}{s}\right)\left(\frac{p_h f_{yt}}{f_y}\right) \cot^2 37.5°$$

When the threshold torsional moment is exceeded, the minimum permissible area of longitudinal reinforcement is given by ACI Sec. 11.5.5.3 as

$$A_{l,\min} = \frac{5\sqrt{f'_c} A_{cp}}{f_y} - \left(\frac{A_t}{s}\right) p_h \left(\frac{f_{yt}}{f_y}\right) \quad \text{[ACI 11-24]}$$

$$\frac{A_t}{s} \geq \frac{25 b_w}{f_{yt}}$$

The bars are distributed around the inside perimeter of the closed stirrups at a maximum spacing of 12 in. A longitudinal bar is required in each corner of a closed stirrup.

The minimum diameter of longitudinal reinforcement is given by ACI Sec. 11.5.6.2 as

$$d_{bl} = \frac{s}{24}$$

$$\geq \text{no. 3 bar}$$

To prevent crushing of the concrete compression diagonals, the combined stress caused by the factored torsion and shear forces is limited by ACI Sec. 11.5.3.1. The dimensions of the section must be such that

$$\sqrt{\left(\frac{V_u}{b_w d}\right)^2 + \left(\frac{T_u p_h}{1.7 A_{oh}^2}\right)^2} \leq \phi\left(\frac{V_c}{b_w d} + 8\sqrt{f'_c}\right)$$

Example 3.7

The prestressed beam described in Ex. 3.5 is simply supported over a span of 30 ft, and has a 28 day concrete strength of 6000 lbf/in^2. The area of the low relaxation prestressing tendons provided is $A_{ps} = 0.765$ in^2 with a specified tensile strength of $f_{pu} = 270$ kips/in^2, a yield strength of $f_{py} = 243$ kips/in^2, and an effective stress of $f_{se} = 150$ kips/in^2 after all losses. The centroid of the prestressing tendons is 10 in above the soffit of the beam at a distance of 15 in from the support. The beam supports a uniformly distributed factored load, including its self-weight, of 8 kips/ft, and an applied factored torsional moment of 150 in-kips. Determine the shear and torsional reinforcement required at the critical section, if the yield strength of the reinforcement is $f_{yt} = f_y = 60$ kips/in^2.

Solution

From Ex. 3.5, the required shear reinforcement at the critical section is

$$\frac{A_v}{s} = 0.293 \text{ in}^2/\text{ft}$$

The threshold torsion is obtained in Ex. 3.6 as

$$T_u = 128 \text{ in-kips}$$

$$< 150 \text{ in-kips}$$

Torsional reinforcement is required.

With 1 in cover to $^1/_2$ in stirrups, the distance between the center lines of the vertical legs of the closed stirrups is

$$x_o = b_w - d_b - 2(\text{clear cover})$$

$$= 12 \text{ in} - 0.5 \text{ in} - (2)(1 \text{ in})$$

$$= 9.5 \text{ in}$$

The distance between the center lines of the horizontal legs of the closed stirrups is

$$y_o = h - d_b - 2(\text{clear cover})$$

$$= 30 \text{ in} - 0.5 \text{ in} - (2)(1 \text{ in})$$

$$= 27.5 \text{ in}$$

The area enclosed by the center line of the closed stirrups is

$$A_{oh} = x_o y_o$$
$$= (9.5 \text{ in})(27.5 \text{ in})$$
$$= 261 \text{ in}^2$$

The gross area enclosed by shear flow is given by ACI Sec. 11.5.3.6 as

$$A_o = 0.85 A_{oh}$$
$$= (0.85)(261 \text{ in}^2)$$
$$= 222 \text{ in}^2$$

The perimeter of the center line of the closed stirrups is

$$p_h = 2(x_o + y_o)$$
$$= (2)(9.5 \text{ in} + 27.5 \text{ in})$$
$$= 74 \text{ in}$$

From ACI Eq. (11-21), assuming that $\theta = 37.5°$, the required area of one leg of a closed stirrup is

$$\frac{A_t}{s} = \frac{T_u}{2\phi A_o f_{yt} \cot 37.5°}$$
$$= \frac{(150 \text{ in-kips})\left(12 \dfrac{\text{in}}{\text{ft}}\right)}{(2)(0.75)(222 \text{ in}^2)\left(60 \dfrac{\text{kips}}{\text{in}^2}\right)(1.303)}$$
$$= 0.069 \text{ in}^2/\text{ft}/\text{leg}$$

The summation of the areas of required shear and torsion reinforcement is

$$\sum \frac{A_{v+t}}{s} = \frac{A_v}{s} + \frac{2A_t}{s} = 0.293 \frac{\text{in}^2}{\text{ft}} + (2)\left(0.069 \frac{\text{in}^2}{\text{ft}}\right)$$
$$= 0.43 \text{ in}^2/\text{ft}$$

The minimum permissible combined area of stirrups for shear and torsion is given by ACI Sec. 11.5.5.2 as the greater of

$$\frac{A_v + 2A_t}{s} = \frac{0.75\sqrt{f'_c} b_w}{f_{yt}}$$
$$= \frac{0.75\sqrt{6000 \dfrac{\text{lbf}}{\text{in}^2}}(12 \text{ in})\left(12 \dfrac{\text{in}}{\text{ft}}\right)}{60{,}000 \dfrac{\text{lbf}}{\text{in}^2}}$$
$$= 0.139 \text{ in}^2/\text{ft} \quad [\text{does not govern}]$$

$$\frac{A_v + 2A_t}{s} = \frac{50 b_w}{f_{yt}}$$
$$= \frac{(50)(12 \text{ in})\left(12 \dfrac{\text{in}}{\text{ft}}\right)}{60{,}000 \dfrac{\text{lbf}}{\text{in}^2}}$$
$$= 0.12 \text{ in}^2/\text{ft} \quad [\text{does not govern}]$$

The governing maximum permissible spacing of the closed stirrups is specified by ACI Sec. 11.5.6.1 as

$$s_{\max} = \frac{p_h}{8}$$
$$= \frac{74 \text{ in}}{8}$$
$$= 9.25 \text{ in}$$

Closed stirrups consisting of two arms of no. 3 bars at 6 in spacing provides an area of

$$\frac{A}{s} = \left(2 \frac{\text{bars}}{\text{ft}}\right)(2)\left(0.11 \frac{\text{in}^2}{\text{bar}}\right) = 0.44 \frac{\text{in}^2}{\text{ft}}$$
$$> 0.43 \text{ in}^2/\text{ft} \quad [\text{satisfactory}]$$

The required area of the longitudinal reinforcement is given by ACI Eq. (11-22), assuming $\theta = 37.5°$, as

$$A_t = \left(\frac{A_t}{s}\right) p_h \left(\frac{f_{yt}}{f_y}\right) \cot^2 \theta$$
$$= \frac{\left(0.069 \dfrac{\text{in}^2}{\text{ft} \cdot \text{leg}}\right)(74 \text{ in})(\cot 37.5°)^2}{12 \dfrac{\text{in}}{\text{ft}}}$$
$$= 0.72 \text{ in}^2$$

$$\frac{25 b_w}{f_{yt}} = \frac{(25)(12 \text{ in})}{60{,}000 \dfrac{\text{kips}}{\text{in}^2}}$$
$$= 0.0050 \text{ in}^2/\text{in-leg}$$
$$< A_t/s \quad [= 0.00576 \text{ in}^2/\text{in-leg}]$$

Use $A_t/s = 0.00576$ in²/in/leg in ACI Eq. (11-24). The minimum permissible area of longitudinal reinforcement is given by ACI Eq. (11-24) as

$$A_{l,\min} = \frac{5\sqrt{f'_c}A_{cp}}{f_y} - \left(\frac{A_t}{s}\right)p_h\left(\frac{f_{yt}}{f_y}\right)$$

$$= \frac{5\sqrt{6000\,\frac{\text{lbf}}{\text{in}^2}}(360\text{ in}^2)}{60{,}000\,\frac{\text{lbf}}{\text{in}^2}}$$

$$- \left(0.00576\,\frac{\frac{\text{in}^2}{\text{in}}}{\text{leg}}\right)(74\text{ in})$$

$$= 1.89\text{ in}^2 \quad [\text{governs}]$$

Use 10 no. 4 bars placed in the corners of the closed stirrups and distributed along the vertical legs to give a longitudinal steel area of

$$A_l = 2.0\text{ in}^2$$
$$> 1.89\text{ in}^2 \quad [\text{satisfactory}]$$

The minimum permitted diameter of longitudinal reinforcement is given by ACI Sec. 11.5.6.2 as

$$d_{bl} = \frac{s}{24}$$
$$= \frac{6\text{ in}}{24}$$
$$= 0.25\text{ in}$$
$$< 0.5\text{ in} \quad [\text{no. 4 bars are satisfactory}]$$

A check on the crushing of the concrete compression diagonals is provided by ACI Sec. 11.5.3.1. The left side of ACI Eq. (11-18) is

$$\sqrt{\left(\frac{V_u}{b_w d}\right)^2 + \left(\frac{T_u p_h}{1.7 A_{oh}^2}\right)^2}$$

$$= \sqrt{\left(\frac{110\text{ in-kips}}{(12\text{ in})(0.8)(30\text{ in})}\right)^2 + \left(\frac{(150\text{ in-kips})(74\text{ in})}{(1.7)(261\text{ in}^2)^2}\right)^2}$$

$$= \sqrt{0.146\,\frac{\text{kips}}{\text{in}^2} + 0.009\,\frac{\text{kips}}{\text{in}^2}}$$

$$= 0.394\text{ kips/in}^2$$

The right side of ACI Eq. (11-18) is

$$\phi\left(\frac{V_c}{b_w d} + 8\sqrt{f'_c}\right)$$

$$= \frac{(0.75)\left(\frac{111{,}540\text{ lbf}}{(12\text{ in})(0.8)(30\text{ in})} + 8\sqrt{6000\,\frac{\text{lbf}}{\text{in}^2}}\right)}{1000\,\frac{\text{lbf}}{\text{kip}}}$$

$$= 0.755\,\frac{\text{kips}}{\text{in}^2}$$

$$> 0.394\text{ kips/in}^2 \quad [\text{ACI Eq. (11-18) is satisfied}]$$

3. PRESTRESS LOSSES

Nomenclature

A_c	area of the gross concrete section	in²
A_{ps}	area of prestressing tendon	in²
b_w	web width, or diameter of circular section	in
C	factor for relaxation losses	–
E_c	modulus of elasticity of concrete at 28 days	kips/in²
E_{ci}	modulus of elasticity of concrete at time of initial prestress	kips/in²
E_p	modulus of elasticity of prestressing tendon	kips/in²
f'_{ci}	concrete strength at transfer	lbf/in²
f_{pd}	compressive stress at level of tendon centroid after elastic losses and including sustained dead load	lbf/in²
f_{pi}	compressive stress at level of tendon centroid after elastic losses	lbf/in²
f_{pp}	compressive stress at level of tendon centroid before elastic losses	lbf/in²
g	sag of prestressing tendon	in
h	overall thickness or height of member	in
H	ambient relative humidity	–
I_c	moment of inertia of the gross concrete section	in⁴
J	factor for relaxation losses	–
K	wobble friction coefficient per foot of prestressing tendon	–
K_{sh}	factor for shrinkage losses accounting for elapsed time between completion of casting and transfer of prestressing force	–
K_{re}	factor for relaxation losses	–
l_{px}	length of prestressing tendon from jacking end to any point x measured along the curve	ft
M_D	bending moment caused by superimposed dead load	ft-kips
M_G	bending moment caused by member self-weight	ft-kips
n	E_p/E_c	–
n_i	E_p/E_{ci}	–

p_{cp}	outside perimeter of the concrete cross section	in
P_i	prestressing tendon force after elastic losses	kips
P_p	prestressing tendon force before elastic losses	kips
P_{pj}	prestressing tendon force at jacking end	kips
P_{px}	prestressing tendon force at a distance of x from the jacking end	kips
$P_{\Delta cr}$	loss of tendon force caused by creep	kips
$P_{\Delta el}$	loss of tendon force caused by elastic shortening	kips
$P_{\Delta re}$	loss of tendon force caused by relaxation	kips
$P_{\Delta sh}$	loss of tendon force caused by shrinkage	kips
R	radius of curvature of tendon profile	ft
w_c	unit weight of concrete	lbf/ft^3

Symbols

α	angular change in radians of tendon profile from jacking end to any point x	–
ϵ_{sh}	basic shrinkage strain	–
μ	curvature friction coefficient	–

Friction Losses

Friction losses, which occur in post-tensioned members because of friction and unintentional out-of-straightness of the ducts, are determined by ACI Sec. 18.6.2.1 as

$$P_{pj} = P_{px} e^{Kl_{px} + \mu_p \alpha_{px}} \quad [\text{ACI 18-1}]$$

For a value of $(Kl_{px} + \mu\alpha)$ not greater than 0.3, ACI Eq. (18-1) can be approximated by

$$P_{pj} = P_{px}(1 + Kl_{px} + \mu_p \alpha_{px}) \quad [\text{ACI 18-2}]$$

Rearranging the terms gives

$$P_{px} = P_{pj}(1 - Kl_{px} - \mu\alpha)$$

Illustration for Example 3.8

Example 3.8

The post-tensioned beam shown in the illustration is simply supported over a span of 30 ft and has a 28 day concrete strength of 6000 lbf/in^2. The area of the low-relaxation prestressing tendon provided is $A_{ps} = 0.765$ in^2, with a specified tensile strength of $f_{pu} = 270$ kips/in^2, and a yield strength of $f_{py} = 243$ kips/in^2. The cable centroid, as shown, is parabolic in shape, and is stressed simultaneously from both ends with a jacking force of $P_{pj} = 160$ kips. The value of the wobble friction coefficient is $K = 0.0010$/ft, and the curvature friction coefficient is $\mu = 0.20$. Determine the cable force at midspan of the member before elastic losses.

Solution

The nominal radius of the cable profile is

$$R = \frac{a^2}{2g}$$

$$= \frac{(15 \text{ ft})^2 \left(12 \dfrac{\text{in}}{\text{ft}}\right)}{(2)(10.0 \text{ in})}$$

$$= 135 \text{ ft}$$

The cable length along the curve, from the jacking end to midspan is

$$l_{px} = a + \frac{g^2}{3a}$$

$$= 15 \text{ ft} + \frac{\left(\dfrac{10 \text{ in}}{12 \dfrac{\text{in}}{\text{ft}}}\right)^2}{(3)(15 \text{ ft})}$$

$$= 15.02 \text{ ft}$$

The angular change of the cable profile over length l_x is

$$\alpha = \frac{l_{px}}{R}$$
$$= \frac{15.02 \text{ ft}}{135 \text{ ft}}$$
$$= 0.111 \text{ radians}$$
$$Kl_{px} + \mu\alpha = (0.0010 \text{ ft}^{-1})(15.02 \text{ ft})$$
$$\qquad + (0.20)(0.111 \text{ radians})$$
$$= 0.037$$
$$< 0.3 \quad [\text{ACI Eq. (18-2) is applicable}]$$

The cable force at midspan is

$$P_{px} = P_{pj}(1 - Kl_{px} - \mu\alpha)$$
$$= (160 \text{ kips})(1 - 0.01502 - 0.0222)$$
$$= 154 \text{ kips}$$

Anchor Seating Losses

Anchor seating losses result from slip or set that occur in the anchorage when prestressing force is transferred to the anchor device. Anchorage slip depends on the prestressing system used, and may be compensated by increasing the jacking force.

Elastic Shortening Losses

Losses occur in a prestressed concrete beam at transfer due to the elastic shortening of the concrete. The concrete stress at the level of the centroid of the prestressing tendons after elastic shortening is

$$f_{pi} = P_i\left(\frac{1}{A_g} + \frac{e^2}{I_g}\right) - \frac{eM_G}{I_g}$$

Conservatively, this may be taken as the concrete stress at the level of the centroid of the prestressing tendons before elastic shortening, given by

$$f_{pi} \approx f_{pp} = P_p\left(\frac{1}{A_g} + \frac{e^2}{I_g}\right) - e\frac{M_G}{I_g}$$

If necessary, a process of iteration may now be used to refine the calculated value of the estimated losses.

In a pretensioned member with transfer occurring simultaneously in all tendons, the loss of prestressing force is

$$P_{\Delta el} = n_i A_{ps} f_{pi}$$

In a post-tensioned member with only one tendon, there is no loss from elastic shortening.

In a post-tensioned member with several tendons stressed sequentially, the maximum loss occurs in the first tendon stressed, and no loss occurs in the last tendon stressed. The total loss is then half the value for a pretensioned member.

$$P_{\Delta el} = \frac{n_i A_{ps} f_{pi}}{2}$$

Example 3.9

The post-tensioned beam described in Ex. 3.8 has a concrete strength at transfer of 4500 lbf/in². The beam is simply supported over a span of 30 ft, and is prestressed with two cables having a combined area of $A_{ps} = 0.765$ in². At the center of the span, the total initial force in the two tendons after friction losses is $P_p = 154$ kips. The modulus of elasticity of the prestressing tendon is $E_p = 28 \times 10^6$ lbf/in². Determine the loss of prestressing force caused by elastic shortening.

Solution

The self-weight of the beam is

$$w = w_c b_w h$$
$$= \left(150 \; \frac{\text{lbf}}{\text{ft}^3}\right)(1.0 \text{ ft})(2.5 \text{ ft})$$
$$= 375 \text{ lbf/ft}$$

The self-weight bending moment at midspan is

$$M_G = \frac{wl^2}{8}$$
$$= \frac{\left(375 \; \frac{\text{lbf}}{\text{ft}}\right)(30 \text{ ft})^2 \left(12 \; \frac{\text{in}}{\text{ft}}\right)}{(8)\left(1000 \; \frac{\text{lbf}}{\text{kip}}\right)}$$
$$= 506 \text{ in-kips}$$

From ACI Sec. 8.5, the modulus of elasticity of the concrete at transfer is

$$E_{ci} = 57{,}000\sqrt{f'_c}$$
$$= 57{,}000\sqrt{4500 \; \frac{\text{lbf}}{\text{in}^2}}$$
$$= 3{,}824{,}000 \; \frac{\text{lbf}}{\text{in}^2}$$
$$n_i = \frac{E_p}{E_{ci}}$$
$$= \frac{28 \times 10^6 \; \frac{\text{lbf}}{\text{in}^2}}{3{,}824{,}000 \; \frac{\text{lbf}}{\text{in}^2}}$$
$$= 7.32$$

The concrete stress at the level of the centroid of the prestressing tendons before elastic shortening is

$$f_{pp} = P_p\left(\frac{1}{A_g} + \frac{e^2}{I_g}\right) - \frac{eM_G}{I_g}$$

$$= (154 \text{ kips})\left(\frac{1}{360 \text{ in}^2} + \frac{(10 \text{ in})^2}{27{,}000 \text{ in}^4}\right)$$

$$- \frac{(10 \text{ in})(506 \text{ in-kips})}{27{,}000 \text{ in}^4}$$

$$= 0.811 \text{ kips/in}^2$$

The loss of prestressing force caused by elastic shortening may be estimated as

$$P_{\Delta el} = \frac{n_i A_{ps} f_{pp}}{2}$$

$$= \frac{(7.32)(0.765 \text{ in}^2)\left(0.811 \dfrac{\text{kips}}{\text{in}^2}\right)}{2}$$

$$= 2.3 \text{ kips}$$

The force in the cables after friction and elastic losses is

$$P_i = 154 \text{ kips} - 2.3 \text{ kips}$$
$$= 151.7 \text{ kips}$$

Creep Losses

Creep is the increasing strain that occurs in a prestressed concrete member caused by the sustained compressive stress. The rate of creep is increased when the initial stress in the member is increased. The earlier the age at which the stress is applied to the member, the greater is the rate of creep. The concrete stress at the level of the centroid of the prestressing tendons after elastic shortening, allowing for sustained dead load, is

$$f_{pd} = P_i\left(\frac{1}{A_g} + \frac{e^2}{I_g}\right) - \frac{eM_G}{I_g} - \frac{eM_D}{I_g}$$

For post-tensioned members, with transfer at 28 days, the creep loss is given by[1,2]

$$P_{\Delta cr} = 1.6 n A_{ps} f_{pd}$$

For pretensioned members, with transfer at 3 days, the creep loss is

$$P_{\Delta cr} = 2.0 n A_{ps} f_{pd}$$

Example 3.10

For the post-tensioned beam of Ex. 3.9, the cable force at midspan after elastic losses is $P_i = 151.7$ kips. The 28 day concrete strength is $f'_c = 6000$ lbf/in^2. The superimposed dead load moment is $M_D = 800$ in-kips. Determine the loss of prestressing force caused by creep.

Solution

From ACI Sec. 8.5, the modulus of elasticity of the concrete at 28 days is

$$E_c = 57{,}000\sqrt{f'_c}$$

$$= 57{,}000\sqrt{6000 \dfrac{\text{lbf}}{\text{in}^2}}$$

$$= 4{,}415{,}000 \text{ lbf/in}^2$$

The modular ratio at 28 days is

$$n = \frac{E_p}{E_c}$$

$$= \frac{28 \times 10^6 \dfrac{\text{lbf}}{\text{in}^2}}{4{,}415{,}000 \dfrac{\text{lbf}}{\text{in}^2}}$$

$$= 6.34$$

The concrete stress at the level of the centroid of the prestressing tendons, after elastic shortening and allowing for sustained dead load, is

$$f_{pd} = P_i\left(\frac{1}{A_g} + \frac{e^2}{I_g}\right) - \frac{eM_G}{I_g} - \frac{eM_D}{I_g}$$

$$= (151.7 \text{ kips})\left(\frac{1}{360 \text{ in}^2} + \frac{(10 \text{ in})^2}{27{,}000 \text{ in}^4}\right)$$

$$- \frac{(10 \text{ in})(506 \text{ in-kips})}{27{,}000 \text{ in}^4} - \frac{(10 \text{ in})(800 \text{ in-kips})}{27{,}000 \text{ in}^4}$$

$$= 0.499 \text{ kips/in}^2$$

The loss of prestressing force caused by creep is

$$P_{\Delta cr} = 1.6 n A_{ps} f_{pd}$$

$$= (1.6)(6.34)(0.765 \text{ in}^2)\left(0.499 \dfrac{\text{kips}}{\text{in}^2}\right)$$

$$= 3.9 \text{ kips}$$

The force in the cable after friction, elastic and creep losses is

$$P = 151.7 \text{ kips} - 3.9 \text{ kips}$$
$$= 147.8 \text{ kips}$$

---

[1]Zia, P., et al., 1979 (See References and Codes)
[2]Portland Cement Association, 2008 (See References and Codes)

Shrinkage Losses

The shrinkage of a prestressed beam produces a shortening of the beam with a corresponding loss of prestress. The basic shrinkage strain is given by[3,4]

$$\epsilon_{sh} = 8.2 \times 10^{-6} \text{ in/in}$$

Allowing for the ambient relative humidity, H, and the ratio of the member's volume to surface area, A_g/p_{cp}, the shrinkage loss for a pretensioned member is

$$P_{\Delta sh} = A_{ps}\epsilon_{sh}E_p\left(1 - \frac{0.06A_g}{p_{cp}}\right)(100 - H)$$

For a post-tensioned member, with transfer after some shrinkage has already occurred, the shrinkage loss is

$$P_{\Delta sh} = K_{sh}A_{ps}\epsilon_{sh}E_p\left(1 - \frac{0.06A_g}{p_{cp}}\right)(100 - H)$$

Example 3.11

The post-tensioned beam of Ex. 3.10 is located in an area with an ambient relative humidity of 60%, and transfer is effected 10 days after the completion of curing, giving a value[5] of 0.73 for K_{sh}. Determine the loss of prestressing force caused by shrinkage.

Solution

The shrinkage loss is

$$\begin{aligned}P_{\Delta sh} &= K_{sh}A_{ps}\epsilon_{sh}E_p\left(1 - \frac{0.06A_g}{p_{cp}}\right)(100 - H) \\ &= (0.73)(0.765 \text{ in}^2)\left(8.2 \times 10^{-6} \frac{\text{in}}{\text{in}}\right) \\ &\quad \times \left(28 \times 10^3 \frac{\text{kips}}{\text{in}^2}\right) \\ &\quad \times \left(1 - \frac{(0.06 \text{ in}^{-1})(360 \text{ in}^2)}{84 \text{ in}}\right)(100 - 60) \\ &= 3.8 \text{ kips}\end{aligned}$$

The force in the cable after friction, elastic, creep, and shrinkage losses is

$$\begin{aligned}P &= 147.8 \text{ kips} - 3.8 \text{ kips} \\ &= 144 \text{ kips}\end{aligned}$$

Relaxation Losses

Relaxation is the reduction in tensile force that occurs in a prestressing tendon under sustained tensile strain. The loss in prestress depends on the tendon properties, the initial force in the tendon, and the losses caused by creep, shrinkage, and elastic shortening. The relaxation loss is given by[5,3]

$$P_{\Delta re} = \left(A_{ps}K_{re} - J(P_{\Delta cr} + P_{\Delta sh} + P_{\Delta el})\right)C$$

Example 3.12

For the post-tensioned beam of Ex. 3.11, the values of the relevant parameters are[4]

$$K_{re} = 5 \text{ kips/in}^2$$
$$J = 0.04$$
$$C = 0.09$$

Determine the loss of prestressing force caused by relaxation.

Solution

The loss caused by relaxation is

$$\begin{aligned}P_{\Delta re} &= \left(A_{ps}K_{re} - J(P_{\Delta cr} + P_{\Delta sh} + P_{\Delta el})\right)C \\ &= \left(\begin{array}{c}(0.765 \text{ in}^2)\left(5 \frac{\text{kips}}{\text{in}^2}\right) \\ -(0.04)(3.9 \text{ kips} + 3.8 \text{ kips} + 2.3 \text{ kips})\end{array}\right) \\ &\quad \times (0.9) \\ &= 3.1 \text{ kips}\end{aligned}$$

The force in the cable after friction, elastic, creep, shrinkage, and relaxation losses is

$$\begin{aligned}P_e &= P - P_{\Delta re} \\ &= 144 \text{ kips} - 3.1 \text{ kips} \\ &= 140.9 \text{ kips}\end{aligned}$$

4. COMPOSITE CONSTRUCTION

Nomenclature

A_c	area of precast surface or area of contact surface for horizontal shear	in^2
A_g	gross area of precast girder	in^2
A_v	area of ties within a distance s	in^2
A_{vf}	area of friction reinforcement	in
b	effective flange width	in
b_f	actual flange width	in
b_v	width of girder at contact surface	in
b_w	width of girder web	in
b/n	transformed flange width	in
d	distance from extreme compression fiber to centroid of prestressed reinforcement (not less than $0.8h$)	in
e	eccentricity of prestressing force	in
E_f	modulus of elasticity of flange concrete	kips/in^2

---

[3]Zia, P., et al., 1979 (See References and Codes)
[4]Portland Cement Association, 2008 (See References and Codes)
[5]Ibid.

E_w	modulus of elasticity of precast girder concrete	kips/in^2
f_b	stress in the bottom fiber of the composite section	lbf/in^2
f_t	stress in the top fiber of the composite section	lbf/in^2
h	depth of composite section	in
h_f	depth of flange of the composite section	in
h_w	depth of precast girder	in
I_c	moment of inertia of composite section	in^4
I_g	moment of inertia of gross concrete section about centroidal axis, neglecting reinforcement	in^4
l	span length	ft
M	applied bending moment	ft-kips
M_F	moment caused by the flange concrete	ft-kips
M_G	moment caused by the precast girder self-weight	ft-kips
M_L	moment caused by the superimposed load	ft-kips
n	modular ratio, E_w/E_f	–
P_e	force in the prestressing tendons at service loads after allowance for all losses	kips
P_Δ	total loss of prestress	kips
s	spacing of ties	in
s_b	spacing of precast girders	ft
S_{cb}	section modulus at bottom of composite section	in^3
S_{ct}	section modulus at top of composite section	in^3
S_g	section modulus at bottom of precast girder	in^3
V_{nh}	nominal horizontal shear strength	kips
V_u	factored shear force at section	kips
w_F	weight of concrete flange	lbf/ft
w_G	weight of precast girder	lbf/ft
w_L	superimposed load	lbf/ft
$\bar{y}_c$	distance from neutral axis of transformed composite section to bottom of precast girder	in

Symbols

λ	correlation factor related to unit weight of concrete	–
μ	coefficient of friction	–
ρ_v	ratio of tie reinforcement area to area of contact surface, A_v/b_vs	–
ϕ	strength reduction factor	–

General Considerations

As shown in Fig. 3.7, a composite beam consists of a concrete flange acting integrally with a precast prestressed concrete girder to resist the applied loads. Transfer of horizontal shear at the interface of the flange and the girder is required to provide composite action between the two components. The compressive stress in the concrete flange is not uniform over the width of the flange, and an effective width is assumed in calculations.

As shown in Fig. 3.7, the effective width of the flange is defined by ACI Sec. 8.12 as the least of

- $l/4$
- $b_w + 16h_f$
- s_b

Figure 3.7 Composite Section

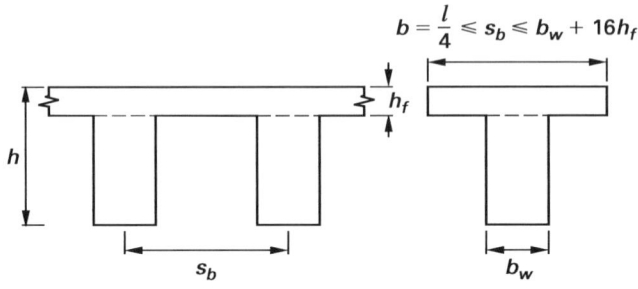

For service load design, composite section properties are calculated by assuming that plane sections remain plane, and that strain varies linearly with distance from the neutral axis of the composite section. When the 28 day compressive strengths of the flange and the precast member are different, the flange area is transformed into an equivalent area corresponding to the properties of the precast member. This equivalent area is obtained by dividing the effective slab width by the modular ratio n.

As shown in Fig. 3.8, the flange transformed area is

$$A_f = \frac{bh_f}{n}$$

Figure 3.8 Transformed Section

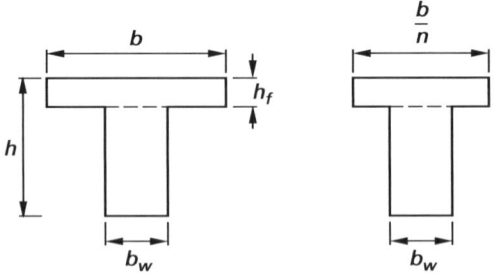

The bending moment applied to the composite section produces a stress in the bottom fiber of the precast girder, given by

$$f_b = \frac{M}{S_{cb}}$$

$$S_{cb} = \frac{I_c}{\bar{y}_c}$$

The stress produced in the top of the flange by the bending moment applied to the composite section is

$$f_t = \frac{M}{nS_{ct}}$$

$$S_{ct} = \frac{I_c}{h - \bar{y}_c}$$

Example 3.13

A composite beam has an effective span of $l = 30$ ft.

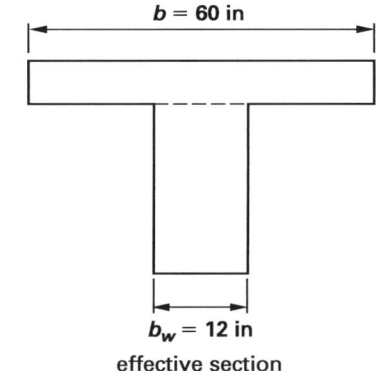

effective section

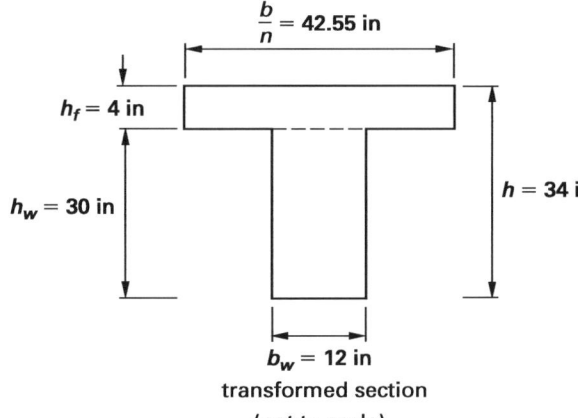

transformed section
(not to scale)

The precast, prestressed girders have a 28 day concrete strength of 6000 lbf/in² and are spaced at $s_b = 5$ ft centers. The flange has a 28 day concrete strength of 3000 lbf/in². Determine the transformed section properties of the composite section.

Solution

The modulus of elasticity of the girder and flange concrete is obtained from ACI Sec. 8.5 as

$$E_w = 57{,}000\sqrt{f'_c} = 57{,}000\sqrt{6000\ \frac{\text{lbf}}{\text{in}^2}}$$

$$= 4{,}415{,}000\ \text{lbf/in}^2$$

$$E_f = 57{,}000\sqrt{f'_c} = 57{,}000\sqrt{3000\ \frac{\text{lbf}}{\text{in}^2}}$$

$$= 3{,}122{,}000\ \text{lbf/in}^2$$

The modular ratio is

$$n = \frac{E_w}{E_f}$$

$$= \frac{4{,}415{,}000\ \frac{\text{lbf}}{\text{in}^2}}{3{,}122{,}000\ \frac{\text{lbf}}{\text{in}^2}}$$

$$= 1.41$$

The effective flange width is limited to the lesser of

$$b = \frac{l}{4}$$

$$= \frac{(30\ \text{ft})\left(12\ \frac{\text{in}}{\text{ft}}\right)}{4}$$

$$= 90\ \text{in}$$

$$b = b_w + 16h_f$$

$$= 12\ \text{in} + (16)(4\ \text{in})$$

$$= 76\ \text{in}$$

$$b = s_b$$

$$= 60\ \text{in} \quad [\text{governs}]$$

The transformed flange width is

$$\frac{b}{n} = \frac{60\ \text{in}}{1.41}$$

$$= 42.55\ \text{in}$$

The relevant properties of the precast girder are

$$A_g = 360\ \text{in}^2$$

$$I_g = 27{,}000\ \text{in}^4$$

$$S_g = 1800\ \text{in}^3$$

The relevant properties of the flange are

$$A_f = 170\ \text{in}^2$$

$$I_f = 227\ \text{in}^4$$

The properties of the transformed section are shown in Table 3.1.

Table 3.1 Transformed Section Moments and Centroids

part	A (in²)	y (in)	I (in⁴)	Ay (in³)	Ay^2 (in⁴)
girder	360	15	27,000	5400	81,000
flange	170	32	227	5440	174,080
total	530	–	27,227	10,840	255,080

The height of the centroid of the composite section is

$$\bar{y}_c = \frac{\sum Ay}{\sum A}$$
$$= \frac{10{,}840 \text{ in}^3}{530 \text{ in}^2}$$
$$= 20.45 \text{ in}$$

The moment of inertia of the composite section is

$$I_c = \sum I + \sum Ay^2 + \bar{y}_c^2 \sum A - 2\bar{y}\sum Ay$$
$$= 27{,}227 \text{ in}^4 + 255{,}080 \text{ in}^4 + (20.45 \text{ in})^2(530 \text{ in}^2)$$
$$\quad - (2)(20.45 \text{ in})(10{,}840 \text{ in}^3)$$
$$= 60{,}598 \text{ in}^4$$

The section modulus of the composite section referred to the top of the flange is

$$S_{ct} = \frac{I_c}{h - y_c}$$
$$= \frac{60{,}598 \text{ in}^4}{34 \text{ in} - 20.45 \text{ in}}$$
$$= 4472 \text{ in}^3$$

The section modulus of the composite section referred to the bottom of the girder is

$$S_{cb} = \frac{I_c}{\bar{y}_c}$$
$$= \frac{60{,}598 \text{ in}^4}{20.45 \text{ in}}$$
$$= 2963 \text{ in}^3$$

Shear Connection

Transfer of the horizontal shear force across the interface is necessary to provide full composite action. This may be effected by intentionally roughening the top surface of the precast girder and providing reinforcement to tie the two components together. ACI Sec. 17.5.3 specifies that the factored shear force at a section must not exceed

$$V_u = \phi V_{nh} \quad \text{[ACI 17-1]}$$

When the interface is intentionally roughened, ACI Sec. 17.5.3.1 specifies that the nominal horizontal shear strength must not be greater than

$$V_{nh} = 80 b_v d$$

When the interface is smooth, with the minimum ties required by ACI Sec. 11.4.6.3 provided across the interface, ACI Sec. 17.5.3.2 specifies that the nominal horizontal shear strength must not be greater than

$$V_{nh} = 80 b_v d$$

When the interface is roughened to 0.25 in amplitude, with the minimum ties required by ACI Sec. 11.4.6.3 provided across the interface, ACI Sec. 17.5.3.3 specifies that the nominal horizontal shear strength is

$$V_{nh} = (260 + 0.6\rho_v f_y)\lambda b_v d$$
$$\leq 500 b_v d$$

The correction factor related to the unit weight of concrete is given by ACI Sec. 11.6.4.3 as

$$\lambda = 1.0 \quad \text{[normal-weight concrete]}$$
$$= 0.85 \quad \text{[sand-lightweight concrete]}$$
$$= 0.75 \quad \text{[all lightweight concrete]}$$

When the factored shear force exceeds $\phi(500 b_v d)$, ACI Sec. 17.5.3.4 requires that the design shall be based on the shear-friction method given in ACI Sec. 11.6.4, with the nominal horizontal shear strength given by

$$V_{nh} = A_{vf} f_y \mu \quad \text{[ACI 11-25]}$$
$$\leq 0.2 f'_c A_c$$
$$\leq 800 A_c$$

The coefficient of friction is given by ACI Sec. 11.6.4.3 as

$$\mu = 1.0\lambda \quad \text{[interface roughened to an amplitude of 0.25 in]}$$
$$= 0.6\lambda \quad \text{[interface not roughened]}$$

In accordance with ACI Sec. 17.6.1, the tie spacing must not exceed four times the least dimension of the supported element, or 24 in.

Example 3.14

The composite section of Ex. 3.13 has a factored shear force of $V_u = 100$ kips at the critical section. Normal-weight concrete is used for both the prestressed girder and the flange. Determine the required tie reinforcement at the interface, which is intentionally roughened to an amplitude of 0.25 in. If ties are needed, use 12 in spacing.

Solution

From ACI Sec. 17.5.2, the effective depth of the prestressing tendons may be taken as

$$d = 0.8h$$
$$= (0.8)(34 \text{ in})$$
$$= 27.2 \text{ in}$$

The limiting value of the factored shear force for the application of the shear-friction method is given by ACI Sec. 17.5.3.4 as

$$500\phi b_v d = \frac{\left(500 \ \frac{\text{lbf}}{\text{in}^2}\right)(0.75)(12 \text{ in})(27.2 \text{ in})}{1000 \ \frac{\text{lbf}}{\text{kip}}}$$

$$= 122 \text{ kips}$$
$$> V_u$$

ACI Sec. 17.5.3.3 applies, and

$$V_u = (100 \text{ kips})\left(1000 \ \frac{\text{lbf}}{\text{kip}}\right)$$
$$= 100{,}000 \text{ lbf}$$
$$= \phi(260 + 0.6\rho_v)\lambda b_v d$$
$$= (0.75)\left(260 \ \frac{\text{lbf}}{\text{in}^2} + 0.6\rho_v\left(60{,}000 \ \frac{\text{lbf}}{\text{in}^2}\right)\right)$$
$$\quad \times (1.0)(12 \text{ in})(27.2 \text{ in})$$
$$\rho_v = 0.0041$$

The required area of vertical ties at a spacing of 12 in is

$$A_v = \rho_v b_v s$$
$$= (0.0041)(12 \text{ in})(12 \text{ in})$$
$$= 0.59 \text{ in}^2$$

Provide no. 5 grade 60 ties (two legs) at a spacing of 12 in to give an area of

$$A_v = 0.62 \text{ in}^2$$
$$> 0.59 \text{ in}^2 \quad [\text{satisfactory}]$$

Design for Flexure

Composite beams may be constructed using either of two methods, shored construction or unshored construction. In shored construction, the precast girder is propped before casting the flange, in order to eliminate stresses in the precast girder caused by the deposited concrete. On removal of the props, the weight of the flange, additional superimposed dead load, and live loads are supported by the composite section. In unshored construction, the flange is cast without propping the precast girder. The precast girder alone must be adequate to support the weight of the flange and all other loads that are applied before the flange concrete has attained its design strength. The composite section supports additional superimposed dead load and live loads.

For both shored and unshored construction, in accordance with ACI Sec. 17.2.4, the design of the composite section for ultimate loads is identical.

Example 3.15

The precast prestressed girder of the composite section described in Ex. 3.13 has a final prestressing force of $P_e = 140$ kips, with an eccentricity of $e = 10$ in at midspan. The loss of prestress may be assumed to occur before the flange is cast. The precast, prestressed girders have a span of $l = 30$ ft, and are spaced at $s_b = 5$ ft centers. The composite section has a span of $l = 30$ ft. The superimposed applied load is $w_L = 1000$ lbf/ft, and the precast section is not propped. Determine the stresses at midspan in the composite section.

Solution

The self-weight bending moment at the midspan of the precast girder is

$$M_G = \frac{w_G l^2}{8}$$
$$= \frac{\left(375 \ \frac{\text{lbf}}{\text{ft}}\right)(30 \text{ ft})^2 \left(12 \ \frac{\text{in}}{\text{ft}}\right)}{(8)\left(1000 \ \frac{\text{lbf}}{\text{kip}}\right)}$$
$$= 506 \text{ in-kips}$$

The bending moment produced at midspan by the weight of the flange concrete is

$$M_F = \frac{w_F l^2}{8}$$
$$= \frac{\left(250 \ \frac{\text{lbf}}{\text{ft}}\right)(30 \text{ ft})^2 \left(12 \ \frac{\text{in}}{\text{ft}}\right)}{(8)\left(1000 \ \frac{\text{lbf}}{\text{kip}}\right)}$$
$$= 338 \text{ in-kips}$$

The bending moment produced at midspan by the applied superimposed load is

$$M_L = \frac{w_L l^2}{8}$$
$$= \frac{\left(1000 \ \frac{\text{lbf}}{\text{ft}}\right)(30 \text{ ft})^2 \left(12 \ \frac{\text{in}}{\text{ft}}\right)}{(8)\left(1000 \ \frac{\text{lbf}}{\text{kip}}\right)}$$
$$= 1350 \text{ in-kips}$$

The stress in the bottom fiber of the prestressed girder caused by the final prestressing force is

$$f_b = P_e\left(\frac{1}{A_g} + \frac{e}{S_g}\right)$$
$$= (140 \text{ kips})\left(\frac{1}{360 \text{ in}^2} + \frac{10 \text{ in}}{1800 \text{ in}^3}\right)$$
$$= 1.167 \text{ kips/in}^2 \quad [\text{compression}]$$

The stress in the bottom fiber of the prestressed girder caused by its self-weight and the weight of the flange concrete is

$$f_b = \frac{M_G + M_F}{S_g}$$
$$= \frac{506 \text{ in-kips} + 338 \text{ in-kips}}{1800 \text{ in}^3}$$
$$= 0.469 \text{ kips/in}^2 \quad [\text{tension}]$$

The stress in the bottom fiber of the prestressed girder caused by the applied superimposed load on the composite section is

$$f_b = \frac{M_L}{S_{cb}}$$
$$= \frac{1350 \text{ in-kips}}{2963 \text{ in}^3}$$
$$= 0.456 \text{ kips/in}^2 \quad [\text{tension}]$$

The final stress in the bottom fiber of the prestressed girder caused by all loads is

$$f_b = 1.167 \frac{\text{kips}}{\text{in}^2} - 0.469 \frac{\text{kips}}{\text{in}^2} - 0.456 \frac{\text{kips}}{\text{in}^2}$$
$$= 0.242 \text{ kips/in}^2 \quad [\text{compression}]$$

The final stress in the top fiber of the flange caused by all loads is

$$f_t = \frac{M_L}{nS_{ct}}$$
$$= \frac{1350 \text{ in-kips}}{(1.41)(4472 \text{ in}^3)}$$
$$= 0.214 \text{ kips/in}^2$$

5. LOAD BALANCING PROCEDURE

Nomenclature

A_g	area of concrete section	in^2
f_c	concrete stress	lbf/in^2
g	sag of prestressing tendon	in
l	span length	ft
M_B	balancing load moment caused by w_B	in-lbf
M_o	out-of-balance moment caused by w_o, $M_W - M_B$	in-lbf
M_W	applied load moment caused by w_W	in-lbf
P	prestressing force	kips
S	section modulus	in^3
w_B	balancing-load produced by prestressing tendon	kips/ft or kips
w_o	out-of-balance load, $w_W - w_B$	kips/ft or kips
w_W	superimposed applied load	kips/ft or kips

The load balancing technique uses the upward pressure of the prestressing tendons on a prestressed beam to balance the downward pressure of the applied loads. As shown in Fig. 3.9, a tendon with a parabolic profile produces a uniformly distributed upward pressure on a beam of w_B. Equating moments in the free-body diagram gives

$$\frac{w_B l^2}{8} = M_B = Pg$$

The balancing load is

$$w_B = \frac{8Pg}{l^2}$$

The applied load w_W on the beam produces a moment

$$M_W = \frac{w_W l^2}{8}$$

When the applied load w_W equals the balancing load w_B, the moment produced by the applied load is

$$M_W = \frac{w_B l^2}{8} = Pg$$

Figure 3.9 Load Balancing Concepts

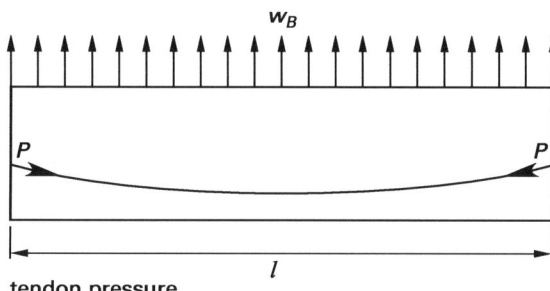
tendon pressure

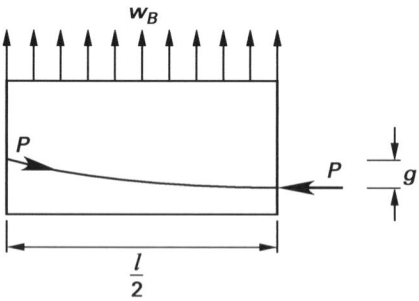
free-body diagram

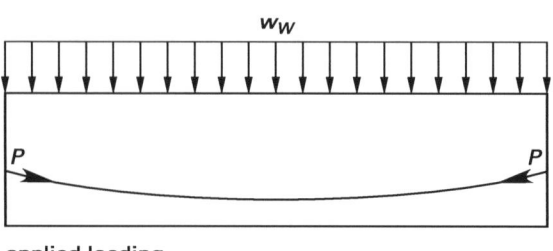
applied loading

Figure 3.10 Alternative Tendon Profiles

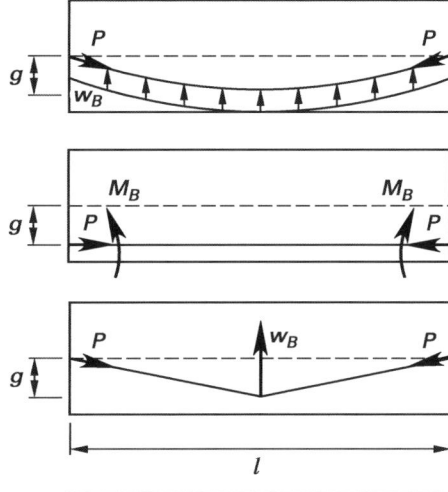

When the applied load w_W equals the balancing load w_B, there is no moment on the beam. At any section, there is a constant compressive stress of

$$f_c = \frac{P}{A_g}$$

If the downward load is not fully balanced by the upward force, the out-of-balance load on the beam is

$$w_o = w_W - w_B$$

This produces an out-of-balance moment of

$$M_o = M_W - M_B$$

The stress in the concrete is then

$$f_c = \frac{P}{A_g} \pm \frac{M_o}{S}$$

The balancing loads produced by other tendon profiles may be derived,[6,7,8] and are shown in Fig. 3.10. Deflections also may be readily calculated using this method.

Example 3.16

The beam shown in the illustration is prestressed with a tendon having a parabolic profile and a final prestressing force of $P_e = 140$ kips. Determine the uniformly distributed load, including the weight of the beam, which will exactly balance the prestressing force, and determine the resulting stress in the beam. Calculate the stresses in the beam if the uniformly distributed load is increased by 20%.

Solution

The sag of the tendon is

$$g = 10 \text{ in}$$

The uniformly distributed load required to exactly balance the prestressing force is

$$\begin{aligned} w_W &= \frac{8P_e g}{l^2} \\ &= \frac{(8)(140 \text{ kips})(10 \text{ in})}{(30 \text{ ft})^2 \left(12 \frac{\text{in}}{\text{ft}}\right)} \\ &= 1.037 \text{ kips/ft} \end{aligned}$$

The uniform compressive stress throughout the beam is then

$$\begin{aligned} f_c &= \frac{P}{A_g} \\ &= \frac{140 \text{ kips}}{360 \text{ in}^2} \\ &= 0.389 \text{ kips/in}^2 \quad \text{[compression]} \end{aligned}$$

---

[6]Lin, T.Y., 1963 (See References and Codes)
[7]Prestressed Concrete Institute, 2004 (See References and Codes)
[8]Freyermuth, C.L., and Schoolbred, R.A., 1967 (See References and Codes)

Illustration for Example 3.16

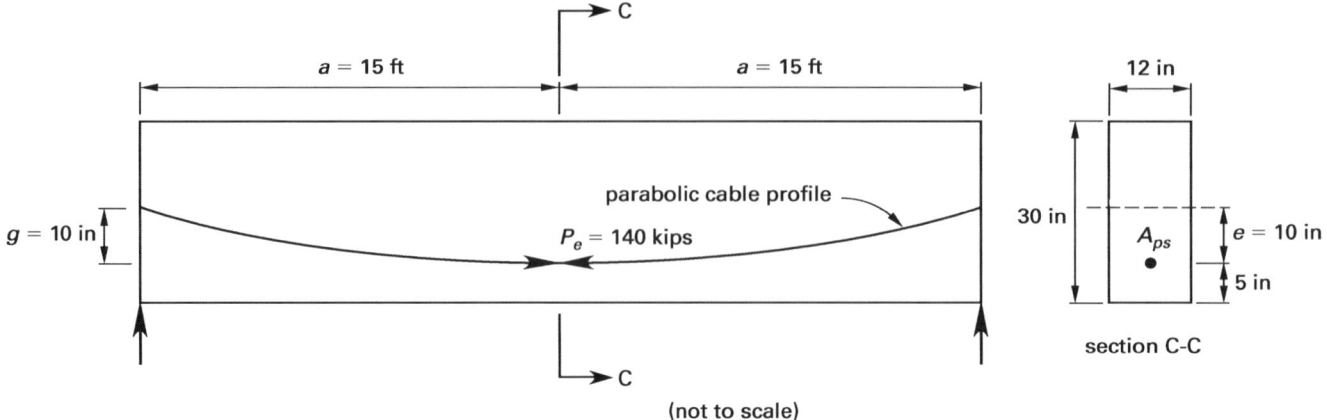

(not to scale)

An increase of 20% in w_W produces an out-of-balance load of

$$w_o = (0.2)\left(1.037 \; \frac{\text{kips}}{\text{ft}}\right)$$
$$= 0.207 \; \text{kips/ft}$$

The out-of-balance moment produced at midspan by w_o is

$$M_o = \frac{w_o l^2}{8}$$
$$= \frac{\left(0.207 \; \frac{\text{kips}}{\text{ft}}\right)(30 \; \text{ft})^2 \left(12 \; \frac{\text{in}}{\text{ft}}\right)}{8}$$
$$= 279 \; \text{in-kips}$$

The resultant stresses at mid-span are

$$f_b = f_c - \frac{M_o}{S}$$
$$= 0.389 \; \frac{\text{kips}}{\text{in}^2} - \frac{279 \; \text{in-kips}}{1800 \; \text{in}^3}$$
$$= 0.234 \; \text{kips/in}^2 \quad [\text{compression}]$$

$$f_b = f_c + \frac{M_o}{S}$$
$$= 0.389 \; \frac{\text{kips}}{\text{in}^2} + \frac{279 \; \text{in-kips}}{1800 \; \text{in}^3}$$
$$= 0.544 \; \text{kips/in}^2 \quad [\text{compression}]$$

6. CONCORDANT CABLE PROFILE

Nomenclature

A_g	area of concrete section transfer	in^2
e	eccentricity of prestressing force	in
e'	resultant cable eccentricity	in
EI	flexural stiffness of compression member	in^2-lbf
m	moment produced by unit value of the redundant	in-kips
M_p	primary moment produced by prestressing force	in-kips
M_r	resultant moment produced by prestressing force and secondary effects, $Pe + M_s$	in-kips
M_s	secondary moment produced by secondary effects	in-kips
P	prestressing force	kips
R	reaction, support restraint	kips
S	section modulus	in^3

Symbols

δ_p	deflection at midspan	in

As shown in Fig. 3.11, prestressing a simply supported beam with a tendon having a constant eccentricity produces a uniform bending moment in the beam that has no effect on the support reactions. This bending moment, known as the *primary bending moment*, is

$$M_p = Pe$$

The primary bending moment is shown in Fig. 3.11, using the sign convention that moment causing compressive stress in the bottom fiber of the beam is positive. The primary bending moment produces an upward deflection in the beam. The deflection at midspan is

$$\delta_p = \frac{M_p(2l)^2}{8EI}$$
$$= \frac{Pe(2l)^2}{8EI}$$

Introducing an additional support 3 to the beam at midspan produces an indeterminate two-span continuous beam. Upward deflection of the beam at the central support is now restrained by the support. The support reaction is obtained from Fig. 3.10 as

$$R_3 = \frac{48\delta_p EI}{(2l)^3}$$
$$= \frac{6Pe}{2l}$$

This reaction produces a secondary bending moment in the beam, with a maximum value at the central support of

$$M_s = \frac{R_3(2l)}{4}$$
$$= \frac{3Pe}{2}$$

Superimposing the primary and secondary moment diagrams gives the combined moment diagram shown in Fig. 3.11. The value of the resultant moment at the central support is

$$M_r = M_p - M_s$$
$$= Pe - \frac{3Pe}{2}$$
$$= \frac{-Pe}{2}$$

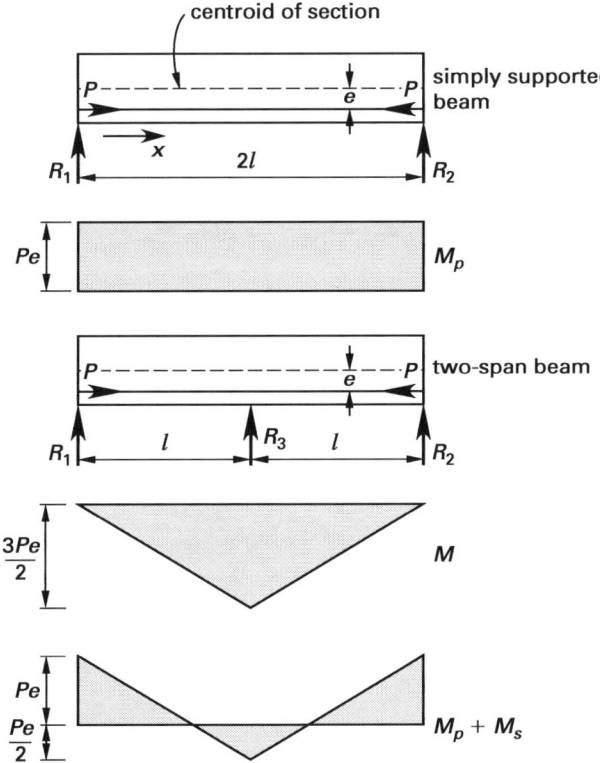

Figure 3.11 Secondary Effects

In general, the equation for the resultant moment is

$$M_r = Pe - \frac{3Pex}{2l}$$

The effective tendon eccentricity is

$$e' = \frac{M_r}{P}$$
$$= e - \frac{3ex}{2l}$$

Hence, a tendon with an initial eccentricity of e' produces no secondary effects in the member and no support reactions. It is termed the *concordant cable*.

External loads applied to a continuous beam on unyielding supports produce a bending moment diagram with magnitude M_x at any section x. To some convenient scale, this is the concordant profile for this system of loads because the effective eccentricity is

$$e' = \frac{M_x}{P}$$

In addition, a concordant profile may be modified as shown in Fig. 3.12, using a linear transformation by varying the location of the tendon at interior supports, without changing the resultant moment.

Figure 3.12 Concordant Cable Profile

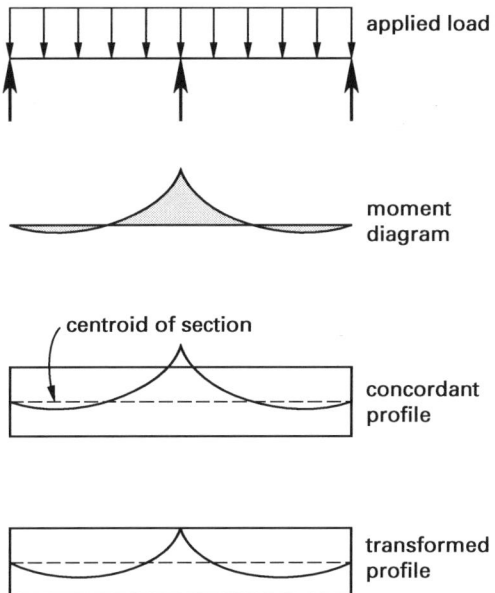

Example 3.17

The beam shown in the illustration is prestressed with a tendon having a final prestressing force of $Pe = 140$ kips. Determine the eccentricity of the concordant cable at support 3 to balance a uniformly distributed load of 1.0 kip/ft applied to the beam, and determine the resulting stress in the beam.

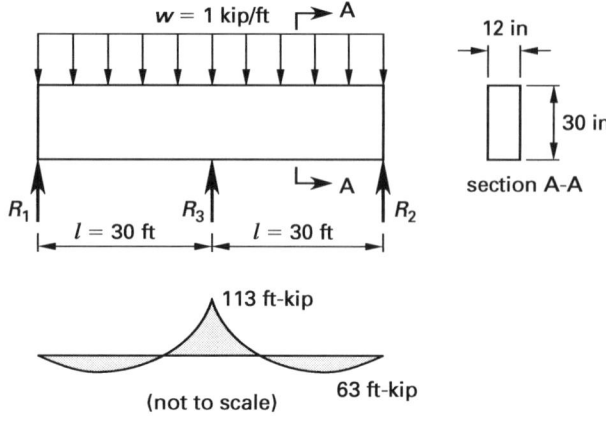

Solution

The bending moment produced at support 3 by the distributed load is

$$M_3 = 0.125 wl^2$$
$$= (0.125)\left(1\ \frac{\text{kip}}{\text{ft}}\right)(30\ \text{ft})^2 \left(12\ \frac{\text{in}}{\text{ft}}\right)$$
$$= 1350\ \text{in-kips}$$

The eccentricity of the concordant cable at support 3 is

$$e' = \frac{M_3}{Pe}$$
$$= \frac{1350\ \text{in-kips}}{140\ \text{kips}}$$
$$= 9.6\ \text{in} \quad \text{[above the centroid of the section]}$$

The resultant stresses at support 3 are

$$f_b = \frac{Pe}{A_g} - \frac{M_3}{S}$$
$$= \frac{140\ \text{kips}}{360\ \text{in}^2} - \frac{1350\ \text{in-kips}}{1800\ \text{in}^3}$$
$$= -0.361\ \text{kips/in}^2 \quad \text{[tension]}$$

$$f_t = \frac{Pe}{A_g} + \frac{M_3}{S}$$
$$= \frac{140\ \text{kips}}{360\ \text{in}^2} + \frac{1350\ \text{in-kips}}{1800\ \text{in}^3}$$
$$= 1.139\ \text{kips/in}^2 \quad \text{[compression]}$$

PRACTICE PROBLEMS

Problems 1–5 refer to the normal weight concrete prestressed beam shown in the illustration. The beam is simply supported over a span of 25 ft, and has a 28 day concrete strength of 6000 lbf/in^2. The area of the low-relaxation prestressing tendons provided is $A_{ps} = 0.612$ in^2, with a specified tensile strength of $f_{pu} = 270$ kips/in^2, a yield strength of $f_{py} = 243$ kips/in^2, and an effective stress of $f_{se} = 150$ kips/in^2 after all losses.

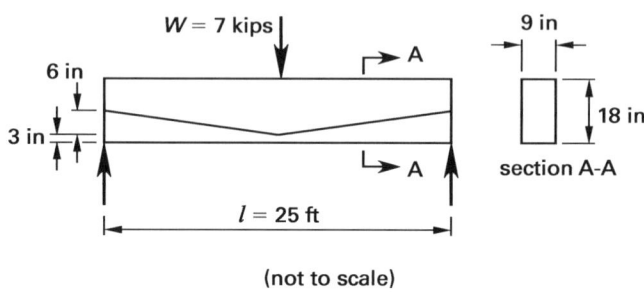

1. What is most nearly the cracking moment of the beam at midspan?

(A) 650.0 in-kips

(B) 863.0 in-kips

(C) 1108 in-kips

(D) 1324 in-kips

2. The post-tensioned unbonded beam has an area of 0.6 in^2 of grade 60 auxiliary reinforcement provided at a height of 3 in above the beam soffit. Which of the given statements is/are true?

I. The design flexural strength of the beam is 1718 in-kips.

II. The design flexural strength of the beam is 1431 in-kips.

III. The beam complies with ACI Sec. 18.8.2.

IV. The beam does not comply with ACI Sec. 18.8.2.

(A) I and III only

(B) I and IV only

(C) II and III only

(D) II and IV only

3. The beam supports a concentrated load of 7 kips at midspan. What is most nearly the prestressing force required in the tendons to exactly balance the applied load?

(A) 44 kips

(B) 88 kips

(C) 130 kips

(D) 180 kips

4. What is most nearly the resulting stress in the beam at midspan?

(A) 270 lbf/in^2

(B) 460 lbf/in^2

(C) 540 lbf/in^2

(D) 810 lbf/in^2

5. The prestressed beam of Prob. 3 supports an additional uniformly distributed load of 200 lbf/ft. What are most nearly the resulting stresses in the beam at midspan?

(A) $f_b = -115$ lbf/in^2, $f_t = 655$ lbf/in^2

(B) $f_b = 154$ lbf/in^2, $f_t = 926$ lbf/in^2

(C) $f_b = 425$ lbf/in^2, $f_t = 1190$ lbf/in^2

(D) $f_b = 695$ lbf/in^2, $f_t = 1470$ lbf/in^2

SOLUTIONS

1. The relevant properties of the beam are

$$A_g = bh$$
$$= (9 \text{ in})(18 \text{ in})$$
$$= 162 \text{ in}^2$$

$$S = \frac{A_g h}{6}$$
$$= \frac{(162 \text{ in}^2)(18 \text{ in})}{6}$$
$$= 486 \text{ in}^3$$

$$e = \frac{h}{2} - 3 \text{ in}$$
$$= \frac{18 \text{ in}}{2} - 3 \text{ in}$$
$$= 6 \text{ in}$$

$$R_b = \frac{1}{A_g} + \frac{e}{S}$$
$$= \frac{1}{162 \text{ in}^2} + \frac{6 \text{ in}}{486 \text{ in}^3}$$
$$= 0.0185 \text{ in}^{-2}$$

After allowance for all losses, the force in the prestressing tendons at service loads is

$$Pe = A_{ps} f_{se}$$
$$= (0.612 \text{ in}^2)\left(150 \ \frac{\text{kips}}{\text{in}^2}\right)$$
$$= 91.80 \text{ kips}$$

The modulus of rupture of the concrete is given by ACI Eq. (9-10) as

$$f_r = 7.5 \lambda \sqrt{f'_c}$$
$$= (7.5)(1.0)\sqrt{6000 \ \frac{\text{lbf}}{\text{in}^2}}$$
$$= 581 \text{ lbf/in}^2$$

The cracking moment is

$$M_{cr} = S(PeR_b + f_r)$$
$$= (486 \text{ in}^3)\begin{pmatrix}(91.80 \text{ kips})(0.0185 \text{ in}^{-2}) \\ + \left(0.581 \ \frac{\text{kips}}{\text{in}^2}\right)\end{pmatrix}$$
$$= 1108 \text{ in-kips}$$

The answer is (C).

2. Because the ratio of the effective stress to the tensile strength f_{se}/f_{pu} is greater than 0.5, the method of ACI Sec. 18.7.2 may be used. The ratio of prestressed reinforcement is

$$\rho_p = \frac{A_{ps}}{bd_p} = \frac{0.612 \text{ in}^2}{(9 \text{ in})(15 \text{ in})}$$
$$= 0.00453$$

For unbonded tendons and a span-to-depth ratio of 35 or less, the stress in the unbonded tendons at nominal strength is given by ACI Sec. 18.7.2 as

$$f_{ps} = f_{se} + 10 \frac{\text{kips}}{\text{in}^2} + \frac{f'_c}{100\rho_p} \quad [\text{ACI 18-4}]$$
$$= 150 \frac{\text{kips}}{\text{in}^2} + 10 \frac{\text{kips}}{\text{in}^2} + \frac{6 \frac{\text{kips}}{\text{in}^2}}{(100)(0.00453)}$$
$$= 173 \text{ kips/in}^2$$
$$< f_{py} \quad [\text{satisfactory}]$$
$$< f_{se} + 60 \text{ kips/in}^2 \quad [\text{satisfactory}]$$

The minimum required area of auxiliary reinforcement is specified by ACI Sec. 18.9.2 as

$$A_s = 0.004 A_{ct} \quad [\text{ACI 18-6}]$$
$$= (0.004)(9 \text{ in})(9 \text{ in})$$
$$= 0.324 \text{ in}^2$$
$$< 0.60 \text{ in}^2 \text{ provided} \quad [\text{satisfactory}]$$

Assuming full utilization of the auxiliary reinforcement, the depth of the stress block is

$$a = \frac{A_{ps}f_{ps} + A_s f_y}{0.85 f'_c b}$$
$$= \frac{(0.612 \text{ in}^2)\left(173 \frac{\text{kips}}{\text{in}^2}\right) + (0.6 \text{ in}^2)\left(60 \frac{\text{kips}}{\text{in}^2}\right)}{(0.85)\left(6 \frac{\text{kips}}{\text{in}^2}\right)(9 \text{ in})}$$
$$= 3.09 \text{ in}$$

The maximum depth of the stress block for a tension-controlled section is given by ACI Sec. 10.3.4 as

$$a_t = 0.375 \beta_1 d_p$$
$$= (0.375)(0.75)(15 \text{ in})$$
$$= 4.22 \text{ in}$$
$$> a$$

The section is tension-controlled and

$$\phi = 0.9$$

The nominal flexural strength of the section is

$$M_n = A_{ps}f_{ps}\left(d_p - \frac{a}{2}\right) + A_s f_y\left(d - \frac{a}{2}\right)$$
$$= (0.612 \text{ in}^2)\left(173 \frac{\text{kips}}{\text{in}^2}\right)\left(15 \text{ in} - \frac{3.09 \text{ in}}{2}\right)$$
$$+ (0.60 \text{ in}^2)\left(60 \frac{\text{kips}}{\text{in}^2}\right)\left(15 \text{ in} - \frac{3.09 \text{ in}}{2}\right)$$
$$= 1909 \text{ in-kips}$$

The design flexural strength of the beam is

$$\phi M_n = (0.9)(1909 \text{ in-kips})$$
$$= 1718 \text{ in-kips}$$

The cracking moment of the beam is determined in Prob. 1 as

$$M_{cr} = 1108 \text{ in-kips}$$
$$\frac{\phi M_n}{M_{cr}} = \frac{1718 \text{ in-kips}}{1108 \text{ in-kips}}$$
$$= 1.55$$
$$> 1.2$$

The beam complies with ACI Sec. 18.8.2.

The answer is (A).

3. The sag of the tendon is

$$g = 6 \text{ in}$$

The prestressing force required to exactly balance the applied load is obtained from Fig. 3.10 as

$$P = \frac{Wl}{4g}$$
$$= \frac{(7 \text{ kips})(300 \text{ in})}{(4)(6 \text{ in})}$$
$$= 87.5 \text{ kips} \quad (88 \text{ kips})$$

The answer is (B).

4. The uniform compressive stress throughout the beam is

$$f_c = \frac{P}{A_g} = \frac{87,500 \text{ lbf}}{162 \text{ in}^2}$$
$$= 540 \text{ lbf/in}^2$$

The answer is (C).

5. The moment produced at midspan by the additional distributed load is

$$M = \frac{wl^2}{8} = \frac{\left(200 \, \frac{\text{lbf}}{\text{ft}}\right)(25 \text{ ft})^2 \left(12 \, \frac{\text{in}}{\text{ft}}\right)}{8}$$
$$= 187,500 \text{ in-lbf}$$

The resultant stresses at midspan are

$$f_b = f_c - \frac{M}{S}$$
$$= 540 \, \frac{\text{lbf}}{\text{in}^2} - \frac{187,500 \text{ in-lbf}}{486 \text{ in}^3}$$
$$= 154 \text{ lbf/in}^2$$
$$f_t = f_c + \frac{M}{S}$$
$$= 540 \, \frac{\text{lbf}}{\text{in}^2} + \frac{187,500 \text{ in-lbf}}{486 \text{ in}^3}$$
$$= 926 \text{ lbf/in}^2$$

The answer is (B).

Structural Steel Design

1. Plastic Design 4-1
2. Eccentrically Loaded Bolt Groups 4-11
3. Eccentrically Loaded Weld Groups 4-19
4. Composite Beams 4-27
   Practice Problems 4-33
   Solutions 4-34

1. PLASTIC DESIGN

Nomenclature

A	gross area of member [ASD]	in^2
A_f	area of flange	in^2
A_g	gross area of member [LRFD]	in^2
BF	tabulated factor used to calculate the design flexural strength for unbraced lengths between L_p and L_r	kips
C_b	bending coefficient	–
C_c	column slenderness ratio separating elastic and inelastic buckling	–
C_m	factor relating actual moment diagram to an equivalent uniform moment diagram	–
D	dead load	kips or lbf
D	degree of indeterminacy of a structure	–
E	earthquake load	kips or lbf
E	modulus of elasticity of steel, 29,000 kips/in^2	–
F_a	allowable axial stress	kips/in^2
F_{cr}	critical stress	kips/in^2
F'_e	factored Euler critical stress tabulated in AISC Table 8	kips/in^2
F_y	yield stress	kips/in^2
g	stiffness ratio	–
H	horizontal force	kips or lbf
I	moment of inertia of section about centroidal axis	in^4
I_c	moment of inertia of column	in^4
I_g	moment of inertia of beam	in^4
K	effective-length factor	–
l	span length	ft
l_{cr}	critical unbraced segment length adjacent to a plastic hinge	in
L	live load due to occupancy	kips or lbf
L_b	unbraced segment length	in
L_c	maximum unbraced segment length at which the allowable bending stress may be taken to be $0.66F_y$	ft
L_p	limiting laterally unbraced length for full plastic bending capacity	ft or in
L_{pd}	limiting laterally unbraced length for plastic analysis	ft or in
L_r	limiting laterally unbraced length for inelastic lateral torsional buckling	ft or in
L_r	roof live load	kips or lbf
L_u	maximum unbraced segment length at which the allowable bending stress may be taken to be $0.60F_y$	ft
m_i	number of independent collapse mechanisms in a structure	–
M/M_p	end moment ratio, positive when the segment is bent in reverse curvature	–
M_m	maximum moment that can be resisted by the member in the absence of axial load	kips/in^2
M_{nx}	nominal flexural strength about the strong axis in the absence of axial load	kips
M_{ny}	nominal flexural strength about the weak axis in the absence of axial load	kips
M_p	plastic moment of resistance of a member, larger of the moments at the ends of the unbraced segment	ft-kips
M'_p	plastic moment of resistance modified for axial compression	ft-kips
M_s	bending moment produced by factored loads acting on the cut-back structure	ft-kips
M_{ux}	required flexural strength about the strong axis, including second-order effects	kips
M_{uy}	required flexural strength about the weak axis, including second-order effects	kips
M_y	moment at which the extreme fibers of a member yield in flexure	ft-kips
M_1	smaller moment at end of unbraced length of beam	ft-kips
M_2	larger moment at end of unbraced length of beam	ft-kips
M_1/M_2	positive when moments cause reverse curvature and negative for single curvature	–
p	number of possible hinge locations in a structure	–
P	axial force	kips or lbf

P	factored axial load	kips
P_{cr}	maximum strength of an axially loaded compression member	kips
P_e	Euler buckling load	kips
P_{max}	maximum load on a column	kips
P_n	nominal axial strength in the absence of bending moment	kips
P_u	required axial strength in compression	kips
P_y	plastic axial load	kips
r	radius of gyration of cross section	in
r_b	corresponding radius of gyration	in
r_T	radius of gyration of compression flange plus one-third of the compression portion of the web	in
r_x	radius of gyration of the member about its strong axis	in
r_y	radius of gyration of the member about its weak axis	in
R	rainwater or ice load	kips or lbf
S	elastic section modulus, M_y/F_y	ft-kips
S	snow load	kips or lbf
V	shear force	kips or lbf
W	applied force	kips or lbf
W	wind load	kips or lbf
Z	plastic section modulus, M_p/F_y	ft-kips

Symbols

θ	plastic hinge rotation	deg
λ	load factor	–
λ_c	column slenderness parameter	–
ϕ	resistance factor	–
ϕ_b	resistance factor for flexure, 0.90	–
ϕ_c	resistance factor for compression, 0.85	–
υ	shape factor, Z/S	–

General Considerations

The plastic method, or inelastic method, of structural analysis[1,2] is used to determine the maximum loads that a structure can support prior to collapse. The plastic analysis method has several advantages over the ASD and LRFD design techniques. The principal advantages of the plastic analysis technique are that it

- produces a more economical structure
- provides a simple and direct design technique
- accurately models the structure at ultimate loads
- realistically predicts the ultimate strength

Plastic Hinges

The plastic method of structural analysis is applicable to structures constructed with a ductile material possessing ideal elastic-plastic characteristics. As shown in Fig. 4.1, such a material initially exhibits a linear relationship between stress and strain, until the yield point

---
[1]Horne, M.R., and L.J. Morris, 1982 (See References and Codes)
[2]Neal, B.G., 1970 (See References and Codes)

is reached. After this point, the stress remains constant at the yield stress F_y, while the strain continues to increase indefinitely as plastic yielding of the material occurs. The plastic hinge formed at the location where yielding occurs has a plastic moment of resistance, M_p. Rotation continues at the hinge without any increase in resisting moment.

Figure 4.1 Elastic-Plastic Material

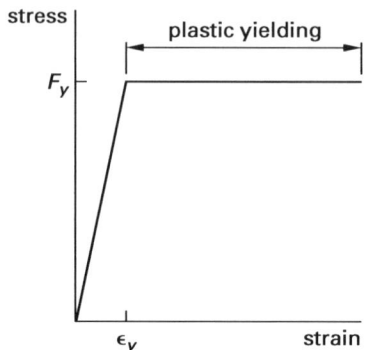

Shape Factor

As shown in Fig. 4.2, an increasing applied bending moment on a steel beam eventually causes the extreme fibers to reach the yield stress. The resisting moment developed in the beam is the yield moment given by

$$M_y = F_y S$$

Figure 4.2 Shape Factor Determination

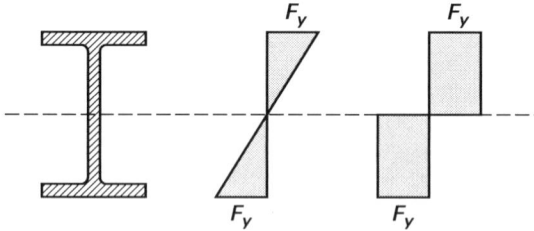

As the moment on the section continues to increase, the yielding at the extreme fibers progresses toward the equal area axis until finally the whole of the section has yielded. The resisting moment developed in the section is the plastic moment given by

$$M_p = F_y Z \quad \text{[AISC F2-1]}$$

The plastic section modulus Z is calculated as the arithmetic sum of the first moments of the area about the neutral axis. A plastic hinge has now formed in the section, and rotation of the hinge continues without increase in the resisting moment. The shape factor is defined as

$$v = \frac{Z}{S} = \frac{M_p}{M_s}$$
$$\approx 1.14 \quad \text{[for an I section]}$$
$$= 1.5 \quad \text{[for a solid rectangular section]}$$

Values of the plastic modulus are tabulated in the AISC Manual. To ensure that adequate plastic rotations can occur, AISC Apps. 1.2 and 1.3 specify that members must be adequately braced compact sections with a yield stress not exceeding 65 kips/in^2.

Example 4.1

A W12 × 65 section has 10 in × $\frac{1}{2}$ in plates welded to both flanges, as shown in the illustration. The yield stress of the W section and the plates is 50 kips/in^2. Determine the plastic moment of resistance and the shape factor for the combination section.

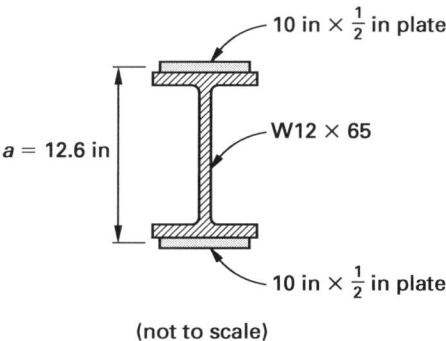

(not to scale)

Solution

From AISC Table 1-1, the properties of the W12 × 65 section are

$$d = 12.1 \text{ in}$$
$$A_W = 19.1 \text{ in}^2$$
$$S_W = 87.9 \text{ in}^3$$
$$I_W = 533 \text{ in}^4$$
$$Z_W = 96.8 \text{ in}^3$$

The area of each plate is

$$A_P = 5 \text{ in}^2$$

The distance between the centroids of the two plates is

$$a = 12.6 \text{ in}$$

Neglecting the moment of inertia of the plates about their individual centroidal axes, the moment of inertia of the combination section is

$$I_C = I_W + \frac{A_P a^2}{2} = 533 \text{ in}^4 + \frac{(5 \text{ in}^2)(12.6 \text{ in})^2}{2}$$
$$= 930 \text{ in}^4$$

The elastic section modulus of the combination section is

$$S_C = \frac{I_C}{\dfrac{d}{2} + 0.5 \text{ in}} = \frac{930 \text{ in}^4}{\dfrac{12.1 \text{ in}}{2} + 0.5 \text{ in}}$$
$$= 142 \text{ in}^3$$

The plastic section modulus of the combination section is

$$Z_C = Z_W + A_P a$$
$$= 96.8 \text{ in}^3 + (5 \text{ in}^2)(12.6 \text{ in})$$
$$= 160 \text{ in}^3$$

The plastic moment of resistance of the combination section is

$$M_p = F_y Z_C$$
$$= \frac{\left(50 \dfrac{\text{kips}}{\text{in}^2}\right)(160 \text{ in}^3)}{12 \dfrac{\text{in}}{\text{ft}}}$$
$$= 667 \text{ ft-kips}$$

The shape factor is

$$v = \frac{Z_C}{S_C} = \frac{160 \text{ in}^3}{142 \text{ in}^3}$$
$$= 1.13$$

Plastic Hinge Formation and Load Factors

A plastic hinge is formed in a structure as the bending moment at a specific location reaches the plastic moment of resistance at that location. As shown in Fig. 4.3, a fixed-ended beam supports a uniformly distributed service load, W.

The bending moments produced in the beam are shown in Fig. 4.3(a). The moments at the ends of the beam are twice that at the center.

As the load W is progressively increased to a value W', plastic hinges are formed simultaneously at both ends of the beam. The bending moments in the beam are as shown in Fig. 4.3(b).

As shown in Fig. 4.3(c), the system is now equivalent to a simply supported beam with an applied load W' and moments M_p at both ends. Progressively increasing the applied load causes the two plastic hinges to rotate, while the moments at both ends remain constant at the value M_p.

Figure 4.3 Formation of Plastic Hinges

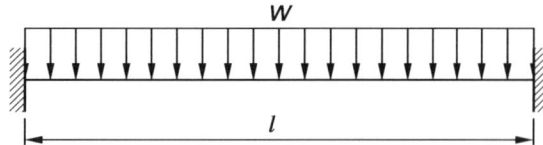
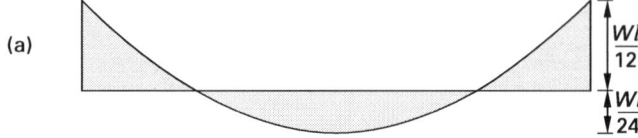
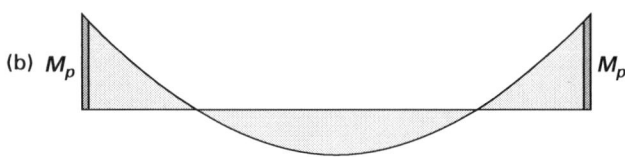
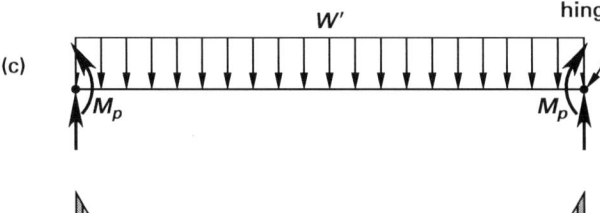
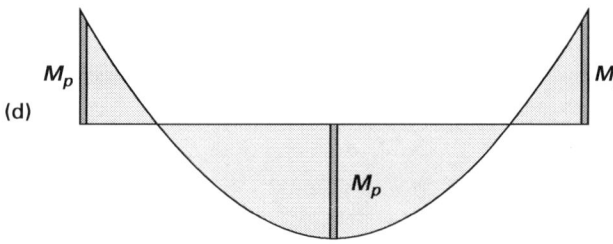

Finally, as the applied load is increased to the value λW, a third plastic hinge forms in the center of the span, giving the distribution of bending moment shown in Fig. 4.3(d).

The system is now an unstable mechanism shown in Fig. 4.3(e), and collapse occurs under the ultimate, or plastic-limit, load λW. Immediately before collapse, the system is statically determinate and the ultimate load may be calculated. Taking moments about the center of the span for the left half of the beam gives

$$2M_p = \left(\frac{\lambda W}{2}\right)\left(\frac{l}{2}\right) - \left(\frac{\lambda W}{2}\right)\left(\frac{l}{4}\right)$$

$$\lambda W = \frac{16 M_p}{l}$$

The ratio of the collapse load to the service load is

$$\frac{\lambda W}{W} = \lambda$$

AISC App. 1.1 specifies that plastic design is not permitted using ASD design methods.

AISC Sec. B2 specifies that load factors and load combinations shall be as stipulated in the appropriate code or in ASCE[3]. The required ultimate loads are given by IBC[4] Sec. 1605.2 as

$\lambda W = 1.4 D$ [IBC 16-1]

$\lambda W = 1.2 D + 1.6 L + 0.5 (L_r \text{ or } S \text{ or } R)$ [IBC 16-2]

$\lambda W = 1.2 D + 1.6 (L_r \text{ or } S \text{ or } R) + (0.5 L \text{ or } 0.8 W)$
[IBC 16-3]

$\lambda W = 1.2 D + 1.6 W + 0.5 L + 0.5 (L_r \text{ or } S \text{ or } R)$
[IBC 16-4]

$\lambda W = 1.2 D \pm 1.0 E + 0.5 L + 0.2 S$ [IBC 16-5]

$\lambda W = 0.9 D \pm 1.6 W$ [IBC 16-6]

$\lambda W = 0.9 D \pm 1.0 E$ [IBC 16-7]

For garages, places of public assembly, and areas where $L > 100 \text{ lbf/ft}^2$, replace $0.5L$ with $1.0L$ in IBC Eqs. (16-4) and (16-5). For roof configurations that do not shed snow, replace $0.2S$ with $0.7S$ in IBC Eq. (16-5).

The plastic modulus of a structure that is required to support a given load combination may be determined by either the statical design method or the mechanism design method.

Statical Design Method

The statical method is a convenient technique to use for continuous beams, as illustrated in Fig. 4.4. The three-span continuous beam of uniform section is first cut back to a statically determinate condition and the factored loads applied, as shown in Fig. 4.4(a). The free bending moment diagram for this condition is shown in Fig. 4.4(b). It consists of three moment envelopes, each with a maximum value of

$$M_s = \frac{\lambda W l}{8}$$

---
[3]American Society of Civil Engineers, 2005 (See References and Codes)
[4]International Code Council, 2009 (See References and Codes)

Figure 4.4 Statical Design Method

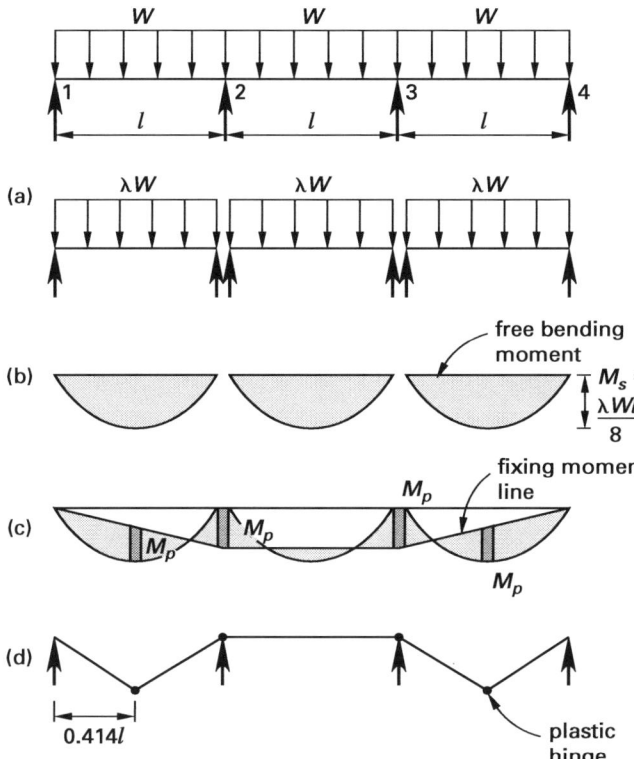

Figure 4.5 Nonuniform Beam

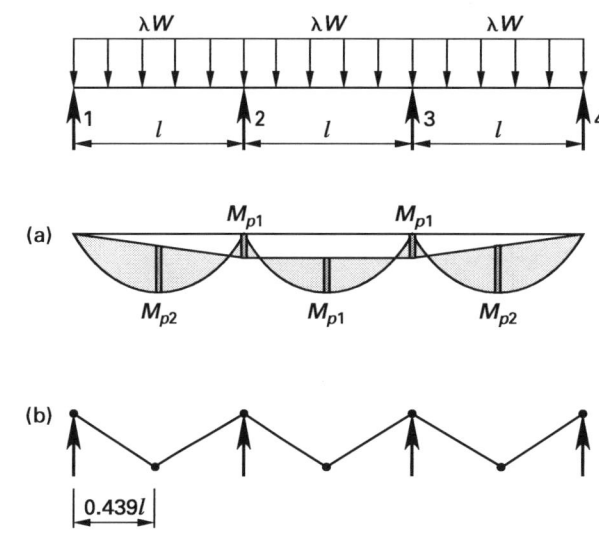

The fixing moment line that represents the support moments in the original structure is now superimposed on the free moment diagram, as shown in Fig. 4.4(c). The fixing moment line is positioned to make the moments at supports 1 and 2 and in spans 12 and 34 equal to M_p, the required plastic moment of resistance. Hence, the collapse mechanism shown in Fig. 4.4(d) is formed, and collapse occurs simultaneously in the two end spans, with the plastic hinges occurring at a distance of $0.414l$ from the end supports. The required plastic moment of resistance is determined from the geometry of the figure as

$$M_p = 0.686 M_s$$
$$= 0.0858 \lambda W l$$

When the continuous beam shown in Fig. 4.5 is of nonuniform section, a complete collapse mechanism is possible, with plastic hinges forming in all spans and at both interior supports. The location of the fixing moment line that produces this condition is shown in Fig. 4.5(a). The required plastic moments of resistance are

$$M_{p1} = 0.5 M_s$$
$$= \frac{\lambda W l}{16}$$
$$M_{p2} = 0.766 M_s$$
$$= 0.0958 \lambda W l$$

The collapse mechanism is shown in Fig. 4.5(b). The plastic hinges occur at a distance of $0.439l$ from the end supports.

Example 4.2

The two-span continuous beam shown in the illustration supports the factored load indicated, including an allowance for the self-weight of the beam. Assuming that adequate lateral support is provided to the beam, determine the lightest W12 section required, using grade 50 steel.

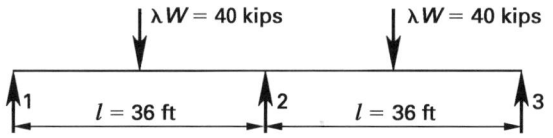

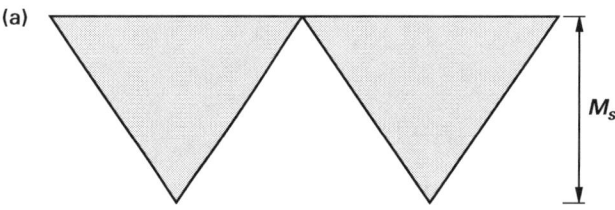

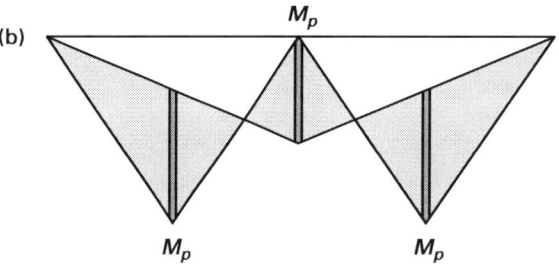

Solution

As indicated in illustration (a), the free bending moment in each span is

$$M_s = \frac{\lambda W l}{4}$$

$$= \frac{(40 \text{ kips})(36 \text{ ft})}{4}$$

$$= 360 \text{ ft-kips}$$

The first plastic hinge forms at the central support. To produce collapse, three plastic hinges are necessary, as shown in illustration (c). The fixing moment line is located as shown in illustration (b). The required plastic moment of resistance is derived from the geometry of the illustration as

$$M_p = \tfrac{2}{3} M_s$$

$$= \left(\tfrac{2}{3}\right)(360 \text{ ft-kips})$$

$$= 240 \text{ ft-kips}$$

From AISC Table 3-6, a W12 × 45 has

$$\phi_b M_p = 241 \text{ ft-kips}$$

$$> 240 \text{ ft-kips} \quad [\text{satisfactory}]$$

Beam Bracing Requirements

To ensure that the necessary hinge rotations can occur before collapse, the maximum unbraced length of a member adjacent to a plastic hinge is restricted by AISC Eq. (A-1-7) to a value of

$$L_{pd} = \left(0.12 + (0.076)\left(\frac{M_1}{M_2}\right)\right)\frac{E r_y}{F_y}$$

Rotation does not occur in the last hinge to form, and the bracing requirements of AISC Sec. F2.2 are applicable. Similarly, AISC Sec. F2.2 applies to segments remote from a plastic hinge.

Example 4.3

The two-span continuous beam of Ex. 4.2 is laterally braced at the supports and at the center of the spans, as shown in the illustration. Determine whether the beam is adequately braced.

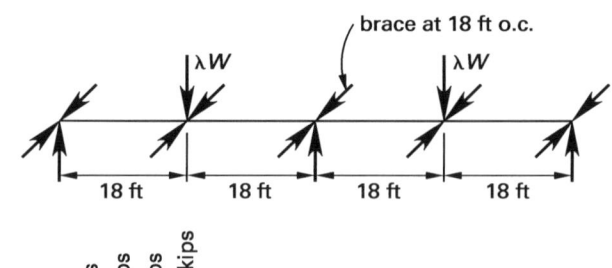

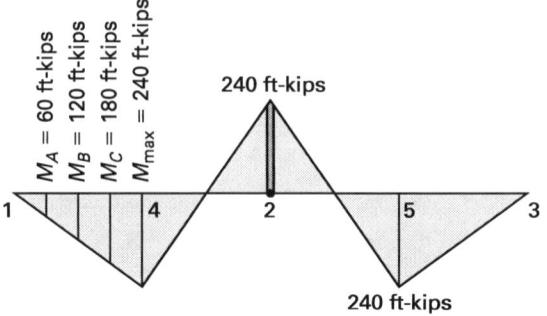

Solution

From AISC Tables 1-1 and 3-6, the relevant properties for a W12 × 45 are

$$r_y = 1.95 \text{ in}$$

$$\phi_b M_p = 241 \text{ ft-kips}$$

$$\phi_b M_r = 151 \text{ ft-kips}$$

$$L_p = 6.89 \text{ ft}$$

$$L_r = 22.4 \text{ ft}$$

$$BF = 5.75 \text{ kips}$$

$$F_y = 50 \, \frac{\text{kips}}{\text{in}^2}$$

The unbraced length in the end span is

$$L_b = L_{14} = L_{24}$$

$$= 0.5 L_{12} = (0.5)(36 \text{ ft})$$

$$= 18 \text{ ft}$$

$$> L_p$$

$$< L_r$$

For section 14, the plastic hinge at 4 is the last hinge to form, and the bracing requirements of AISC Sec. F2.2 are applicable. The bending moments acting at the quarter points of section 14 are shown in the moment diagram. The relative values of these moments are

$$M_{\max} = 4$$

$$M_A = 1$$

$$M_B = 2$$

$$M_C = 3$$

The bending coefficient is given by AISC Eq. (F1-1) where R_M is 1.0, as

$$C_b = \frac{12.5 M_{\max}}{2.5 M_{\max} + 3 M_A + 4 M_B + 3 M_C} R_M$$
$$= \left(\frac{(12.5)(4)}{(2.5)(4) + (3)(1) + (4)(2) + (3)(3)}\right)(1.0)$$
$$= 1.67$$

The nominal flexural strength for an unbraced length between the full plastic bending capacity length L_p and the inelastic lateral torsional buckling length L_r is given by AISC Sec. F2.2b as

$$M_n = C_b\left(M_p - (M_p - 0.7 F_y S_x)\left(\frac{L_b - L_p}{L_r - L_p}\right)\right)$$
[AISC F2-2]

The flexural design strength is derived as

$$\phi_b M_n = C_b\big(\phi_b M_p - (BF)(L_b - L_p)\big)$$
$$= (1.67)\binom{(241 \text{ ft-kips})}{-(5.75 \text{ kips})(18 \text{ ft} - 6.89 \text{ ft})}$$
$$\leq \phi_b M_p$$
$$= 241 \text{ ft-kips} \quad [\text{maximum}]$$
$$> M_{\max} \quad [\text{satisfactory}]$$

For section 24, the bracing requirements of AISC App. 1.7 are applicable.

$$\frac{M_{24}}{M_{42}} = \frac{+M_p}{M_p} \quad [\text{moments cause reverse curvature}]$$
$$= 1.0$$

The maximum allowable unbraced length is given by AISC Eq. (A-1-7) as

$$L_{pd} = \left(0.12 + (0.076)\left(\frac{M_{24}}{M_{42}}\right)\right)\frac{E r_y}{F_y}$$
$$= \big(0.12 + (0.076)(1.0)\big)$$
$$\times \left(\frac{\left(29{,}000 \frac{\text{kips}}{\text{in}^2}\right)(1.95 \text{ in})}{\left(50 \frac{\text{kips}}{\text{in}^2}\right)\left(12 \frac{\text{in}}{\text{ft}}\right)}\right)$$
$$= 18.5 \text{ ft}$$
$$> L_{25} \quad [\text{satisfactory}]$$

The beam is adequately braced.

Mechanism Design Method

The mechanism design method may be used to determine the required plastic moment of resistance in rigid frames and beams. As shown in Fig. 4.6, plastic hinges may form at the point of application of a concentrated load, at the ends of members, and at the location of zero shear in a prismatic beam. In the case of two members meeting at a joint, a plastic hinge forms in the weaker member. In the case of three or more members meeting at a joint, a plastic hinge may form at the ends of each of the members.

Figure 4.6 Hinge Locations

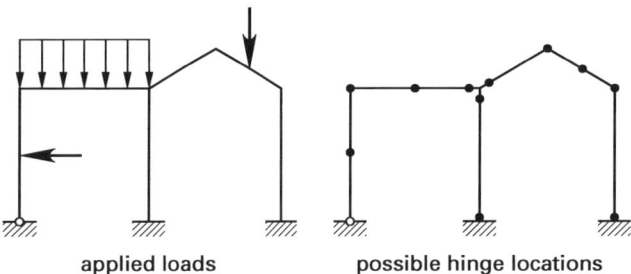

applied loads   possible hinge locations

An independent collapse mechanism corresponds to a condition of unstable equilibrium in a structure. A structure that is indeterminate to the degree D becomes determinate when D plastic hinges have formed. The formation of one more plastic hinges produces a collapse mechanism. Hence, in a structure that has p possible hinge locations, the number of possible independent mechanisms is

$$m_i = p - D$$

In addition, the independent mechanisms may be combined to form combined mechanisms.

The structure shown in Fig. 4.6 is five degrees indeterminate, with 11 possible hinge locations. The number of possible independent mechanisms is

$$m_i = p - D = 11 - 5$$
$$= 6$$

As shown in Fig. 4.7, the independent mechanisms consist of the beam mechanisms B_1, B_2, and B_3; the sway mechanism S; the gable mechanism G; and the joint mechanism J. Any of these independent mechanisms may be combined to form a combined mechanism. Two such combinations, $(B_2 + S + J)$ and $(B_2 + G + J)$, are shown in Fig. 4.8. In combining mechanisms, the objective is to eliminate hinges in order to produce a more critical loading condition. If the correct mechanism has been selected, the bending moment diagram of the structure at collapse must show that the plastic moment of resistance is not anywhere exceeded.

Figure 4.7 Independent Mechanisms

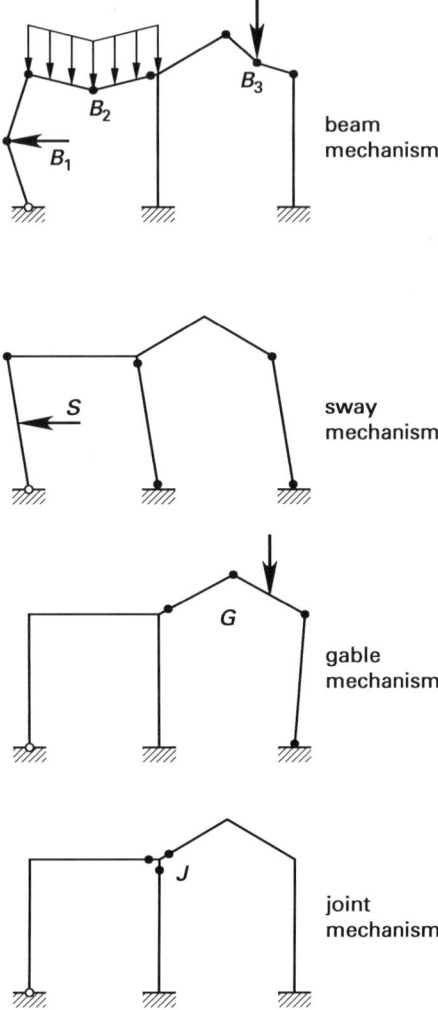

Figure 4.8 Combined Mechanisms

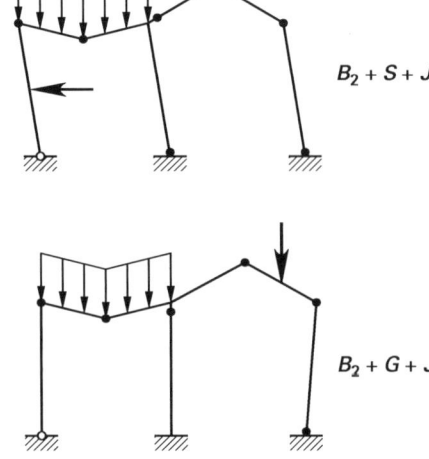

The virtual work principle, which is used in the mechanism design method, entails establishing equilibrium relationships resulting from real forces undergoing virtual displacements. The principle states that if a structure in equilibrium under a system of applied forces is subjected to a system of virtual displacements compatible with the external restraints and the geometry of the structure, the total external work done by the applied forces equals the work done by the internal forces (caused by the internal deformations corresponding to the external displacements). The expression *virtual displacement* implies an extremely small imaginary displacement, and the external work done is the product of a real loading system and imaginary displacements. The internal work, or internal strain energy, is the product of the internal forces in the system and the deformations produced by the imaginary displacements. The equilibrium relationship is: Internal work equals external work.

The application of the method is illustrated in Ex. 4.4.

Example 4.4

The rigid frame shown in the example illustration is fabricated from members of a uniform section in grade 50 steel. For the factored loading indicated, ignoring the member self-weight and assuming adequate lateral support is provided, determine the plastic moment of resistance required.

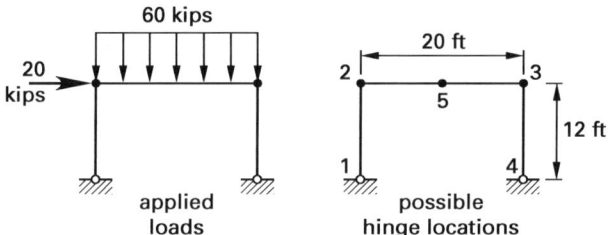

Solution

The three possible collapse mechanisms are the independent beam mechanism, the independent sway mechanism, and the combined mechanism. These are shown in the solution illustration. Applying a virtual displacement to each of these mechanisms in turn, and equating internal and external work provides three equations. A value of M_p may be obtained from each of the three equations. The largest value of M_p governs.

For the beam mechanism, applying a virtual vertical displacement at midspan of the beam, and setting internal work equal to external work,

$$4M_p(\theta) = \frac{(2)\left(3\ \frac{\text{kips}}{\text{ft}}\right)(10\ \text{ft})(10\ \text{ft})(\theta)}{2}$$
$$= (300\ \text{ft-kips})(\theta)$$
$$M_p = 75\ \text{ft-kips}$$

For the sway mechanism, applying a virtual horizontal displacement at the top of the columns, and setting internal work equal to external work,

$$2M_p(\theta) = (20 \text{ kips})(12 \text{ ft})(\theta)$$

$$= (240 \text{ ft-kips})(\theta)$$

$$M_p = 120 \text{ ft-kips}$$

For the combined mechanism, adding internal and external work of both independent mechanisms, and setting internal work equal to external work gives

$$4M_p(\theta) = (300 \text{ ft-kips})(\theta) + (240 \text{ ft-kips})(\theta)$$

$$= (540 \text{ ft-kips})(\theta)$$

$$M_p = 135 \text{ ft-kips} \quad \text{[governs]}$$

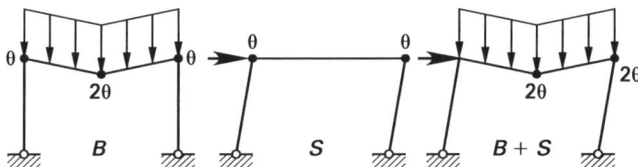

Example 4.5

For the rigid frame described in Ex. 4.4, determine the lightest W10 section required using grade 50 steel.

Solution

From LRFD Table 3-6, a W10 × 30 has

$$\phi_b M_p = 137 \text{ ft-kips}$$

$$> 135 \text{ ft-kips} \quad \text{[satisfactory]}$$

Static Equilibrium Check

The mechanism method provides an upper bound on the collapse load. Using static equilibrium methods, a bending moment diagram of the assumed collapse mode may be constructed. If the correct collapse mode has been selected, the value of M_p should nowhere be exceeded.

Example 4.6

Calculate the member forces and draw the bending moment diagram for the assumed collapse mechanism of Ex. 4.4.

Solution

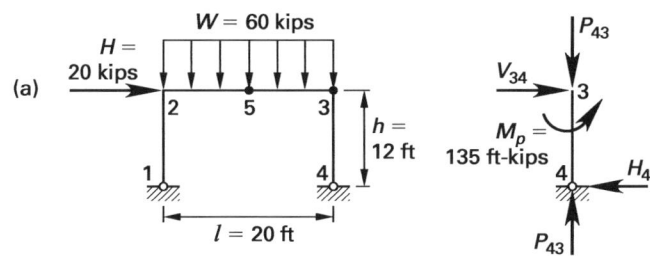

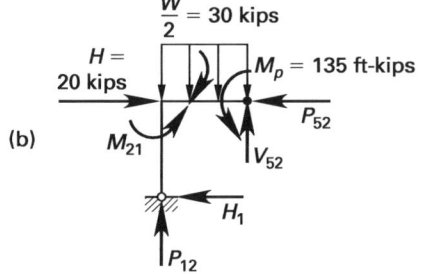

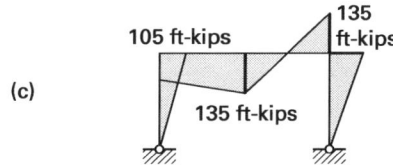

For the free-body diagram of member 34, as shown in illustration (a), taking moments about point 3 gives the horizontal reaction at support 4 as

$$H_4 = \frac{M_p}{h} = \frac{135 \text{ ft-kips}}{12 \text{ ft}}$$

$$= 11.25 \text{ kips} \quad \text{[acting to the left as shown]}$$

$$= V_{34}$$

$$= P_{52}$$

For the free-body diagram of member 125, as shown in illustration (b), resolving forces gives the horizontal reaction at support 1 as

$$H_1 = H - P_{52} = 20 \text{ kips} - 11.25 \text{ kips}$$

$$= 8.75 \text{ kips} \quad \text{[acting to the left as shown]}$$

Taking moments about point 5 gives the axial force in member 12 as

$$P_{12} = \frac{M_p + \left(\dfrac{W}{2}\right)\left(\dfrac{l}{4}\right) - H_1 h}{\dfrac{l}{2}}$$

$$= \frac{135 \text{ ft-kips} + (30 \text{ kips})\left(\dfrac{20 \text{ ft}}{4}\right)}{\dfrac{20 \text{ ft}}{2}}$$

$$= 18 \text{ kips}$$

Taking moments about joint 2 for member 12, gives the moment at joint 2 as

$$M_{21} = H_1 h = (8.75 \text{ kips})(12 \text{ ft})$$
$$= 105 \text{ ft-kips}$$

Resolving forces gives the shear force at point 5 as

$$V_{52} = \frac{W}{2} - P_{12} = 30 \text{ kips} - 18 \text{ kips}$$
$$= 12 \text{ kips}$$

For the whole structure, resolving forces gives the axial force in member 43 as

$$P_{43} = W - P_{12} = 60 \text{ kips} - 18 \text{ kips}$$
$$= 42 \text{ kips}$$

The bending moment diagram is shown in illustration (c). Because $M_p = 135$ ft-kips is not exceeded at any point in the frame, the combined mechanism is the correct failure mode.

Column Design Requirements

The column maximum unbraced length, L, for plastic design is limited by AISC App 1.6 to

$$\frac{L}{r} = 4.71 \sqrt{\frac{E}{F_y}}$$
$$= 113 \quad [\text{for } F_y = 50 \text{ kips/in}^2]$$

The actual unbraced length in the plane of bending is L. The corresponding radius of gyration is r.

In accordance with AISC App. 1.5, the axial load on a column shall not exceed the value

$$P_{\max} = 0.85 \phi_c A_g F_y \quad [\text{braced frame}]$$
$$P_{\max} = 0.75 \phi_c A_g F_y \quad [\text{sway frame}]$$

AISC Comm. Sec. 1.5 permits low rise frames to be designed without consideration of second-order (P-delta) effects.

As for beams, the maximum unbraced length of a column segment must not exceed the value

$$L_{pd} = \left(0.12 + (0.076)\left(\frac{M_1}{M_2}\right)\right)\frac{Er_y}{F_y} \quad [\text{AISC A-1-7}]$$

For combined axial force and flexure, the interaction expressions from LRFD Sec. H1 apply, and

$$\frac{P_u}{\phi_c P_n} + \left(\frac{8}{9}\right)\left(\frac{M_{ux}}{\phi_b M_{nx}} + \frac{M_{uy}}{\phi_b M_{ny}}\right) \leq 1.0 \quad \left[\frac{P_u}{\phi_c P_n} \geq 0.2\right]$$

$$\frac{P_u}{2\phi_c P_n} + \left(\frac{M_{ux}}{\phi_b M_{nx}} + \frac{M_{uy}}{\phi_b M_{ny}}\right) \leq 1.0 \quad \left[\frac{P_u}{\phi_c P_n} < 0.2\right]$$

Example 4.7

Determine whether column 34 of the sway frame in Ex. 4.4 is satisfactory. The column and girder consist of a grade 50, W10 × 30 section, and the column is laterally braced at 4 ft centers. It may be assumed that the plastic hinge located at point 5 is the last to form.

Solution

The relevant properties of a W10 × 30 section are obtained from AISC Tables 1-1 and 3-6 as

$$A_g = 8.84 \text{ in}^2$$
$$I = 170 \text{ in}^4$$
$$r_x = 4.38 \text{ in}$$
$$r_y = 1.37 \text{ in}$$
$$\phi_b M_p = 137 \text{ ft-kips}$$
$$L_p = 4.84 \text{ ft}$$

The maximum unbraced length of a column segment shall not exceed the value given by AISC Eq. (A-1-7) as

$$L_{pd} = \left(0.12 - (0.076)\left(\frac{M_1}{M_2}\right)\right)\frac{Er_y}{F_y}$$
$$= \left(0.12 + (0.076)\left(\frac{\frac{2M_p}{3}}{M_p}\right)\right)$$
$$\times \left(\frac{\left(29{,}000 \frac{\text{kips}}{\text{in}^2}\right)(1.37 \text{ in})}{\left(50 \frac{\text{kips}}{\text{in}^2}\right)}\right)$$
$$= 4.6 \text{ ft}$$
$$> 4 \text{ ft} \quad [\text{satisfactory}]$$

$$\frac{L_x}{r_x} = \frac{144 \text{ in}}{4.38 \text{ in}} = 32.9$$
$$< 113 \quad [\text{satisfies AISC App. 1.6}]$$

$$\frac{L_y}{r_y} = \frac{48 \text{ in}}{1.37 \text{ in}} = 35.0$$
$$< 113 \quad [\text{satisfies AISC App. 1.6}]$$

For the pinned connection at joint 4, AISC Comm. Sec. C2 specifies a stiffness ratio of $G_A = 10$. At joint 3,

$$G_B = \frac{\sum\left(\frac{I_c}{l_c}\right)}{\sum\left(\frac{I_g}{l_g}\right)} = \frac{\frac{I}{12}}{\frac{I}{20}}$$

$$= 1.7$$

From the alignment chart for sway frames in AISC Comm. Fig. C-C2.4, the effective length factor is

$$K_{34} = 2.0$$

The slenderness ratio about the x-axis is

$$\frac{K_{34}L_x}{r_x} = \frac{(2.0)(12\text{ ft})\left(12\,\frac{\text{in}}{\text{ft}}\right)}{4.38\text{ in}}$$

$$= 66$$

$$< 200 \quad [\text{satisfies AISC Sec. E2}]$$

The frame is braced in the lateral direction and the effective length factor about the y-axis is given by AISC Table C-C2.2 as

$$K_{34} = 1.0$$

The slenderness ratio about the y-axis is

$$\frac{K_{34}L_y}{r_y} = \frac{(1.0)(4\text{ ft})\left(12\,\frac{\text{in}}{\text{ft}}\right)}{1.37\text{ in}}$$

$$= 35.0$$

The slenderness ratio about the x-axis governs.

In accordance with AISC App. 1.5, the maximum, axial load in the column of a sway frame is restricted to

$$P_{\max} = 0.75\phi_c A_g F_y$$

$$= (0.75)(0.90)(8.84\text{ in}^2)\left(50\,\frac{\text{kips}}{\text{in}^2}\right)$$

$$= 298\text{ kips}$$

$$> P_{43} \quad [\text{satisfactory}]$$

From AISC Table 3-6, for a W10 × 30,

$$L_p = 4.84\text{ ft}$$

$$> 4.0\text{ ft} \quad [\text{full plastic bending capacity available}]$$

$$\phi_b M_{nx} = \phi_b M_p$$

$$= 137\text{ ft-kips}$$

From AISC Table 4-22, for a $K_{34}L_x/r_x$ value of 66, the design stress for axial load is

$$\phi_c F_{cr} = 32.7\text{ kips/in}^2$$

The design axial strength is

$$\phi_c P_n = \phi_c F_{cr} A_g \quad [\text{AISC E3-1}]$$

$$= \left(32.7\,\frac{\text{kips}}{\text{in}^2}\right)(8.84\text{ in}^2)$$

$$= 289\text{ kips}$$

$$\frac{P_{34}}{\phi_c P_n} = \frac{42\text{ kips}}{289\text{ kips}}$$

$$= 0.15$$

$$< 0.20 \quad [\text{AISC Eq. (H1-1b) governs}]$$

Secondary effects may be neglected and AISC Eq. (H1-1b) reduces to

$$\frac{P_{34}}{2\phi_c P_n} + \frac{M_p}{\phi_b M_{nx}} \leq 1.0$$

$$\frac{42\text{ kips}}{(2)(289\text{ kips})} + \frac{135\text{ ft-kips}}{137\text{ ft-kips}} = 1.06$$

$$\approx 1.0 \quad [\text{satisfactory}]$$

The column is satisfactory.

2. ECCENTRICALLY LOADED BOLT GROUPS

Nomenclature

A_b	nominal unthreaded body area of bolt	in²
C	coefficient for eccentrically loaded bolt and weld groups	–
d_b	nominal bolt diameter	in
d_m	moment arm between resultant tensile and compressive forces caused by an eccentric force	in
D	lateral distance between bolts	in
e	eccentricity of applied load	in
f_v	computed shear stress	kips/in²
F_{nt}	nominal tensile stress of bolt	kips/in²
F_{nv}	nominal shear stress of bolt	kips/in²
F_u	specified minimum tensile strength	kips/in²
n	number of bolts in one vertical row	–
n'	number of bolts above the neutral axis (in tension)	–
N	total number of bolts in a group	–
P	externally applied eccentric load	kips
P_u	factored load on connection	kips
P_v	allowable shear capacity of bolt	kips
R	resultant force on bolt	kips
R_n	nominal strength	kips
s	center-to-center pitch of two consecutive bolts	in
T	unfactored tensile force	kips
T_b	minimum pre-tension force	kips
T_u	factored tensile force	kips
V_P	shear force on bolt	kips

Symbols

ϕ	resistance factor	–

Bolt Group Eccentrically Loaded in the Plane of the Faying Surface, LRFD Method

Eccentrically loaded bolt groups of the type shown in Fig. 4.9 may be conservatively designed by means of the elastic unit area method. The moment of inertia of the bolt group about the x-axis is

$$I_x = \sum y^2$$

Figure 4.9 Eccentrically Loaded Bolt Group, LRFD Method

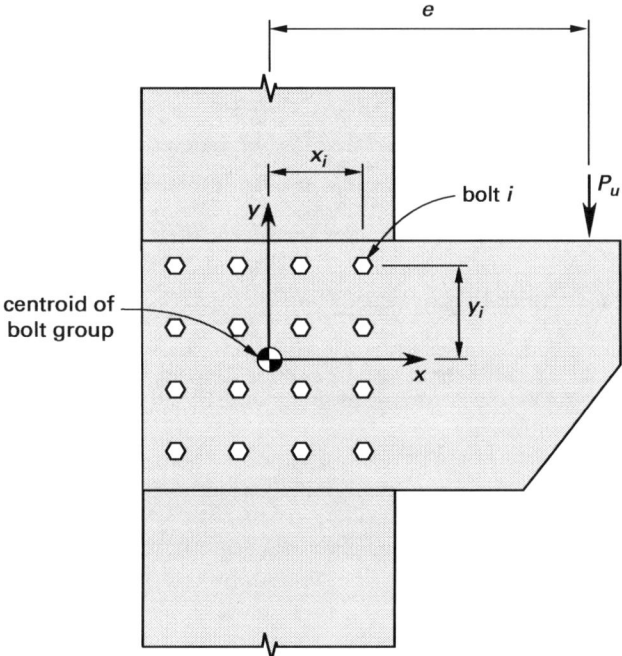

The moment of inertia of the bolt group about the y-axis is

$$I_y = \sum x^2$$

The polar moment of inertia of the bolt group about the centroid is

$$I_o = I_x + I_y$$

The vertical force on bolt i, caused by the applied load P_u, is

$$V_P = \frac{P_u}{N}$$

The vertical force on bolt i, caused by the eccentricity e, is

$$V_e = \frac{P_u e x_i}{I_o}$$

The horizontal force on bolt i, caused by the eccentricity e, is

$$H_e = \frac{P_u e y_i}{I_o}$$

The resultant force on bolt i is

$$R = \sqrt{(V_P + V_e)^2 + H_e^2}$$

Example 4.8

Using the LRFD method, determine the diameter of the A325 bearing-type bolts required in the bolted bracket shown in the illustration. Threads are included in the shear plane.

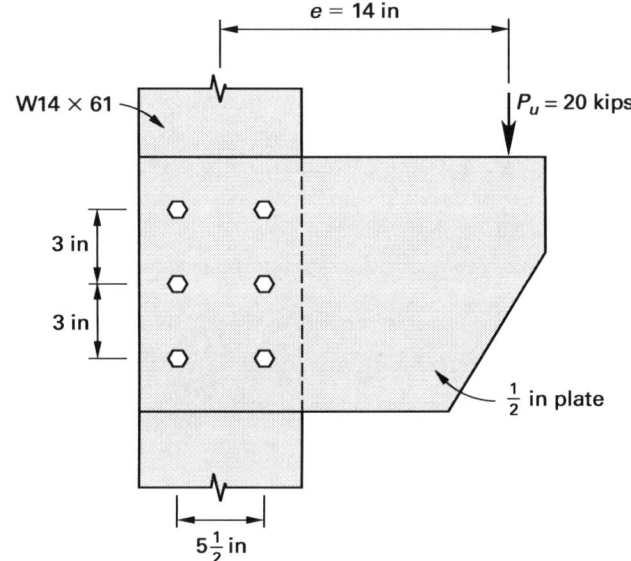

Solution

The geometrical properties of the bolt group are obtained by applying the unit area method. The moment of inertia about the x-axis is

$$I_x = \sum y^2 = (4)(3 \text{ in})^2$$
$$= 36 \text{ in}^4/\text{in}^2$$

The moment of inertia about the y-axis is

$$I_y = \sum x^2 = (6)(2.75 \text{ in})^2$$
$$= 45.38 \text{ in}^4/\text{in}^2$$

The polar moment of inertia about the centroid is

$$I_o = I_x + I_y = 36 \frac{\text{in}^4}{\text{in}^2} + 45.38 \frac{\text{in}^4}{\text{in}^2}$$
$$= 81.38 \text{ in}^4/\text{in}^2$$

The top right bolt is the most heavily loaded, and the coexistent forces on this bolt are the following. The vertical force caused by the applied load is

$$V_P = \frac{P_u}{N} = \frac{20 \text{ kips}}{6}$$
$$= 3.33 \text{ kips}$$

The vertical force caused by the eccentricity is

$$V_e = \frac{P_u e x_i}{I_o}$$
$$= \frac{(20 \text{ kips})(14 \text{ in})(2.75 \text{ in})}{81.38 \frac{\text{in}^4}{\text{in}^2}}$$
$$= 9.46 \text{ kips}$$

The horizontal force caused by the eccentricity is

$$H_e = \frac{P_u e y_i}{I_o}$$
$$= \frac{(20 \text{ kips})(14 \text{ in})(3 \text{ in})}{81.38 \frac{\text{in}^4}{\text{in}^2}}$$
$$= 10.32 \text{ kips}$$

The resultant force is

$$R = \sqrt{(V_P + V_e)^2 + H_e^2}$$
$$= \sqrt{(3.33 \text{ kips} + 9.46 \text{ kips})^2 + (10.32 \text{ kips})^2}$$
$$= 16.4 \text{ kips}$$

Shear controls. From AISC Table 7-1, the design shear strength of a $^7/_8$ in diameter A325N bolt in a standard hole, in single shear with threads included in the shear plane, is

$$\phi R_n = 21.6 \text{ kips}$$
$$> 16.4 \text{ kips} \quad [\text{satisfactory}]$$

Bolt Group Eccentrically Loaded in the Plane of the Faying Surface, ASD Method

Eccentrically loaded bolt groups of the type shown in Fig. 4.10 may be conservatively designed by means of the elastic unit area method. The moment of inertia of the bolt group about the x-axis is

$$I_x = \sum y^2$$

The moment of inertia of the bolt group about the y-axis is

$$I_y = \sum x^2$$

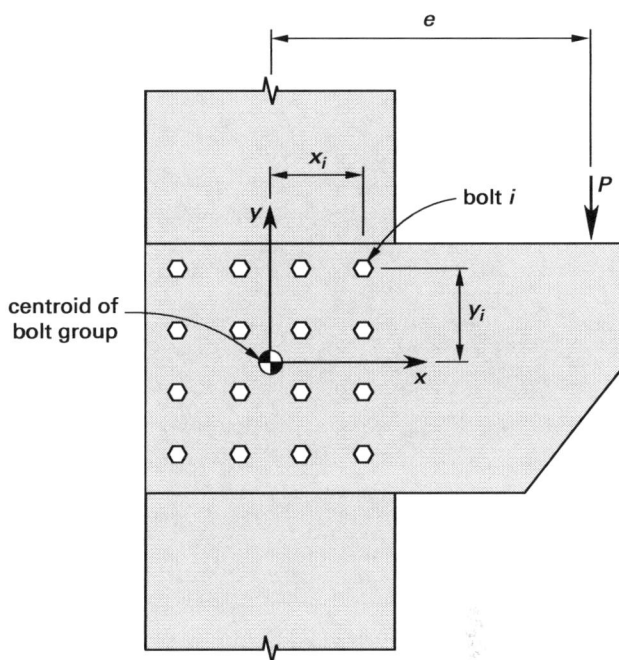

Figure 4.10 Eccentrically Loaded Bolt Group, ASD Method

The polar moment of inertia of the bolt group about the centroid is

$$I_o = I_x + I_y$$

The vertical force on bolt i, caused by the applied load P, is

$$V_P = \frac{P}{N}$$

The vertical force on bolt i, caused by the eccentricity e, is

$$V_e = \frac{P e x_i}{I_o}$$

The horizontal force on bolt i, caused by the eccentricity e, is

$$H_e = \frac{P e y_i}{I_o}$$

The resultant force on bolt i is

$$R = \sqrt{(V_P + V_e)^2 + H_e^2}$$

Example 4.9

Using the ASD method, determine the diameter of the A325 bearing-type bolts required in the bolted bracket shown in the illustration. Threads are included in the shear plane.

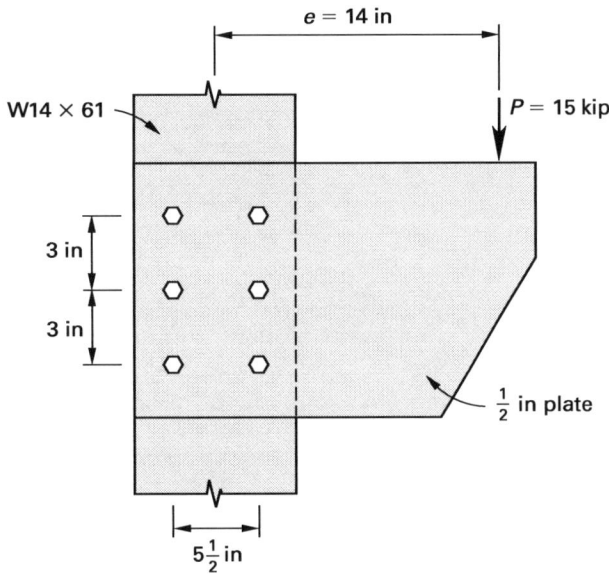

(not to scale)

Solution

The geometrical properties of the bolt group are obtained by applying the unit area method. The moment of inertia about the x-axis is

$$I_x = \sum y^2 = (4)(3 \text{ in})^2$$
$$= 36 \text{ in}^4/\text{in}^2$$

The moment of inertia about the y-axis is

$$I_y = \sum x^2 = (6)(2.75 \text{ in})^2$$
$$= 45.38 \text{ in}^4/\text{in}^2$$

The polar moment of inertia about the centroid is

$$I_o = I_x + I_y = 36 \frac{\text{in}^4}{\text{in}^2} + 45.38 \frac{\text{in}^4}{\text{in}^2}$$
$$= 81.38 \text{ in}^4/\text{in}^2$$

The top right bolt is the most heavily loaded, and the coexistent forces on this bolt are the following. The vertical force caused by the applied load is

$$V_P = \frac{P}{N} = \frac{15 \text{ kips}}{6}$$
$$= 2.50 \text{ kips}$$

The vertical force caused by the eccentricity is

$$V_e = \frac{Pex_i}{I_o}$$
$$= \frac{(15 \text{ kips})(14 \text{ in})(2.75 \text{ in})}{81.38 \frac{\text{in}^4}{\text{in}^2}}$$
$$= 7.10 \text{ kips}$$

The horizontal force caused by the eccentricity is

$$H_e = \frac{Pey_i}{I_o}$$
$$= \frac{(15 \text{ kips})(14 \text{ in})(3 \text{ in})}{81.38 \frac{\text{in}^4}{\text{in}^2}}$$
$$= 7.74 \text{ kips}$$

The resultant force is

$$R = \sqrt{(V_P + V_e)^2 + H_e^2}$$
$$= \sqrt{(2.50 \text{ kips} + 7.10 \text{ kips})^2 + (7.74 \text{ kips})^2}$$
$$= 12.3 \text{ kips}$$

Shear controls. From AISC Table 7-1, the allowable shear strength of a $7/8$ in diameter A325N bolt in a standard hole, in single shear with threads included in the shear plane, is

$$P_v = 14.4 \text{ kips}$$
$$> 12.3 \text{ kips} \quad [\text{satisfactory}]$$

Instantaneous Center of Rotation, LRFD Method

The instantaneous center of rotation method for analyzing eccentrically loaded bolt groups affords a more realistic estimate of a bolt group's capacity. The eccentrically loaded member is assumed to rotate about a point designated the instantaneous center of rotation.

As shown in Fig. 4.11, the bolt that is farthest from the instantaneous center will experience the greatest deformation, and will be the first bolt to fail as the applied load is increased. The deformations in the remaining bolts in the group are proportional to their distances from the instantaneous center.

Figure 4.11 Instantaneous Center of Rotation, LRFD Method

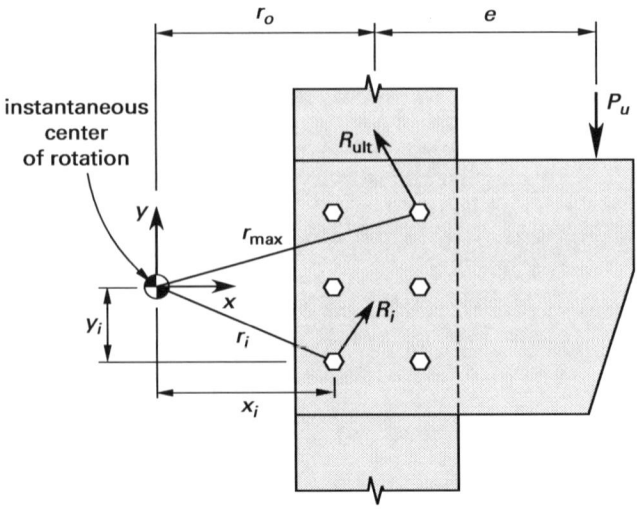

For common bolt group patterns, AISC Tables 7-7–7-14 tabulate the coefficient C, given by

$$C = \frac{P_u}{\phi R_n}$$

The method may be applied to bearing-type and slip-critical bolts, in single or double shear connections.

Example 4.10

Using the instantaneous center of rotation method, determine the diameter of the A325 bearing-type bolts required in the bolted bracket design in Ex. 4.8. Threads are included in the shear plane.

Solution

The relevant details obtained from Ex. 4.8 are the following. The number of bolts in one vertical row is

$$n = 3$$

The bolt pitch is

$$s = 3 \text{ in}$$

The lateral distance between bolts is

$$D = 5\tfrac{1}{2} \text{ in}$$

The eccentricity is

$$e = 14 \text{ in}$$

From AISC Table 7-9, the coefficient C is

$$C = 1.36$$

The required design strength of an individual bolt, based on the instantaneous center of rotation method, is

$$\phi R_n = \frac{P_u}{C} = \frac{20 \text{ kips}}{1.36}$$
$$= 14.7 \text{ kips}$$

Shear controls. From AISC Table 7-1, the design shear strength of a $^3/_4$ in diameter A325N bolt in a standard hole, in single shear with threads included in the shear plane, is

$$\phi R_n = 15.9 \text{ kips}$$
$$> 14.7 \text{ kips} \quad \text{[satisfactory]}$$

Instantaneous Center of Rotation, ASD Method

The instantaneous center of rotation method for analyzing eccentrically loaded bolt groups affords a more realistic estimate of a bolt group's capacity. The eccentrically loaded member is assumed to rotate about a point designated the instantaneous center of rotation.

As shown in Fig. 4.12, the bolt that is farthest from the instantaneous center will experience the greatest deformation, and will be the first bolt to fail as the applied load is increased. The deformations in the remaining bolts in the group are proportional to their distances from the instantaneous center.

Figure 4.12 Instantaneous Center of Rotation, ASD Method

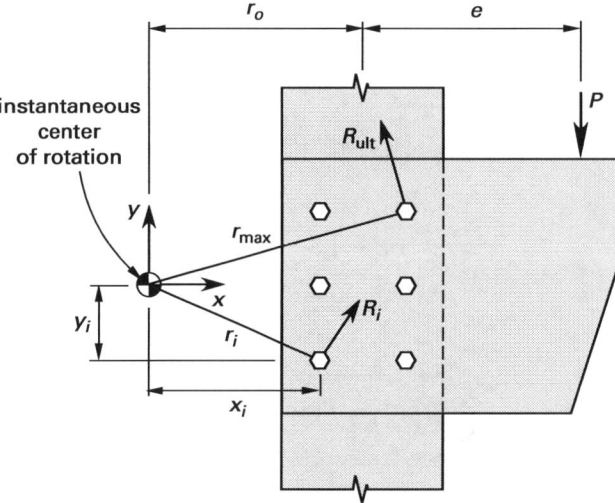

AISC Part 7, Tables 7-7–7-14 tabulate the coefficient C for common bolt group patterns. Here, the coefficient C is

$$C = \frac{P}{P_v}$$

The method may be applied to bearing-type and slip-critical bolts, in single or double shear connections.

Example 4.11

Using the instantaneous center of rotation method, determine the diameter of the A325 bearing-type bolts required in the bolted bracket designed in Ex. 4.9. Threads are included in the shear plane.

Solution

The relevant details from Ex. 4.9 are the following. The number of bolts in one vertical row is

$$n = 3$$

The bolt pitch is

$$b = 3 \text{ in}$$

The lateral distance between bolts is

$$D = 5\tfrac{1}{2} \text{ in}$$

The eccentricity is

$$e = 14 \text{ in}$$

From AISC Table 7-9, the coefficient C is

$$C = 1.36$$

The required strength of an individual bolt, based on the instantaneous center of rotation method, is

$$P_v = \frac{P}{C} = \frac{15 \text{ kips}}{1.36}$$
$$= 11.0 \text{ kips}$$

Shear controls. From AISC Table 7-1, the allowable shear strength of a $^7\!/_8$ in diameter A325N bolt in a standard hole, in single shear with threads included in the shear plane, is

$$P_v = 14.4 \text{ kips}$$
$$> 11.0 \text{ kips} \quad \text{[satisfactory]}$$

Bolt Group Eccentrically Loaded Normal to the Faying Surface, LRFD Method

Eccentrically loaded bolt groups of the type shown in Fig. 4.13 may be conservatively designed by assuming that the neutral axis is located at the centroid of the bolt group, and that a plastic stress distribution is produced in the bolts. The tensile force in each bolt above the neutral axis caused by the eccentricity is

$$T_u = \frac{P_u e}{n' d_m}$$

The shear force in each bolt caused by the factored applied load is

$$V_P = \frac{P_u}{N}$$

Bearing-type connectors may now be selected for combined shear and tensile forces based on the requirements of AISC Sec. J3.7.

Example 4.12

Determine whether the $^7\!/_8$ in diameter A325N bearing-type bolts in the bolted bracket shown in the illustration are adequate. Prying action may be neglected.

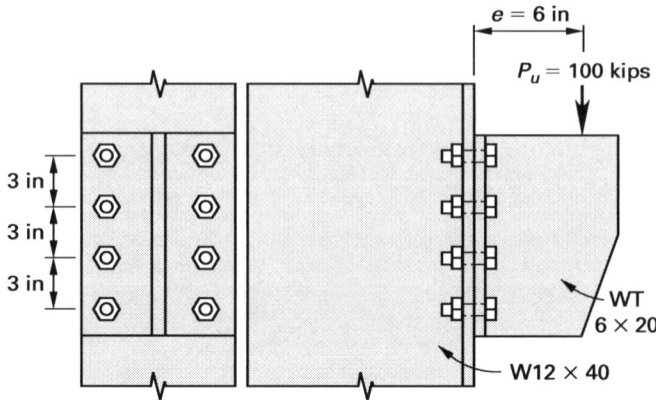

Solution

The tensile force in each bolt above the neutral axis caused by the eccentricity is

$$T_u = \frac{P_u e}{n' d_m} = \frac{(100 \text{ kips})(6 \text{ in})}{(4)(6 \text{ in})}$$
$$= 25 \text{ kips}$$

The shear force in each bolt caused by the applied load is

$$V_P = \frac{P_u}{N} = \frac{100 \text{ kips}}{8}$$
$$= 12.50 \text{ kips}$$

The calculated shear stress on each bolt is

$$f_v = \frac{V_P}{A_b} = \frac{12.5 \text{ kips}}{0.601 \text{ in}^2}$$
$$= 20.8 \text{ kips/in}^2$$

Figure 4.13 Bolt Group Eccentrically Loaded Normal to the Faying Surface, LRFD Method

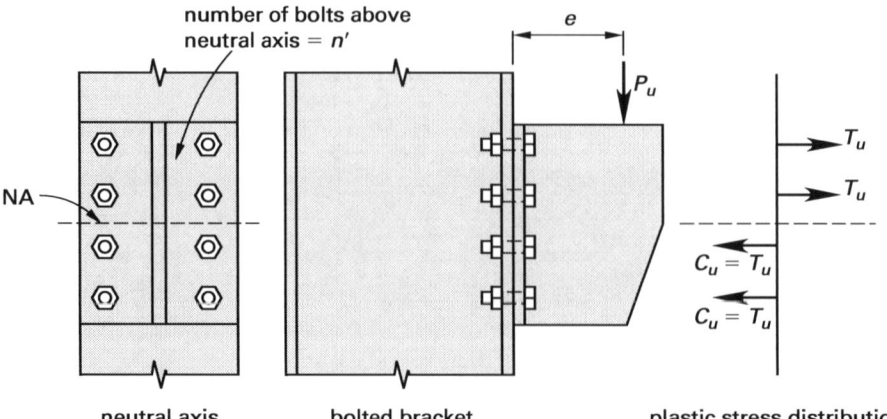

The design shear stress for grade A325 bolts, with threads included in the shear plane, is obtained from AISC Table J3.2 as

$$\phi F_{nv} = (0.75)\left(48\ \frac{\text{kips}}{\text{in}^2}\right)$$
$$= 36\ \frac{\text{kips}}{\text{in}^2}$$
$$> f_v \quad [\text{satisfactory}]$$
$$f_v > 0.2\phi F_{nv}$$

The design tensile stress for grade A325 bolts is obtained from AISC Table J3.2 as

$$\phi F_{nt} = (0.75)\left(90\ \frac{\text{kips}}{\text{in}^2}\right)$$
$$= 67.5\ \frac{\text{kips}}{\text{in}^2}$$

The factored tensile stress in each $7/8$ in diameter bolt is

$$f_t = \frac{T_u}{A_b} = \frac{25\ \text{kips}}{0.601\ \text{in}^2}$$
$$= 41.60\ \frac{\text{kips}}{\text{in}^2}$$
$$< \phi F_{nt} \quad [\text{satisfactory}]$$
$$> 0.2\phi F_{nt}$$

It is necessary to investigate the effects of the combined shear and tensile stress. The nominal tensile stress F'_{nt} of a bolt that is subjected to combined shear and tension is given by AISC Eq. (J3-3a) as

$$F'_{nt} = 1.3 F_{nt} - \frac{f_v F_{nt}}{\phi F_{nv}}$$
$$= (1.3)\left(90\ \frac{\text{kips}}{\text{in}^2}\right) - \frac{\left(20.8\ \frac{\text{kips}}{\text{in}^2}\right)\left(90\ \frac{\text{kips}}{\text{in}^2}\right)}{36\ \frac{\text{kips}}{\text{in}^2}}$$
$$= 65\ \text{kips/in}^2$$

The design tensile stress $\phi F'_{nt}$ of a bolt that is subjected to combined shear and tension is given by AISC Eq. (J3-2) as

$$\phi F'_{nt} = (0.75)\left(65\ \frac{\text{kips}}{\text{in}^2}\right)$$
$$= 48.75\ \frac{\text{kips}}{\text{in}^2}$$
$$> f_t \quad [\text{satisfactory}]$$

Bolt Group Eccentrically Loaded Normal to the Faying Surface, ASD Method

Eccentrically loaded bolt groups of the type shown in Fig. 4.14 may be conservatively designed by assuming that the neutral axis is located at the centroid of the bolt group, and that an elastic stress distribution is produced in the bolts. The moment of inertia of the bolt group about the x-axis is

$$I_x = \sum y^2$$

The total tensile force on the bolts a distance y_i from the neutral axis, caused by the eccentricity e, is

$$T_i = \frac{Pey_i}{I_x}$$

The shear force on each bolt caused by the applied load P is

$$V = \frac{P}{n}$$

Bearing-type connectors may now be selected for combined shear and tensile forces, based on the requirements of AISC Sec. J3.7.

Figure 4.14 Bolt Group Eccentrically Loaded Normal to Faying Surface, ASD Method

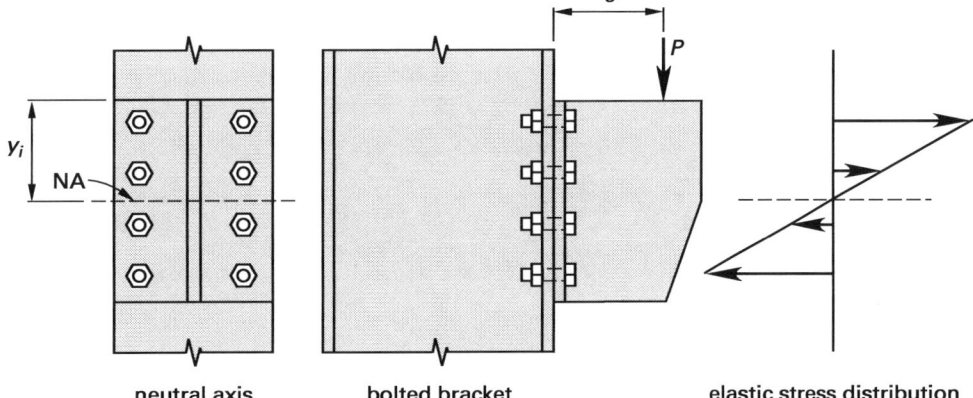

neutral axis   bolted bracket   elastic stress distribution

Example 4.13

Determine whether the $^7/_8$ in diameter A325N bearing-type bolts in the bolted bracket shown in the illustration are adequate. Prying action may be neglected.

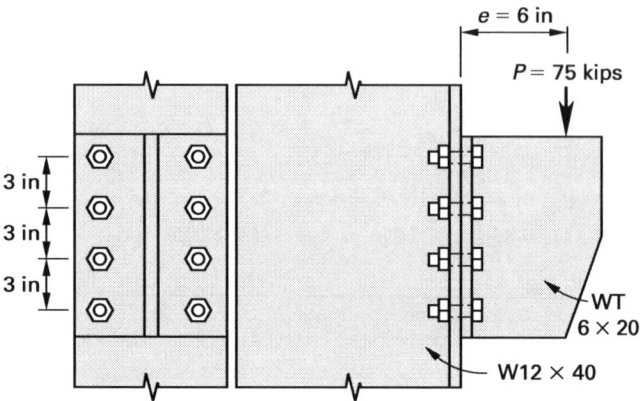

Solution

Assuming the neutral axis occurs at the centroid of the bolt group, and applying the unit area method, the inertia about the x-axis is

$$I_x = \sum y^2 = (4)(1.5 \text{ in})^2 + (4)(4.5 \text{ in})^2$$
$$= 90 \text{ in}^2$$

The applied moment on the bolt group is

$$M = Pe = (75 \text{ kips})(6 \text{ in})$$
$$= 450 \text{ in-kips}$$

The tensile force on each of the top bolts caused by the moment is

$$T = \frac{My}{2I_x} = \frac{(450 \text{ in-kips})(4.5 \text{ in})}{(2)(90 \text{ in}^2)}$$
$$= 11.25 \text{ kips}$$

The shear force on each bolt caused by the vertical load is

$$V = \frac{P}{8} = \frac{75 \text{ kips}}{8}$$
$$= 9.38 \text{ kips}$$

Bearing is not critical. From AISC Table J3.2, the allowable shear stress on a $^7/_8$ in A325N bolt in a standard hole in single shear is

$$\frac{F_{nv}}{\Omega} = \frac{48 \frac{\text{kips}}{\text{in}^2}}{2}$$
$$= 24 \text{ kips/in}^2$$

The shear stress on a top bolt is

$$f_v = \frac{V}{A_b} = \frac{9.38 \text{ kips}}{0.601 \text{ in}^2}$$
$$= 15.61 \text{ kips/in}^2$$
$$< \frac{F_{nv}}{\Omega} \quad [\text{satisfactory}]$$
$$> \frac{0.2 F_{nv}}{\Omega}$$

The allowable tensile stress for grade A325N bolts is obtained from AISC Table J3.2 as

$$\frac{F_{nt}}{\Omega} = \frac{90 \frac{\text{kips}}{\text{in}^2}}{2}$$
$$= 45 \text{ kips/in}^2$$

The tensile stress in each of the top $^7/_8$ in diameter bolt is

$$f_t = \frac{T}{A_b} = \frac{11.25 \text{ kips}}{0.601 \text{ in}^2}$$
$$= 18.72 \frac{\text{kips}}{\text{in}^2}$$
$$< \frac{F_{nt}}{\Omega} \quad [\text{satisfactory}]$$
$$> \frac{0.2 F_{nt}}{\Omega}$$

It is necessary to investigate the effects of the combined shear and tensile stress. The nominal tensile stress F'_{nt} of a bolt subjected to combined shear and tension is given by AISC Eq. (J3-3b) as

$$F'_{nt} = 1.3 F_{nt} - \frac{\Omega f_v F_{nt}}{F_{nv}}$$
$$= (1.3)\left(90 \frac{\text{kips}}{\text{in}^2}\right) - \frac{(2)\left(15.61 \frac{\text{kips}}{\text{in}^2}\right)\left(90 \frac{\text{kips}}{\text{in}^2}\right)}{48 \frac{\text{kips}}{\text{in}^2}}$$
$$= 58.46 \text{ kips/in}^2$$

The allowable tensile stress F'_{nt}/Ω of a bolt subjected to combined shear and tension, is given by AISC Eq. (J3-2) as

$$\frac{F'_{nt}}{\Omega_t} = \frac{58.46 \frac{\text{kips}}{\text{in}^2}}{2}$$
$$= 29.23 \text{ kips/in}^2$$
$$> f_t \quad [\text{satisfactory}]$$

3. ECCENTRICALLY LOADED WELD GROUPS

Nomenclature

a	coefficient for eccentrically loaded weld group	–
A_w	effective area of the weld	in^2
C	coefficient for eccentrically loaded weld group	–
C_1	correction factor for electrode strength given in Tables 4.1 and 4.2	–
D	number of sixteenths of an inch in the weld size	–
F_{EXX}	classification of weld metal	–
F_w	nominal strength of weld electrode	kips/in^2
k	coefficient for eccentrically loaded weld group	–
l	characteristic length of weld group used in tabulated values of instantaneous center method	in
$\bar{l}$	total length of weld	in
M	maximum unfactored moment caused by service loads	in-lbf
P	externally applied load	kips
P_u	externally applied factored load	kips
q	allowable strength of a $1/16$ in fillet weld per in run	kips/in per $1/16$ in
q_u	design strength of a $1/16$ in fillet weld per in run	kips/in per $1/16$ in
R_n	nominal strength	kips
t_e	effective throat thickness of fillet weld	in
w	fillet weld size	in

Weld Group Eccentrically Loaded in the Plane of the Faying Surface, LRFD Method

Eccentrically loaded weld groups of the type shown in Fig. 4.15 may be conservatively designed by means of the elastic vector analysis technique, assuming unit size of weld. The polar moment of inertia of the bolt group about the centroid is

$$I_o = I_x + I_y$$

For a total length of weld $\bar{l}$, the vertical force per linear in of weld caused by the applied load P_u, is

$$V_P = \frac{P_u}{\bar{l}}$$

The vertical force at point i, caused by the eccentricity e, is

$$V_e = \frac{P_u e x_i}{I_o}$$

The horizontal force at point i, caused by the eccentricity e, is

$$H_e = \frac{P_u e y_i}{I_o}$$

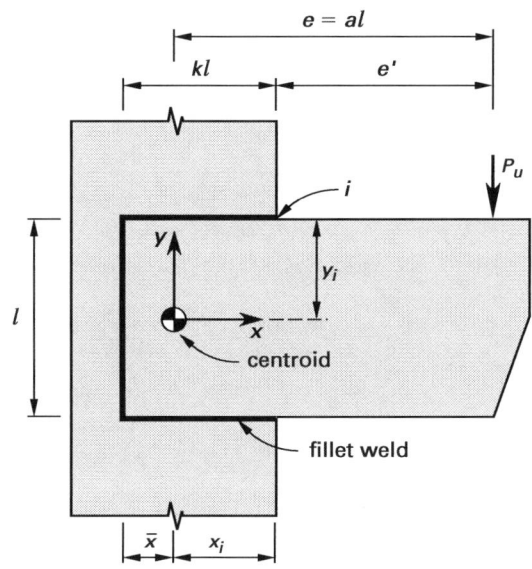

Figure 4.15 Eccentrically Loaded Weld Group, LRFD Method

The resultant force at point i is

$$R = \sqrt{(V_P + V_e)^2 + H_e^2}$$

For design purposes, it is convenient to determine the design strength of a $1/16$ in fillet weld per in run of E70XX grade electrodes, which is

$$\begin{aligned} q_u &= \phi F_w t_e \\ &= (0.75)(0.6)\left(70\ \frac{\text{kips}}{\text{in}^2}\right)(0.707)\left(\frac{1}{16}\ \text{in}\right) \\ &= 1.39\ \text{kips/in per 1/16 in} \end{aligned}$$

Example 4.14

Determine the size of E70XX fillet weld required in the welded bracket shown in the illustration. Use the elastic unit area method.

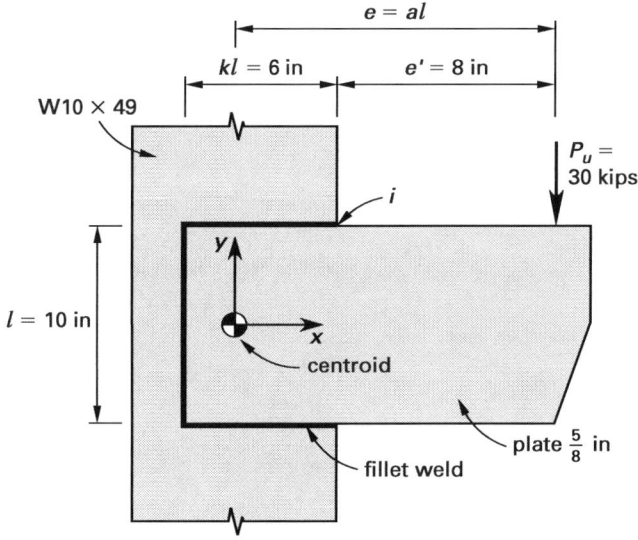

Solution

Assuming unit size of weld, the properties of the weld group are obtained by applying the elastic vector technique. The total length of the weld is

$$\bar{l} = l + 2kl = 10 \text{ in} + (2)(6 \text{ in})$$
$$= 22 \text{ in}$$

The centroid location, given by AISC Table 8-8 for a value of the coefficient $k = 0.60$, for an eccentrically loaded weld group, is

$$\bar{x} = xl = (0.164)(10 \text{ in})$$
$$= 1.64 \text{ in}$$

The moment of inertia about the x-axis is

$$I_x = \frac{l^3}{12} + 2kl\left(\frac{l}{2}\right)^2$$
$$= \frac{(10 \text{ in})^3}{12} + (2)(6 \text{ in})\left(\frac{10 \text{ in}}{2}\right)^2$$
$$= 383 \text{ in}^4/\text{in}$$

The moment of inertia about the y-axis is

$$I_y = \frac{2(kl)^3}{12} + 2kl\left(\frac{kl}{2} - \bar{x}\right)^2 + l\bar{x}^2$$
$$= \frac{(2)(6 \text{ in})^3}{12} + (2)(6 \text{ in})\left(\frac{6 \text{ in}}{2} - 1.64 \text{ in}\right)^2$$
$$+ (10 \text{ in})(1.64 \text{ in})^2$$
$$= 85 \text{ in}^4/\text{in}$$

The polar moment of inertia is

$$I_o = I_x + I_y$$
$$= 383 \, \frac{\text{in}^4}{\text{in}} + 85 \, \frac{\text{in}^4}{\text{in}}$$
$$= 468 \text{ in}^4/\text{in}$$

The eccentricity of the applied load about the centroid of the weld profile is

$$e = e' + kl - \bar{x}$$
$$= 8 \text{ in} + 6 \text{ in} - 1.64 \text{ in}$$
$$= 12.36 \text{ in}$$

The top right-hand corner of the weld profile is the most highly stressed, and the coexistent forces acting at this point in the x-direction and y-direction are the following.

The vertical force caused by the applied load is

$$V_P = \frac{P_u}{\bar{l}} = \frac{30 \text{ kips}}{22 \text{ in}}$$
$$= 1.36 \text{ kips/in}$$

The vertical force caused by the eccentricity is

$$V_e = \frac{P_u e x_i}{I_o}$$
$$= \frac{(30 \text{ kips})(12.36 \text{ in})(4.36 \text{ in})}{468 \, \frac{\text{in}^4}{\text{in}}}$$
$$= 3.45 \text{ kips/in}$$

The horizontal force caused by the eccentricity is

$$H_e = \frac{P_u e y_i}{I_o}$$
$$= \frac{(30 \text{ kips})(12.36 \text{ in})(5 \text{ in})}{468 \, \frac{\text{in}^4}{\text{in}}}$$
$$= 3.96 \text{ kips/in}$$

The resultant force is

$$R = \sqrt{(V_P + V_e)^2 + H_e^2}$$
$$= \sqrt{\left(1.36 \, \frac{\text{kips}}{\text{in}} + 3.45 \, \frac{\text{kips}}{\text{in}}\right)^2 + \left(3.96 \, \frac{\text{kips}}{\text{in}}\right)^2}$$
$$= 6.23 \text{ kips/in}$$

The required fillet weld size per $1/16$ in is

$$D = \frac{R}{q_u} = \frac{6.23 \, \frac{\text{kips}}{\text{in}}}{1.39 \, \frac{\text{kips}}{\text{in}} \text{ per } \frac{1}{16} \text{ in}}$$
$$= 4.5 \text{ sixteenths}$$

Use a weld size of

$$w = \frac{5}{16} \text{ in}$$

The flange thickness of the W10 × 49 is

$$t_f = 0.560 \text{ in}$$
$$\approx \frac{9}{16} \text{ in}$$

From AISC Table J2.4, the minimum size of fillet weld for $9/16$ in thick material is

$$w_{\min} = \frac{1}{4} \text{ in}$$
$$< \frac{5}{16} \text{ in} \quad [\text{satisfactory}]$$

From AISC Sec. J2.2b, the maximum size of fillet weld for the $5/8$ in plate is

$$w_{max} = \frac{5}{8} \text{ in} - \frac{1}{16} \text{ in}$$
$$= \frac{9}{16} \text{ in}$$
$$> \frac{5}{16} \text{ in} \quad [\text{satisfactory}]$$

Weld Group Eccentrically Loaded in the Plane of the Faying Surface, ASD Method

Eccentrically loaded weld groups of the type shown in Fig. 4.16 may be conservatively designed by means of the elastic vector analysis technique, assuming unit size of weld. The polar moment of inertia of the bolt group about the centroid is

$$I_o = I_x + I_y$$

Figure 4.16 Eccentrically Loaded Weld Group, ASD Method

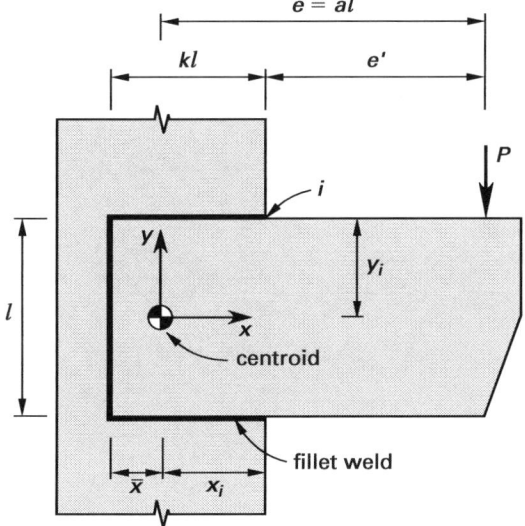

For a total length of weld $\bar{l}$, the vertical force per linear inch of weld, caused by the applied load P, is

$$V_P = \frac{P}{\bar{l}}$$

The vertical force at point i, caused by the eccentricity e, is

$$V_e = \frac{Pex_i}{I_o}$$

The horizontal force at point i, caused by the eccentricity e, is

$$H_e = \frac{Pey_i}{I_o}$$

The resultant force at point i is

$$R = \sqrt{(V_p + V_e)^2 + H_e^2}$$

For design purposes, it is convenient to determine the strength of a $1/16$ in fillet weld per in run of E70XX grade electrodes, which is

$$q = 0.3 F_{u,\text{weld}} t_e$$
$$= (0.3)\left(70 \; \frac{\text{kips}}{\text{in}^2}\right)(0.707)\left(\frac{1}{16} \text{ in}\right)$$
$$= 0.928 \text{ kips/in per } 1/16 \text{ in}$$

Example 4.15

Determine the size of E70XX fillet weld required in the welded bracket shown in the illustration. Use the elastic unit area method.

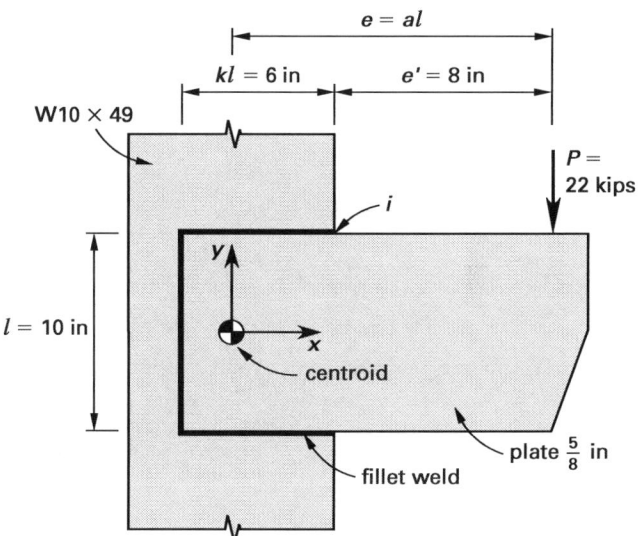

Solution

Assuming unit size of weld, the properties of the weld group are obtained by applying the elastic vector technique. The total length of the weld is

$$\bar{l} = l + 2kl = 10 \text{ in} + (2)(6 \text{ in})$$
$$= 22 \text{ in}$$

The centroid location is given by AISC Table 8-8, for a value of $k = 0.60$, as

$$\bar{x} = xl = (0.163)(10 \text{ in})$$
$$= 1.63 \text{ in}$$

The moment of inertia about the x-axis is

$$I_x = \frac{l^3}{12} + 2kl\left(\frac{l}{2}\right)^2$$
$$= \frac{(10 \text{ in})^3}{12} + (2)(6 \text{ in})\left(\frac{10 \text{ in}}{2}\right)^2$$
$$= 383 \text{ in}^4/\text{in}$$

The moment of inertia about the y-axis is

$$I_y = \frac{(2kl)^3}{12} + 2kl\left(\frac{kl}{2} - \bar{x}\right)^2 + l\bar{x}^2$$
$$= \frac{(2)(6 \text{ in})^3}{12} + (2)(6 \text{ in})\left(\frac{6 \text{ in}}{2} - 1.63 \text{ in}\right)^2$$
$$+ (10 \text{ in})(1.63 \text{ in})^2$$
$$= 85 \text{ in}^4/\text{in}$$

The polar moment of inertia is

$$I_o = I_x + I_y$$
$$= 383 \frac{\text{in}^4}{\text{in}} + 85 \frac{\text{in}^4}{\text{in}}$$
$$= 468 \text{ in}^4/\text{in}$$

The eccentricity of the applied load about the centroid of the weld profile is

$$e = e' + kl - \bar{x}$$
$$= 8 \text{ in} + 6 \text{ in} - 1.63 \text{ in}$$
$$= 12.37 \text{ in}$$

The top right-hand corner of the weld profile is the most highly stressed, and the coexistent forces acting at this point in the x-direction and y-direction are the following.

The vertical force caused by the applied load is

$$V_P = \frac{P}{l} = \frac{22 \text{ kips}}{22 \text{ in}}$$
$$= 1.00 \text{ kips/in}$$

The vertical force caused by the eccentricity is

$$V_e = \frac{Pex_i}{I_o}$$
$$= \frac{(22 \text{ kips})(12.37 \text{ in})(4.37 \text{ in})}{468 \frac{\text{in}^4}{\text{in}}}$$
$$= 2.54 \text{ kips/in}$$

The horizontal force caused by the eccentricity is

$$H_e = \frac{Pey_i}{I_o}$$
$$= \frac{(22 \text{ kips})(12.37 \text{ in})(5 \text{ in})}{468 \frac{\text{in}^4}{\text{in}}}$$
$$= 2.91 \text{ kips/in}$$

The resultant force is

$$R = \sqrt{(V_P + V_e)^2 + H_e^2}$$
$$= \sqrt{\left(1.00 \frac{\text{kips}}{\text{in}} + 2.54 \frac{\text{kips}}{\text{in}}\right)^2 + \left(2.91 \frac{\text{kips}}{\text{in}}\right)^2}$$
$$= 4.58 \text{ kips/in}$$

The required fillet weld size per $1/16$ in is

$$D = \frac{R}{q}$$
$$= \frac{4.58 \frac{\text{kips}}{\text{in}}}{0.928 \frac{\text{kips}}{\text{in}} \text{ per } \frac{1}{16} \text{ in}}$$
$$= 4.94 \text{ sixteenths}$$

Use a weld size of

$$w = \frac{5}{16} \text{ in}$$

The flange thickness of the W10 × 49 is

$$t_f = 0.560 \text{ in}$$
$$\approx \frac{9}{16} \text{ in}$$

From AISC Table J2.4, the minimum size of fillet weld for $9/16$ in thick material is

$$w_{\min} = \frac{1}{4} \text{ in}$$
$$< \frac{5}{16} \text{ in} \quad [\text{satisfactory}]$$

From AISC Sec. J2.2b, the maximum size of fillet weld for the $5/8$ in plate is

$$w_{\max} = \frac{5}{8} \text{ in} - \frac{1}{16} \text{ in}$$
$$= \frac{9}{16} \text{ in}$$
$$> \frac{5}{16} \text{ in} \quad [\text{satisfactory}]$$

Instantaneous Center of Rotation, LRFD Method

The instantaneous center of rotation method for analyzing eccentrically loaded weld groups affords a more realistic estimate of a weld group's capacity. The eccentrically loaded member is assumed to rotate about a point designated the instantaneous center of rotation.

As shown in Fig. 4.17, the weld that is farthest from the instantaneous center will experience the greatest deformation. It will reach the nominal strength first as the applied load is increased. The deformations of other weld elements are proportional to their distances from the instantaneous center.

Figure 4.17 *Instantaneous Center of Rotation, LRFD Method*

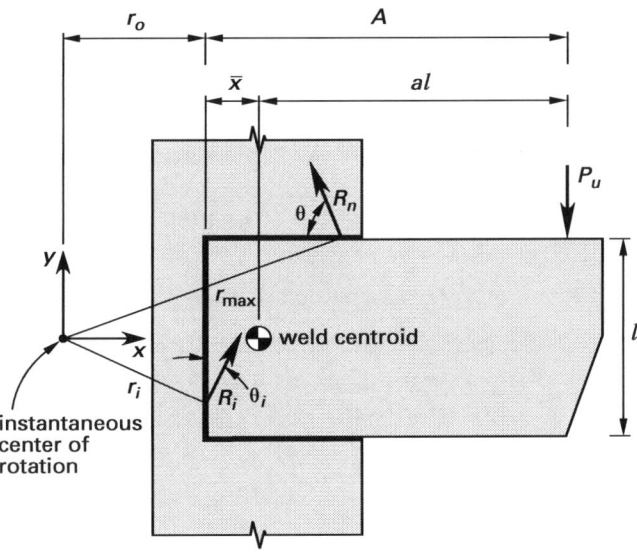

AISC Tables 8-4–8-11 tabulate the coefficient C for common weld group patterns given by

$$C = \frac{R_n}{C_1 D l}$$

$$R_n = \frac{P_u}{\phi}$$

Correction factors C_1 are given in Table 4.1.

Table 4.1 *Correction Factor for Electrode Strength, LRFD*

electrode	E60	E70	E80	E90	E100
C_1	0.857	1.0	1.03	1.16	1.21

Example 4.16

Using the instantaneous center of rotation method, determine the size of E70XX fillet weld required in the welded bracket designed in Ex. 4.14.

Solution

The relevant details obtained from Ex. 4.14 are

$$l = 10 \text{ in}$$
$$e' = 8 \text{ in}$$
$$kl = 6 \text{ in}$$
$$k = 0.6$$

From AISC Table 8-8,

$$\bar{x} = 0.164l$$
$$= (0.164)(10 \text{ in})$$
$$= 1.64 \text{ in}$$
$$a = \frac{e' + kl - \bar{x}}{l}$$
$$= \frac{8 \text{ in} + 6 \text{ in} - 1.64 \text{ in}}{10 \text{ in}}$$
$$= 1.24 \text{ in}$$

From AISC Table 8-8, for values of $a = 1.24$ and $k = 0.6$, the coefficient C is given as

$$C = 1.59$$

For E70XX electrodes, the correction factor for electrode strength is obtained from Table 4.1 as

$$C_1 = 1.0$$

The required weld size is

$$D = \frac{P_u}{\phi C C_1 l}$$
$$= \frac{30 \text{ kips}}{(0.75)(1.59)(1.0)(10 \text{ in})}$$
$$= 2.5 \text{ sixteenths}$$

The flange thickness of the W10 × 49 is

$$t_f = 0.560 \text{ in}$$
$$\approx \frac{9}{16} \text{ in}$$

From AISC Table J2.4, the minimum size of fillet weld for $9/16$ in thick material is

$$w_{\min} = \frac{1}{4} \text{ in}$$

Use a weld size of

$$w = \frac{1}{4} \text{ in}$$

Instantaneous Center of Rotation, ASD Method

The instantaneous center of rotation method for analyzing eccentrically loaded weld groups affords a more realistic estimate of a weld group's capacity. The eccentrically loaded member is assumed to rotate about a point designated the instantaneous center of rotation.

As shown in Fig. 4.18, the weld that is farthest from the instantaneous center will experience the greatest deformation. It will reach maximum strength first as the applied load is increased. The deformations of other weld elements are proportional to their distances from the instantaneous center.

Figure 4.18 Instantaneous Center of Rotation, ASD Method

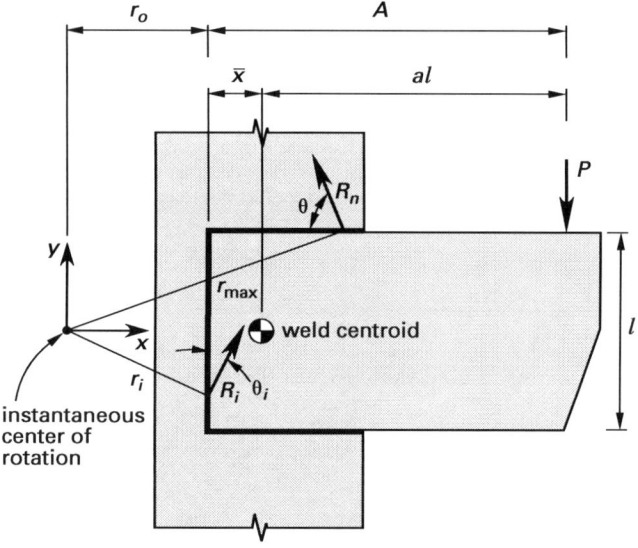

AISC Tables 8-4–8-11 tabulate the coefficient C for common weld group patterns given by

$$C = \frac{R_n}{C_1 D l}$$

$$R_n = \Omega P$$

Correction factors C_1 are given in Table 4.2.

Table 4.2 Correction Factor for Electrode Strength, ASD

electrode	E60	E70	E80	E90	E100
C_1	0.857	1.0	1.03	1.16	1.21

Example 4.17

Using the instantaneous center of rotation method, determine the size of E70XX fillet weld required in the welded bracket designed in Ex. 4.15.

Solution

The relevant details from Ex. 4.15 are

$$l = 10 \text{ in}$$
$$e' = 8 \text{ in}$$
$$kl = 6 \text{ in}$$
$$k = 0.6$$

From AISC Table 8-8,

$$\overline{x} = 0.164l$$
$$= (0.164)(10 \text{ in})$$
$$= 1.64 \text{ in}$$
$$a = \frac{e' + kl - \overline{x}}{l}$$
$$= \frac{8 \text{ in} + 6 \text{ in} - 1.64 \text{ in}}{10 \text{ in}}$$
$$= 1.24 \text{ in}$$

From AISC Table 8-8 for values of $a = 1.24$ and $k = 0.6$, the coefficient C is given as

$$C = 1.59$$

For E70XX electrodes, the correction factor for electrode strength is obtained from Table 4.2 as

$$C_1 = 1.0$$

The required weld size is

$$D = \frac{\Omega P}{C C_1 l}$$
$$= \frac{(2)(22 \text{ kips})}{(1.59)(1.0)(10 \text{ in})}$$
$$= 2.8 \text{ sixteenths}$$

Use a weld size of

$$w = \frac{1}{4} \text{ in}$$

Weld Group Eccentrically Loaded Normal to the Faying Surface, LRFD Method

Eccentrically loaded weld groups of the type shown in Fig. 4.19 may be conservatively designed by means of the elastic vector analysis technique, assuming unit size of weld. For a total length of weld $\bar{l}$, the vertical force per linear inch of weld caused by the applied load P_u is

$$V_P = \frac{P_u}{\bar{l}}$$
$$= P_u/2l$$

Figure 4.19 Weld Group Eccentrically Loaded Normal to the Faying Surface, LRFD Method

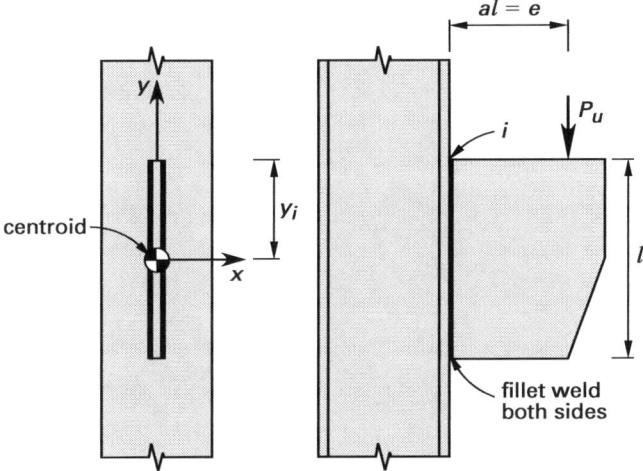

The moment of inertia about the x-axis is

$$I_x = \frac{2l^3}{12}$$
$$= l^3/6$$

The horizontal force at point i, caused by the eccentricity e, is

$$H_e = \frac{P_u e y_i}{I_x}$$
$$= 3P_u e/l^2$$

The resultant force at point i is

$$R = \sqrt{V_P^2 + H_e^2}$$

The instantaneous center of rotation method may also be used to analyze weld groups eccentrically loaded normal to the faying surface. AISC Table 8-4 provides a means of designing weld groups by this method.

Example 4.18

Determine the required size of E70XX fillet weld in the welded gusset plate shown in the illustration. Use the elastic unit area method, and compare with the instantaneous center of rotation method.

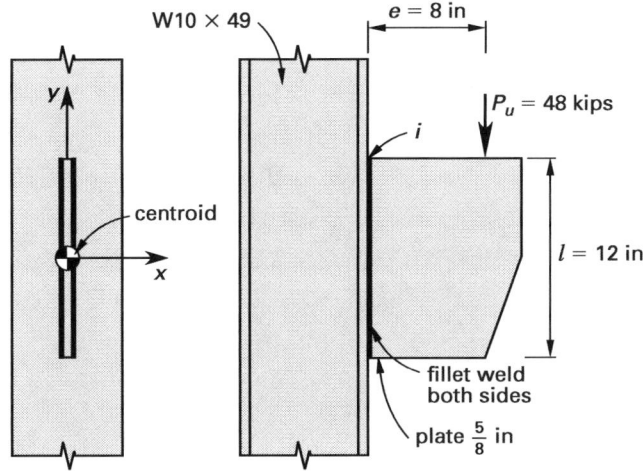

Solution

Assuming unit size of weld, the properties of the weld group are obtained by applying the elastic vector technique. The total length of the weld is

$$\bar{l} = 2l = (2)(12 \text{ in})$$
$$= 24 \text{ in}$$

The vertical force per linear inch of weld, caused by the applied load P_u, is

$$V_P = \frac{P_u}{2l} = \frac{48 \text{ kips}}{(2)(12 \text{ in})}$$
$$= 2 \text{ kips/in}$$

The moment of inertia about the x-axis is

$$I_x = \frac{2l^3}{12} = \frac{(2)(12 \text{ in})^3}{12}$$
$$= 288 \text{ in}^4/\text{in}$$

The horizontal force at point i, caused by the eccentricity e, is

$$H_e = \frac{P_u e y_i}{I_x} = \frac{(48 \text{ kips})(8 \text{ in})(6 \text{ in})}{288 \dfrac{\text{in}^4}{\text{in}}}$$
$$= 8 \text{ kips/in}$$

The resultant force at point i is

$$R = \sqrt{V_P^2 + H_e^2}$$
$$= \sqrt{\left(2.0 \dfrac{\text{kips}}{\text{in}}\right)^2 + \left(8.0 \dfrac{\text{kips}}{\text{in}}\right)^2}$$
$$= 8.25 \text{ kips/in}$$

The required fillet weld size per $1/16$ inch is

$$D = \frac{R}{q_u}$$

$$= \frac{5.46 \; \frac{\text{kips}}{\text{in}}}{1.39 \; \frac{\text{kips}}{\text{in}} \; \text{per} \; \frac{1}{16} \; \text{in}}$$

$$= 3.9 \; \text{sixteenths}$$

Use a weld size of

$$w = \frac{3}{8} \; \text{in}$$

From AISC Table 8-4, for values of $a = 0.67$ and $k = 0$, the coefficient C is given as 1.83. The required fillet weld size per $1/16$ in, based on the instantaneous center of rotation method, is

$$D = \frac{P_u}{\phi C l}$$

$$= \frac{48 \; \text{kips}}{(0.75)\left(1.83 \; \frac{\text{kips}}{\text{in}} \; \text{per} \; \frac{1}{16} \; \text{in}\right)(12 \; \text{in})}$$

$$= 2.9 \; \text{sixteenths}$$

The flange thickness of the W10 × 49 is

$$t_f = 0.560 \; \text{in}$$

$$\approx \frac{9}{16} \; \text{in}$$

From AISC Table J2.4, the minimum size of fillet weld for $9/16$ in thick material is

$$w_{\min} = \frac{1}{4} \; \text{in}$$

Use a weld size of

$$w_{\min} = \frac{1}{4} \; \text{in}$$

Weld Group Eccentrically Loaded Normal to the Faying Surface, ASD Method

Eccentrically loaded bolt groups of the type shown in Fig. 4.20 may be conservatively designed by means of the elastic vector analysis technique, assuming unit size of weld. For a total length of weld $\bar{l}$, the vertical force per linear inch of weld caused by the applied load P is

$$V_P = \frac{P}{\bar{l}}$$

$$= \frac{P}{2l}$$

Figure 4.20 Weld Group Eccentrically Loaded Normal to the Faying Surface, ASD Method

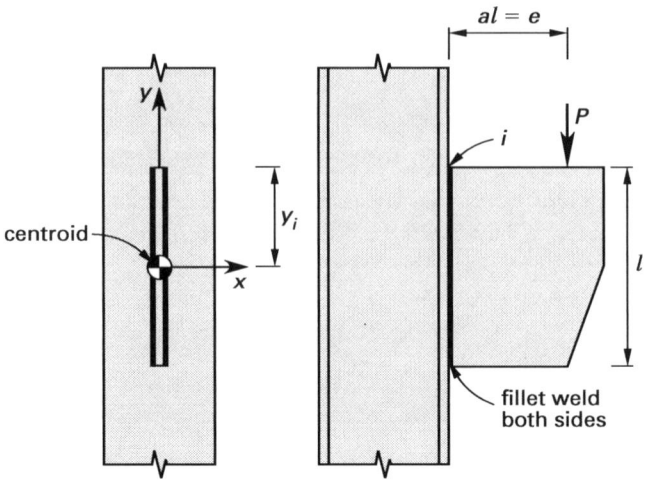

The moment of inertia about the x-axis is

$$I_x = \frac{2l^3}{12}$$

$$= l^3/6$$

The horizontal force at point i, caused by the eccentricity e, is

$$H_e = \frac{Pey_i}{I_x}$$

$$= \frac{3Pe}{l^2}$$

The resultant force at point i is

$$R = \sqrt{V_P^2 + H_e^2}$$

The instantaneous center of rotation method may also be used to analyze weld groups eccentrically loaded normal to the faying surface. AISC Table 8-4 provides a means of designing weld groups by this method.

Example 4.19

Determine the required size of E70XX fillet weld in the welded gusset plate shown in the illustration. Use the elastic unit area method, and compare with the instantaneous center of rotation method.

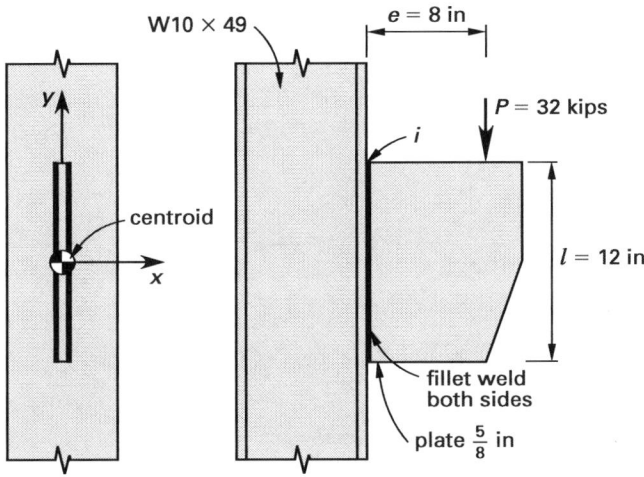

Solution

Assuming unit size of weld, the properties of the weld group are obtained by applying the elastic vector technique. The total length of the weld is

$$\bar{l} = 2l$$
$$= 24 \text{ in}$$

The vertical force per linear inch of weld, caused by the applied load P, is

$$V_P = \frac{P}{2l} = \frac{32 \text{ kips}}{24 \text{ in}}$$
$$= 1.3 \text{ kips/in}$$

The moment of inertia about the x-axis is

$$I_x = \frac{2l^3}{12} = \frac{(2)(12 \text{ in})^3}{12}$$
$$= 288 \text{ in}^4/\text{in}$$

The horizontal force at point i, caused by the eccentricity e, is

$$H_e = \frac{Pey_i}{I_x} = \frac{(32 \text{ kips})(8 \text{ in})(6 \text{ in})}{288 \frac{\text{in}^4}{\text{in}}}$$
$$= 5.3 \text{ kips/in}$$

The resultant force at point i is

$$R = \sqrt{V_P^2 + H_e^2}$$
$$= \sqrt{\left(1.3 \frac{\text{kips}}{\text{in}}\right)^2 + \left(5.3 \frac{\text{kips}}{\text{in}}\right)^2}$$
$$= 5.46 \text{ kips/in}$$

The required fillet weld size per $1/16$ inch is

$$D = \frac{R}{q}$$
$$= \frac{5.46 \frac{\text{kips}}{\text{in}}}{0.928 \frac{\text{kips}}{\text{in}} \text{ per } \frac{1}{16} \text{ in}}$$
$$= 5.9 \text{ sixteenths}$$

Use a weld size of

$$w = \frac{3}{8} \text{ in}$$

From AISC Table 8-4, for values of $a = 0.67$ and $k = 0$, the coefficient C is given as

$$C = 1.83$$

For E70XX electrodes, the correction factor for electrode strength is obtained from Table 4.2 as

$$C_1 = 1.0$$

The required weld size is

$$D = \frac{\Omega P}{CC_1 l}$$
$$= \frac{(2)(32 \text{ kips})}{(1.83)(1.0)(12 \text{ in})}$$
$$= 2.9 \text{ sixteenths}$$

The flange thickness of the W10 × 49 is

$$t_f = 0.560 \text{ in}$$
$$\approx \frac{9}{16} \text{ in}$$

From AISC Table J2.4, the minimum size of fillet weld for $9/16$ in thick material is

$$w_{\min} = \frac{1}{4} \text{ in}$$

Use a weld size of

$$w_{\min} = \frac{1}{4} \text{ in}$$

4. COMPOSITE BEAMS

Nomenclature

a	depth of compression block	in
A_c	area of concrete slab within the effective width	in^2
A_{ctr}	concrete transformed area in compression	in^2
A_s	cross-sectional area of structural steel	in^2
b	effective concrete flange width	in

C_{con}	compressive force in a concrete slab at ultimate load	kips
d	depth of steel beam	in
f'_c	specified compressive strength of the concrete	kips/in^2
f_u	compressive stress in the concrete stress block	kips/in^2
F_y	specified minimum yield stress of the structural steel section	kips/in^2
h	overall height of the transformed composite section	in
L	span length	ft
M	applied bending moment	in-lbf
M_n	nominal flexural strength of member	in-kips or ft-kips
n	number of shear connectors between point of maximum positive moment and point of zero moment	–
Q_n	nominal shear strength of single shear connector	kips
s	beam spacing	ft or in
t_c	actual slab thickness	in
T_{stl}	tensile force in steel at ultimate load	kips
V'	total factored horizontal shear between point of maximum moment and point of zero moment	kips
y	moment arm between centroids of tensile force and compressive force	in
Y_{con}	distance from top of steel beam to top of concrete	in
Y_1	distance from top of steel beam to plastic neutral axis	in
Y_2	distance from top of steel beam to concrete flange force	in

Symbols

$\sum Q_n$	summation of Q_n between point of maximum moment and point of zero moment on either side	kips

General Considerations

As shown in Fig. 4.21, a composite beam consists of a concrete slab acting integrally with a steel beam to resist the applied loads. Shear connectors are welded to the top flange of the beam to provide composite action between the slab and the beam. Because the compressive stress in the concrete slab is not uniform over the width of the slab, an effective width is assumed in calculations. In accordance with AISC Sec. I3.1, the effective width of the concrete slab on either side of the beam center line is defined as the lesser of

- one-eighth of the beam span
- one-half of the beam spacing
- the distance to the edge of the slab

Composite Beam Design, LRFD Method

In accordance with AISC Sec. I1.1a, a rectangular stress block with depth a is formed in the concrete slab at the ultimate load. The stress in the stress block is

$$f_u = 0.85 f'_c$$

The compressive force in the concrete slab is

$$C_{\text{con}} = b a f_u$$
$$= 0.85 f'_c b a$$

As shown in Fig. 4.22, when the stress block depth is less than the depth of the slab, the plastic neutral axis is located at the top of the steel beam. The beam has fully yielded in tension at the yield stress F_y. The force in the steel beam is

$$T_{\text{stl}} = F_y A_s$$

Equating horizontal forces gives

$$T_{\text{stl}} = C_{\text{con}}$$
$$F_y A_s = 0.85 f'_c b a$$

The depth of the stress block, when sufficient shear connectors are provided to ensure full composite action, is

$$a = \frac{F_y A_s}{0.85 f'_c b}$$

Figure 4.21 Composite Section

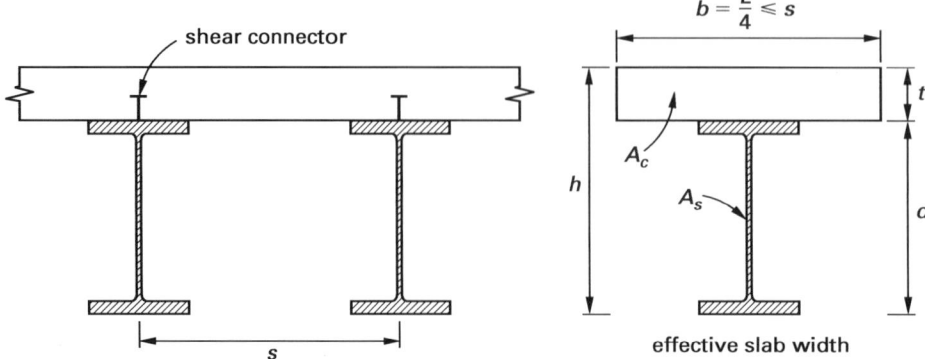

Figure 4.22 Fully Composite Beam at Ultimate Load

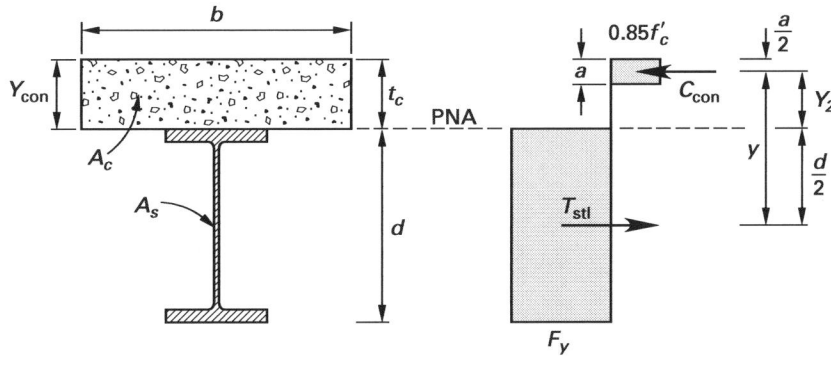

The distance from top of steel beam to the centroid of the compressive force is

$$Y_2 = Y_{\text{con}} - \frac{a}{2}$$

The moment arm between the centroids of the tensile and compressive forces is

$$y = Y_2 + \frac{d}{2}$$

The nominal flexural capacity of the fully composite section is

$$M_n = T_{\text{stl}} y$$
$$= T_{\text{stl}} \left(Y_2 + \frac{d}{2} \right)$$

The design flexural capacity is

$$\phi M_n = 0.9 M_n$$

For given values of Y_2 and $\sum Q_n$, AISC Table 3-19 provides values of ϕM_n for a range of W sections and for the three cases

- the plastic neutral axis located in the steel beam
- the plastic neutral axis located at the top of the steel beam
- shear connectors inadequate to provide full composite action

For all three cases the depth of the stress block is

$$a = \frac{\sum Q_n}{0.85 f'_c b}$$

$\sum Q_n$ is the smaller of

- $0.85 f'_c A_c$, plastic neutral axis is located in the steel beam
- $F_y A_s$, plastic neutral axis is located at the top of the steel beam
- $n Q_n$, shear connectors are inadequate to provide full composite action

Composite beams may be constructed using either of two methods. In shored construction, the steel beam is propped before casting the slab, to eliminate stresses in the steel beam caused by the deposited concrete. In unshored construction, the slab is cast without propping the steel beam. For both shored and unshored beams, the composite section is designed to support the total factored loads, coming from all dead and live loads. For unshored construction, the steel beam alone must be adequate to support all loads applied before the concrete has attained 75% of its required strength.

Example 4.20

A simply supported composite beam consists of a concrete slab with a depth of $t_c = Y_{\text{con}} = 4$ in, directly supported on a W21 × 57 grade 50 steel beam. The beam spacing is $s = 9$ ft, and the span is $L = 30$ ft. The slab consists of normal weight concrete with a strength of $f'_c = 4000$ lbf/in^2. Determine the design moment capacity of the beam if full composite action is provided.

Solution

The effective width b of the concrete slab, in accordance with AISC Sec. I3.1, is the lesser of

$$s = (9 \text{ ft}) \left(12 \frac{\text{in}}{\text{ft}} \right) = 108 \text{ in}$$

$$\frac{L}{4} = \frac{(30 \text{ ft}) \left(12 \frac{\text{in}}{\text{ft}} \right)}{4}$$
$$= 90 \text{ in} \quad [\text{governs}]$$

For full composite action with the plastic neutral axis located at the top of the steel beam

$$\sum Q_n = F_y A_s$$
$$= \left(50 \frac{\text{kips}}{\text{in}^2} \right) (16.7 \text{ in}^2)$$
$$= 835 \text{ kips}$$

The distance from the top of the steel beam to the plastic neutral axis is

$$Y_1 = 0 \text{ in}$$

The depth of the stress block is

$$a = \frac{F_y A_s}{0.85 f'_c b}$$

$$= \frac{835 \text{ kips}}{(0.85)\left(4 \ \frac{\text{kips}}{\text{in}^2}\right)(90 \text{ in})}$$

$$= 2.73 \text{ in}$$

The distance from the top of the steel beam to the line of action of the concrete slab force is

$$Y_2 = Y_{\text{con}} - \frac{a}{2} = 4 \text{ in} - \frac{2.73 \text{ in}}{2}$$

$$= 2.64 \text{ in}$$

By interpolation from AISC Table 3-19, for $Y_1 = 0$ and $Y_2 = 2.64$ in,

$$\phi M_n = 827 \text{ ft-kips}$$

Composite Beam Design, ASD Method

In accordance with AISC Sec. I1.1a, a rectangular stress block with depth a is formed in the concrete slab at the ultimate load. The stress in the stress block is

$$f_u = 0.85 f'_c$$

The compressive force in the concrete slab is

$$C_{\text{con}} = ba f_u$$

$$= 0.85 f'_c ba$$

As shown in Fig. 4.22, when the stress block depth is less than the depth of the slab, the plastic neutral axis is located at the top of the steel beam. The beam has fully yielded in tension at the yield stress F_y. The force in the steel beam is

$$T_{\text{stl}} = F_y A_s$$

Equating horizontal forces gives

$$T_{\text{stl}} = C_{\text{con}}$$

$$F_y A_s = 0.85 f'_c ba$$

The depth of the stress block, when sufficient shear connectors are provided to ensure full composite action, is

$$a = \frac{F_y A_s}{0.85 f'_c b}$$

The distance from top of steel beam to the centroid of the compressive force is

$$Y_2 = Y_{\text{con}} - \frac{a}{2}$$

The moment arm between the centroids of the tensile and compressive forces is

$$y = Y_2 + \frac{d}{2}$$

The nominal flexural capacity of the fully composite section is

$$M_n = T_{\text{stl}} y$$

$$= T_{\text{stl}} \left(Y_2 + \frac{d}{2}\right)$$

The allowable flexural capacity is

$$\frac{M_n}{\Omega} = \frac{M_n}{1.67}$$

For given values of Y_2 and $\sum Q_n$, AISC Table 3-19 provides values of M_n/Ω for a range of W sections and for the three cases

- the plastic neutral axis located in the steel beam
- the plastic neutral axis located at the top of the steel beam
- shear connectors inadequate to provide full composite action

For all three cases the depth of the stress block is

$$a = \frac{\sum Q_n}{0.85 f'_c b}$$

$\sum Q_n$ is the smaller of

- $0.85 f'_c A_c$, plastic neutral axis is located in the steel beam
- $F_y A_s$, plastic neutral axis is located at the top of the steel beam
- nQ_n, shear connectors are inadequate to provide full composite action

Composite beams may be constructed using either of two methods. In shored construction, the steel beam is propped before casting the slab, to eliminate stresses in the steel beam caused by the deposited concrete. In unshored construction, the slab is cast without propping the steel beam. For both shored and unshored beams, the composite section is designed to support the total factored loads, coming from all dead and live loads. For unshored construction, the steel beam alone must be adequate to support all loads applied before the concrete has attained 75% of its required strength.

Example 4.21

A simply supported composite beam consists of a concrete slab with a depth of $t_c = Y_{con} = 4$ in, directly supported on a W21 × 57 grade 50 steel beam. The beam spacing is $s = 9$ ft, and the span is $L = 30$ ft. The slab consists of normal weight concrete with a strength of $f'_c = 4000$ lbf/in². Determine the allowable moment capacity of the beam if full composite action is provided.

Solution

The effective width b of the concrete slab is given by AISC Sec. I3.1 as the lesser of

$$s = (9 \text{ ft})\left(12 \frac{\text{in}}{\text{ft}}\right)$$
$$= 108 \text{ in}$$
$$\frac{L}{4} = \frac{(30 \text{ ft})\left(12 \frac{\text{in}}{\text{ft}}\right)}{4}$$
$$= 90 \text{ in} \quad [\text{governs}]$$

For full composite action with the plastic neutral axis located at the top of the steel beam

$$\sum Q_n = F_y A_s$$
$$= \left(50 \frac{\text{kips}}{\text{in}^2}\right)(16.7 \text{ in}^2)$$
$$= 835 \text{ kips}$$

The distance from the top of the steel beam to the plastic neutral axis is

$$Y_1 = 0 \text{ in}$$

The depth of the stress block is

$$a = \frac{F_y A_s}{0.85 f'_c b}$$
$$= \frac{835 \text{ kips}}{(0.85)\left(4 \frac{\text{kips}}{\text{in}^2}\right)(90 \text{ in})}$$
$$= 2.73 \text{ in}$$

The distance from the top of the steel beam to the line of action of the concrete slab force is

$$Y_2 = Y_{con} - \frac{a}{2} = 4 \text{ in} - \frac{2.73 \text{ in}}{2}$$
$$= 2.64 \text{ in}$$

By interpolation from AISC Table 3-19, for $Y_1 = 0$ and $Y_2 = 2.64$ in, the allowable moment capacity is

$$M_n/\Omega = 550 \text{ ft-kips}$$

Shear Connection Design, LRFD Method

Shear connectors are provided to transfer the horizontal shear force across the interface. The nominal shear strengths Q_n of shear stud connectors are given in AISC Table 3-21. The required number of connectors may be uniformly distributed between the point of maximum moment and the support on either side, with the total horizontal shear being determined by the lesser value of

$$V' = 0.85 f'_c A_c$$
$$V' = F_y A_s$$

To provide complete shear connection and full composite action, the required number of connectors on either side of the point of maximum moment is

$$n = \frac{V'}{Q_n}$$

If a smaller number of connectors is provided, only partial composite action can be achieved, and the nominal flexural strength of the composite member is reduced. The number of shear connectors placed between a concentrated load and the nearest support shall be sufficient to develop the moment required at the load point.

Example 4.22

Determine the number of ⅝ in diameter, shear stud connectors required in the composite beam of Ex. 4.20 to provide full composite action. The beam carries a uniformly distributed load.

Solution

The total horizontal shear resisted by the connectors is the lesser value obtained from AISC Sec. I3.2d as

$$V' = 0.85 f'_c A_c$$
$$= (0.85)\left(\frac{4000 \frac{\text{lbf}}{\text{in}^2}}{1000 \frac{\text{lbf}}{\text{kip}}}\right)(90 \text{ in})(4 \text{ in})$$
$$= 1224 \text{ kips}$$
$$V' = F_y A_s$$
$$= 835 \text{ kips} \quad [\text{from Ex. 4.20; governs}]$$

The nominal shear strength of a 2½ in long, ⅝ in diameter stud connector in normal weight 4000 lbf/in² concrete is obtained from AISC Table 3-21 as

$$Q_n = 18.1 \text{ kips}$$

For full composite action, the total number of studs required on the beam is

$$2n = \frac{2V_h}{Q_n} = \frac{(2)(835 \text{ kips})}{18.1 \text{ kips}}$$
$$= 92$$

Provide 92 studs in pairs at $7\frac{3}{4}$ in centers with end distances of $5\frac{5}{8}$ in.

The minimum longitudinal spacing requirement is given by AISC Sec. I3.2d as

$$s = 6d = (6)\left(\frac{5}{8} \text{ in}\right)$$
$$= 3.75 \text{ in}$$
$$< 7\frac{3}{4} \text{ in} \quad [\text{satisfactory}]$$

The maximum longitudinal spacing requirement is given by AISC Sec. I3.2d as

$$s = 8t = (8)(4 \text{ in})$$
$$= 32 \text{ in}$$
$$< 7\frac{3}{4} \text{ in} \quad [\text{satisfactory}]$$

The minimum transverse spacing requirement is given by AISC Sec. I3.2d as

$$s = 4d = (4)\left(\frac{5}{8} \text{ in}\right)$$
$$= 2.5 \text{ in}$$

Space the connectors 4 in apart transversely.

Shear Connection Design, ASD Method

Shear connectors are provided to transfer the horizontal shear force across the interface. The strengths, Q_n, of shear stud connectors are given in AISC Table 3-21. The required number of connectors may be uniformly distributed between the point of maximum moment and the support on either side, with the total horizontal shear being determined by the lesser value of

$$V' = 0.85f'_c A_c$$
$$V' = F_y A_s$$

To provide complete shear connection and full composite action, the required number of connectors on either side of the point of maximum moment is

$$n = \frac{V'}{Q_n}$$

If a smaller number of connectors is provided, only partial composite action can be achieved, and the nominal flexural strength of the composite member is reduced. The number of shear connectors placed between a concentrated load and the nearest support must be sufficient to develop the moment required at the load point.

Example 4.23

Determine the number of $5/8$ in diameter shear stud connectors required in the composite beam of Ex. 4.21 to provide full composite action. The beam carries a uniformly distributed load.

Solution

The total horizontal shear resisted by the connectors is the lesser value obtained from AISC Sec. I3.2d as

$$V' = 0.85f'_c A_c$$
$$= \frac{(0.85)\left(4000 \frac{\text{lbf}}{\text{in}^2}\right)(4 \text{ in})(90 \text{ in})}{1000 \frac{\text{lbf}}{\text{kip}}}$$
$$= 1224 \text{ kips}$$
$$V' = F_y A_s$$
$$= 835 \text{ kips} \quad [\text{from Ex. 4.21; governs}]$$

The nominal shear strength of a $2\frac{1}{2}$ in long, $5/8$ in diameter stud connector in normal weight 4000 lbf/in^2 concrete is obtained from AISC Table 3-21 as

$$Q_n = 18.1 \text{ kips}$$

For full composite action, the total number of studs required on the beam is

$$2n = \frac{2V_h}{Q_n} = \frac{(2)(835 \text{ kips})}{18.1 \text{ kips}}$$
$$= 92$$

Provide 92 studs in pairs at $7\frac{3}{4}$ in centers with end distances of $5\frac{5}{8}$ in.

The minimum longitudinal spacing requirement is given by AISC Sec. I3.2d as

$$s = 6d = (6)\left(\frac{5}{8} \text{ in}\right)$$
$$= 3.75 \text{ in}$$
$$< 7\frac{3}{4} \text{ in} \quad [\text{satisfactory}]$$

The maximum longitudinal spacing requirement is given by AAISC Sec. I3.2d as

$$s = 8t = (8)(4 \text{ in})$$
$$= 32 \text{ in}$$
$$> 7\frac{3}{4} \text{ in} \quad [\text{satisfactory}]$$

The minimum transverse spacing requirement is given by AISC Sec. I3.2d as

$$s = 4d = (4)\left(\frac{5}{8} \text{ in}\right)$$
$$= 2.5 \text{ in}$$

Space the connectors 4 in apart transversely.

PRACTICE PROBLEMS

1. The rigid frame shown in the illustration is fabricated from members of a uniform section in grade 50 steel. For the factored loading indicated, ignoring the member self weight and assuming adequate lateral support is provided, what is most nearly the required plastic moment of resistance?

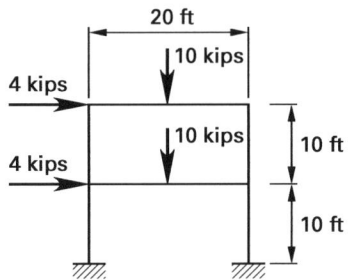

- (A) 24 ft-kips
- (B) 32 ft-kips
- (C) 40 ft-kips
- (D) 48 ft-kips

2. The bolts in the bracket shown are $3/4$ in diameter A325 bearing-type with threads included in the shear plane. Using the ASD method, what is most nearly the maximum allowable load P that the bracket can support?

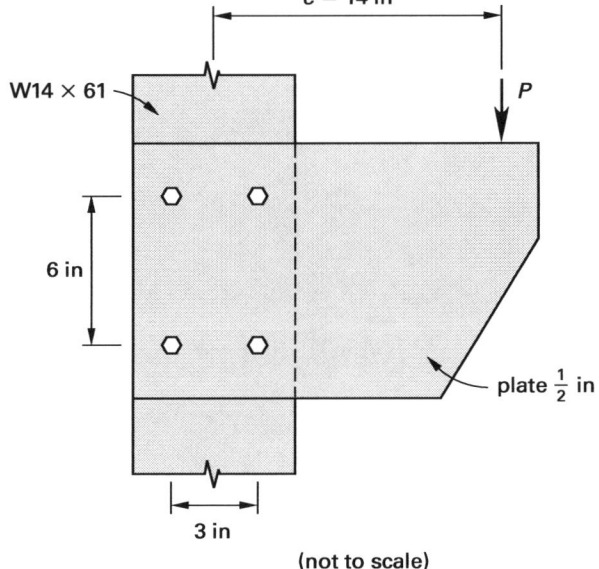

- (A) 9.54 kips
- (B) 10.1 kips
- (C) 11.3 kips
- (D) 12.5 kips

3. In the welded bracket shown, the size of E70XX fillet weld indicated is $1/4$ in. Using the instantaneous center of rotation method, what is most nearly the maximum factored force P_u that the bracket can support?

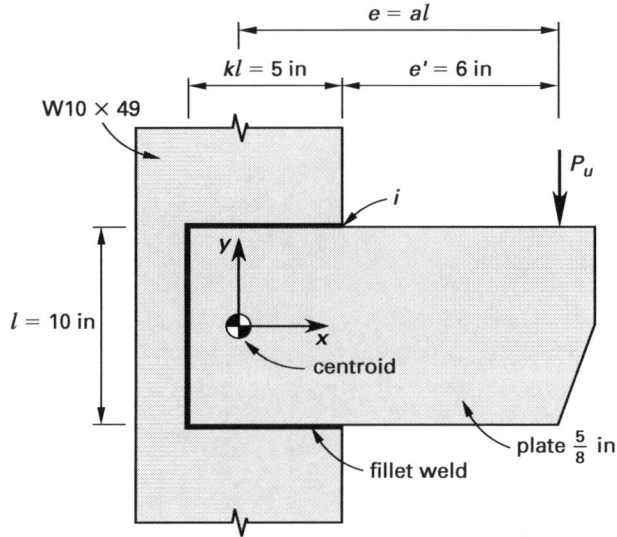

(not to scale)

- (A) 48 kips
- (B) 51 kips
- (C) 54 kips
- (D) 60 kips

4. A simply supported composite beam consists of a concrete slab with a depth of $t_c = Y_{con} = 4$ in, directly supported on a W21 × 57 grade 50 steel beam. The beam spacing is $s = 9$ ft, and the span is $L = 30$ ft. The slab consists of normal weight concrete with a strength of $f'_c = 4000$ lbf/in^2. Using the LRFD method, what is most nearly the design moment capacity of the beam if 60 number $5/8$ in diameter, shear stud connectors are provided along the length of the beam?

- (A) 736 ft-kips
- (B) 749 ft-kips
- (C) 768 ft-kips
- (D) 783 ft-kips

SOLUTIONS

1. The number of possible independent collapse mechanisms is

$$m_i = p - D = 12 - 6$$
$$= 6$$

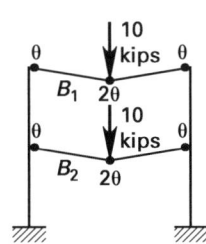

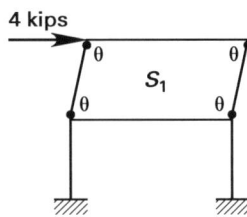

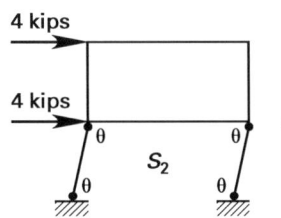

independent mechanisms

The six possible independent collapse mechanisms shown in the illustration are B_1, B_2, S_1, S_2, J_1, and J_2 mechanisms. Applying a virtual displacement to each of these mechanisms in turn, and equating internal and external work, provides the following equations.

For beam mechanism B_1,

$$4M_p = (10 \text{ kips})(10 \text{ ft})$$
$$M_p = 25 \text{ ft-kips}$$

For beam mechanism B_2,

$$M_p = 25 \text{ ft-kips}$$

For sway mechanism S_1,

$$4M_p = (4 \text{ kips})(10 \text{ ft})$$
$$M_p = 10 \text{ ft-kips}$$

For sway mechanism S_2,

$$4M_p = (8 \text{ kips})(10 \text{ ft})$$
$$M_p = 20 \text{ ft-kips}$$

For joint mechanism J_1,

$$3M_p = 0$$

For joint mechanism J_2,

$$3M_p = 0$$

For the combined mechanism $B_1 + B_2 + S_1 + S_2 + J_1 + J_2$ shown in the illustration,

$$10M_p = 100 \text{ ft-kips} + 100 \text{ ft-kips} + 40 \text{ ft-kips}$$
$$+ 80 \text{ ft-kips} + 0 + 0$$
$$= 320 \text{ ft-kips}$$
$$M_p = 32 \text{ ft-kips} \quad \text{[governs]}$$

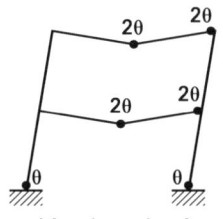

combined mechanism

The answer is (B).

2. The relevant details obtained from the illustration follow. The number of bolts in one vertical row is

$$n = 2$$

The bolt pitch is

$$b = 6 \text{ in}$$

The lateral distance between bolts is

$$D = 3 \text{ in}$$

The eccentricity is

$$e = 14 \text{ in}$$

From AISC Table 7-8, the coefficient C is

$$C = 0.90$$

Shear controls. From AISC Table 7-1, the design shear strength of a $^3/_4$ in diameter, A325N bolt in a standard hole, in single shear with threads included in the shear plane, is

$$P_v = 10.6 \text{ kips}$$

The allowable load, based on the instantaneous center of rotation method, is

$$P = P_v C = (10.6 \text{ kips})(0.90)$$
$$= 9.54 \text{ kips}$$

The answer is (A).

3. Relevant details obtained from the illustration are

$$l = 10 \text{ in}$$
$$e' = 6 \text{ in}$$
$$kl = 5 \text{ in}$$
$$k = 0.5$$

From AISC Table 8-8,

$$\bar{x} = 0.125l = (0.125)(10 \text{ in})$$
$$= 1.25 \text{ in}$$
$$a = \frac{e' + kl - \bar{x}}{l}$$
$$= \frac{6 \text{ in} + 5 \text{ in} - 1.25 \text{ in}}{10 \text{ in}}$$
$$= 0.98 \text{ in}$$

From AISC Table 8-8, for values of $a = 0.98$ and $k = 0.5$, the coefficient C is given as

$$C = 1.70$$

For E70XX electrodes, the correction factor for electrode strength is obtained from AISC Table 8-3 as

$$C_1 = 1.0$$

The maximum factored load that the bracket can support is

$$P_u = \phi C C_1 l D$$
$$= (0.75)(1.70)(1.0)(10 \text{ in})(4 \text{ sixteenths})$$
$$= 51 \text{ kips}$$

The answer is (B).

4. The effective width b of the concrete slab, in accordance with AISC Sec. I3.1, is the lesser of

$$s = (9 \text{ ft})\left(12 \ \frac{\text{in}}{\text{ft}}\right)$$
$$= 108 \text{ in}$$
$$\frac{L}{4} = \frac{(30 \text{ ft})\left(12 \ \frac{\text{in}}{\text{ft}}\right)}{4}$$
$$= 90 \text{ in} \quad [\text{governs}]$$

If the plastic neutral axis is located at the top of the steel beam

$$\sum Q_n = F_y A_s$$
$$= \left(50 \ \frac{\text{kips}}{\text{in}^2}\right)(16.7 \text{ in}^2)$$
$$= 835 \text{ kips}$$

The nominal shear strength of a $2^1/_2$ in long, $^5/_8$ in diameter shear stud connector in normal weight 4000 lbf/in² concrete is obtained from AISC Table 3-21 as

$$Q_n = 18.1 \text{ kips}$$

For 60 shear stud connectors, the shear capacity is

$$\sum Q_n = 30 Q_n$$
$$= 543 \text{ kips} \quad [\text{governs}]$$

The depth of the stress block is

$$a = \frac{\sum Q_n}{0.85 f'_c b}$$
$$= \frac{543 \text{ kips}}{(0.85)\left(4 \ \frac{\text{kips}}{\text{in}^2}\right)(90 \text{ in})}$$
$$= 1.77 \text{ in}$$

The distance from the top of the steel beam to the line of action of the concrete slab force is

$$Y_2 = Y_{\text{con}} - \frac{a}{2} = 4 \text{ in} - \frac{1.77 \text{ in}}{2}$$
$$= 3.12 \text{ in}$$

Interpolate from AISC Table 3-19, for $Y_2 = 3.12$ in and $\sum Q_n = 543$ kips,

$$\phi M_n = 783 \text{ ft-kips}$$

The answer is (D).

5 Design of Wood Structures

1. Design Principles 5-1
2. Design for Flexure 5-6
3. Design for Compression 5-10
4. Design for Tension 5-15
5. Design for Shear 5-18
6. Design of Connections 5-22
   Practice Problems 5-32
   Solutions 5-33

Glossary

The NDS Supplement (Supp.)[1] provides reference design values for the various species and types of wood products that are available. The types of wood products covered in the supplement are the following.

- Table 4A: visually graded dimension lumber, all species except Southern Pine
- Table 4B: visually graded Southern Pine dimension lumber
- Table 4C: mechanically graded dimension lumber
- Table 4D: visually graded timbers
- Table 4E: visually graded decking
- Table 4F: non-North American visually graded dimension lumber
- Table 5A: structural glued laminated softwood timber, stressed primarily in bending
- Table 5B: structural glued laminated softwood timber, stressed primarily in axial tension or compression
- Table 5C: structural glued laminated hardwood timber, stressed primarily in bending
- Table 5D: structural glued laminated hardwood timber, stressed primarily in axial tension or compression

A description and terminology for the available wood products follows.

Decking. Decking consists of solid sawn lumber or glued laminated members, 2 in to 4 in nominal thickness and 4 in or more wide. It is usually single tongue and groove, 2 in thick, and it may be double tongue and groove, 3 in or 4 in thick.

Dimension lumber. Dimension lumber consists of solid sawn lumber members, 2 in to 4 in nominal thickness and 2 in or more wide.

Dressed size. The dimensions of a lumber member after surfacing with a planing machine are usually $1/2$ in to $3/4$ in less than the nominal size.

Grade. Wood products are classified with respect to strength in accordance with specific grading rules.

Joist. A joist is a lumber member, 2 in to 4 in nominal thickness, and 5 in or wider. A joist is typically loaded

1. DESIGN PRINCIPLES

Nomenclature

b	breadth of rectangular bending member	in
C_b	bearing area factor	–
C_c	curvature factor for structural glued laminated members	–
C_D	load duration factor	–
C_F	size factor for sawn lumber	–
C_{fu}	flat use factor	–
C_i	incising factor	–
C_L	beam stability factor	–
C_M	wet service factor	–
C_P	column stability factor	–
C_r	repetitive member factor for dimension lumber	–
C_t	temperature factor	–
C_V	volume factor for structural glued laminated timber	–
d	beam depth	in
E	reference modulus of elasticity	lbf/in^2
E'	adjusted modulus of elasticity	lbf/in^2
$E_{\min}$	reference modulus of elasticity for beam stability	lbf/in^2
$E'_{\min}$	adjusted modulus of elasticity for beam stability	lbf/in^2
F_b	reference bending design value	lbf/in^2
F'_b	adjusted bending design value	lbf/in^2
F_c	reference compression design value parallel to grain	lbf/in^2
F'_c	adjusted compression design value parallel to grain	lbf/in^2
l_b	length of bearing parallel to the grain of the wood	in
L	span length of bending member	ft or in
R	radius of curvature of inside face of laminations	in
t	thickness of laminations	in
x	species parameter for volume factor	–

---
[1]American Forest and Paper Association, 2005 (See References and Codes)

Table 5.1 Applicability of Adjustment Factors

adjustment factor	sawn lumber	glued laminated	F_b	F_t	F_v	$F_{c\perp}$	F_c	E	$E_{\min}$
C_F size	yes	no	√	√	–	–	√	–	–
C_r repetitive member	yes	no	√	–	–	–	–	–	–
C_i incising	yes	no	√	√	√	√	√	√	√
C_V volume[a]	no	yes	√	–	–	–	–	–	–
C_c curvature	no	yes	√	–	–	–	–	–	–
C_D load duration	yes	yes	√	√	√	–	√	–	–
C_M wet service	yes	yes	√	√	√	√	√	√	√
C_b bearing area	yes	yes	–	–	–	√	–	–	–
C_t temperature	yes	yes	√	√	√	√	√	√	√
C_{fu} flat use	yes	yes	√	–	–	–	–	–	–
C_L beam stability[a]	yes	yes	√	–	–	–	–	–	–
C_P column stability	yes	yes	–	–	–	–	√	–	–
C_T buckling	yes	no	–	–	–	–	–	–	√

[a]When applied to glued laminated members, only the lesser value of C_L or C_V is applicable.

on the narrow face, and used as framing in floors and roofs.

Lumber. A lumber member is cut to size in the sawmill and surfaced in a planing machine, and not further processed.

Mechanically graded lumber. Mechanically graded lumber is dimension lumber that has been individually evaluated in a testing machine. Load is applied to the piece of lumber, the deflection measured, and the modulus of elasticity calculated. The strength characteristics of the lumber are directly related to the modulus of elasticity, and can be determined. A visual check is also made on the lumber to detect visible flaws.

Nominal size. Lumber is specified by its nominal, or undressed, size. The finished size of a member after dressing is normally $1/2$ in to $3/4$ in smaller than the original size. Thus, a 2 in nominal × 4 in nominal member has actual dimensions of $1\frac{1}{2}$ in × $3\frac{1}{2}$ in.

Structural glued laminated timber. Glued laminated members, or glulams, are built up from wood laminations bonded together with adhesives. The grain of all laminations is parallel to the length of the beam and the laminations are typically $1\frac{1}{2}$ in thick.

Timbers. Timbers are lumber members of nominal 5 in × 5 in or larger.

Visually stress-graded lumber. Lumber members that have been graded visually to detect flaws and defects, and to assess the inherent strength of the member.

Wood structural panel. A wood structural panel is manufactured from veneers or wood strands bonded together with adhesives. Examples are plywood, oriented strand board, and composite panels.

[2]American Forest and Paper Association, 2005 (See References and Codes)

General Design Requirements

The allowable design values for a wood member depend on the application of the member and on the service conditions under which it is utilized. The reference design values for sawn lumber are tabulated in NDS Supp. Tables 4A–F. For glued laminated members, design values are found in NDS Supp. Tables 5A–D. These reference design values are applicable to normal conditions of service and normal load duration. To determine the relevant design values for other conditions of service, the reference design values are multiplied by adjustment factors, specified in NDS[2] Sec. 2.3. A summary of adjustment factors follows. The applicability of each factor to the reference design values and to sawn lumber or glued laminated timber is given in NDS Tables 4.3.1 and 5.3.1, and summarized in Table 5.1.

Adjustment Factors Applicable Only to Sawn Lumber

Size Factor, C_F

The size factor is applicable to visually graded dimension lumber, visually graded timbers, and visually graded decking. It is not applied to mechanically graded dimension lumber. The reference design values for bending, tension, and compression are multiplied by the size factor, C_F, to give the appropriate design values. For visually graded dimension lumber, 2 in to 4 in thick, values of the size factor are given in NDS Supp. Tables 4A, 4B, and 4F. For visually graded decking, values of the size factor are given in NDS Supp. Table 4E. For visually graded timbers exceeding 12 in depth and 5 in thickness, the size factor is specified in NDS Sec. 4.3.6 as

$$C_F = \left(\frac{12}{d}\right)^{1/9} \quad \text{[NDS 4.3-1]}$$
$$\leq 1.0$$

Repetitive Member Factor, C_r

The repetitive member factor is applicable to visually graded dimension lumber and mechanically graded dimension lumber. It is not applied to visually graded timbers. The design values for visually graded decking in NDS Supp. Table 4E already incorporate the applicable repetitive member factor. The reference design value for bending is multiplied by the repetitive member factor, C_r, when three or more sawn lumber elements, not more than 4 in thick and spaced not more than 24 in apart, are joined by a transverse load distributing element. The value of the repetitive member factor is given in NDS Supp. Tables 4A–C and 4F, and in NDS Sec. 4.3.9, as

$$C_r = 1.15$$

Incising Factor, C_i

Values of the incising factor, C_i, for a prescribed incising pattern are provided in NDS Sec. 4.3.8. These values are applicable to all reference design values for all sawn lumber. The prescribed incising pattern consists of incisions made parallel to the grain at a maximum depth of 0.4 in, a maximum length of $3/8$ in, and at a density of $1100/\text{ft}^2$.

Example 5.1

Selected Douglas Fir-Larch, visually graded 3×6 decking is incised with the prescribed incising pattern. Determine the allowable bending stress. The repetitive member factor is applicable.

Solution

From NDS Table 4.3.8, the applicable incising factor is

$$C_i = 0.80$$

From NDS Supp. Table 4E, the applicable size factor for a 3 in thick member is

$$C_F = 1.04$$

From NDS Supp. Table 4E, the reference design value for bending, including the appropriate repetitive member factor, is

$$F_b C_r = 2000 \; \frac{\text{lbf}}{\text{in}^2}$$

The allowable bending design value is

$$\begin{aligned} F'_b &= C_i C_F F_b C_r \\ &= (0.8)(1.04)\left(2000 \; \frac{\text{lbf}}{\text{in}^2}\right) \\ &= 1700 \; \text{lbf/in}^2 \end{aligned}$$

Adjustment Factors Applicable Only to Glued Laminated Timber

Volume Factor, C_V

The volume factor, C_V, is applicable to the reference design value for bending for glued laminated timber. The volume factor is defined in NDS Sec. 5.3.6 as

$$C_V = \left(\frac{1291.5}{bdL}\right)^{1/x} \quad \text{[NDS 5.3-1]}$$
$$\leq 1.0$$

L is in feet, and b and d are in inches.

$$x = 20 \quad \text{[Southern Pine]}$$
$$= 10 \quad \text{[all other species]}$$

The volume factor is not applied simultaneously with the beam stability factor, C_L. The lesser of these two factors is applicable.

Curvature Factor, C_c

The curvature factor, C_c, is applicable to the reference design value for bending for glued laminated timber. This factor accounts for residual stresses in curved, glued laminated members. The curvature factor is specified in NDS Sec. 5.3.8 as

$$C_c = 1 - (2000)\left(\frac{t}{R}\right)^2 \quad \text{[NDS 5.3-2]}$$
$$\frac{t}{R} \leq \frac{1}{100} \quad \text{[hardwoods and Southern Pine]}$$
$$\leq \frac{1}{125} \quad \text{[other softwoods]}$$

The curvature factor shall not apply to design values in the straight portion of the member, regardless of curvature elsewhere.

Example 5.2

A glued laminated curved beam of stress class 24F-1.7E western species with 1.5 in thick laminations has a radius of curvature of 30 ft, a width of $6^3/4$ in, and a depth of 30 in. The beam has continuous lateral support, and is simply supported over a span of 40 ft. Determine the allowable design value in bending.

Solution

The basic design value for bending, tabulated in NDS Supp. Table 5A, is

$$F_b = 2400 \; \frac{\text{lbf}}{\text{in}^2}$$

The beam is provided with continuous lateral support, giving a stability factor of

$$C_L = 1.0$$

The volume factor governs, and it is

$$C_V = \left(\frac{1291.5}{bdL}\right)^{1/x} \quad \text{[NDS 5.3-1]}$$

$$= \left(\frac{1291.5 \text{ in}^2\text{-ft}}{(6.75 \text{ in})(30 \text{ in})(40 \text{ ft})}\right)^{1/10}$$

$$= 0.832$$

The curvature factor is

$$C_c = 1 - 2000\left(\frac{t}{R}\right)^2 \quad \text{[NDS 5.3-2]}$$

$$= 1 - (2000)\left(\frac{1.5 \text{ in}}{(30 \text{ ft})\left(12 \frac{\text{in}}{\text{ft}}\right)}\right)^2$$

$$= 0.965$$

The adjusted bending stress is

$$F'_b = F_b C_V C_c$$

$$= \left(2400 \frac{\text{lbf}}{\text{in}^2}\right)(0.832)(0.965)$$

$$= 1900 \text{ lbf/in}^2$$

Adjustment Factors Applicable to Both Sawn Lumber and Glued Laminated Members

Load Duration Factor, C_D

Normal load duration is equivalent to applying the maximum allowable load to a member for a period of 10 years. For loads of shorter duration, a member has the capacity to sustain higher loads, and the basic design values are multiplied by the load duration factor, C_D. The load duration factor is applicable to all basic design values with the exception of compression perpendicular to the grain and modulus of elasticity. The load duration factor for the shortest duration load in a combination of loads is applicable for that load combination. Values of the load duration factor are given in NDS Table 2.3.2, and are summarized in Table 5.2.

Table 5.2 Load Duration Factors

design load	load duration	C_D
dead load	permanent	0.90
occupancy live load	10 years	1.00
snow load	2 months	1.15
construction load	7 days	1.25
wind or earthquake load	10 minutes	1.60
impact load	impact	2.0

Example 5.3

The axial stresses produced in a Douglas Fir-Larch, select structural, 6 × 14 visually graded timber post for the applicable load combinations follow.

dead load	700 lbf/in²
dead load + floor live load	1000 lbf/in²
dead load + 0.75 × (floor live load + snow load)	1350 lbf/in²

The post is fully braced in all directions. Determine whether the post is adequate.

Solution

The reference design value for compression parallel to the grain is obtained from NDS Supp. Table 4D as

$$F_c = 1150 \text{ lbf/in}^2$$

The load duration factor from NDS Table 2.3.2 is

$$C_D = 0.9 \quad \text{[dead load]}$$
$$= 1.0 \quad \text{[floor live load]}$$
$$= 1.15 \quad \text{[snow load]}$$

Applying the appropriate adjustment factors, the allowable axial stress for the dead load condition is

$$F'_c = C_D F_c$$

$$= (0.9)\left(1150 \frac{\text{lbf}}{\text{in}^2}\right)$$

$$= 1035 \frac{\text{lbf}}{\text{in}^2}$$

$$> 700 \text{ lbf/in}^2 \quad \text{[satisfactory]}$$

Applying the appropriate adjustment factors, the allowable axial stress for the dead load + floor live load condition is

$$F'_c = C_D F_c$$

$$= (1.0)\left(1150 \frac{\text{lbf}}{\text{in}^2}\right)$$

$$= 1150 \frac{\text{lbf}}{\text{in}^2}$$

$$> 1000 \text{ lbf/in}^2 \quad \text{[satisfactory]}$$

Applying the appropriate adjustment factors, the allowable axial stress for the dead load + 0.75 (floor live load + snow load) condition is

$$F'_c = C_D F_c$$

$$= (1.15)\left(1150 \frac{\text{lbf}}{\text{in}^2}\right)$$

$$= 1323 \frac{\text{lbf}}{\text{in}^2}$$

$$< 1350 \text{ lbf/in}^2 \quad \text{[unsatisfactory]}$$

The post is inadequate.

Wet Service Factor, C_M

The basic design values for sawn lumber and glued laminated timber are applicable to members that are used under dry conditions of service, irrespective of the moisture content at the time of construction. Dry conditions of service are defined as a moisture content not exceeding 19% for sawn lumber or 16% for glued laminated timber. When the moisture content of sawn lumber exceeds 19%, the basic design values are multiplied by the wet service factors, C_M, given in NDS Supp. Tables 4A–F. When the moisture content of glued laminated members exceeds 16%, the wet service factors, C_M, given in NDS Supp. Tables 5A–D are applicable.

Bearing Area Factor C_b

Reference compression design values perpendicular to the grain are applicable to bearings at the end of a member, and to bearings not less than 6 in long at any other location. For bearings less than 6 in long and not less than 3 in from the end of a member, the reference design values for compression perpendicular to the grain are multiplied by the bearing area factor, C_b. This is specified in NDS Sec. 3.10.4 as

$$C_b = \frac{l_b + 0.375}{l_b} \quad \text{[NDS 3.10-2]}$$

The bearing length, l_b, measured parallel to the grain is the diameter of the washer for tightened bolts and lag screws.

Temperature Factor, C_t

The temperature stability factor is applicable to all reference design values for members exposed to sustained temperatures exceeding 100°F. It is specified by NDS Sec. 2.3.3 and tabulated in NDS Table 2.3.3.

Example 5.4

A 4×8 visually graded, select structural, Southern Pine, sawn lumber purlin has a moisture content exceeding 19%, and is subjected to sustained temperatures between 125° and 150°F. The governing load combination is dead load plus live load plus wind load. The purlin is incised with the prescribed incising pattern. Determine the allowable design value in shear.

Solution

The reference design value for shear is tabulated in NDS Supp. Table 4B as

$$F_v = 175 \; \frac{\text{lbf}}{\text{in}^2}$$

The applicable adjustment factors for shear stress are the following. The load duration factor is

$$C_D = 1.6 \quad \text{[NDS Table 2.3.2]}$$

The wet service factor is

$$C_M = 0.97 \quad \text{[NDS Supp. Table 4B]}$$

The temperature factor for wet conditions is

$$C_t = 0.5 \quad \text{[NDS Table 2.3.3]}$$

The incising factor is

$$C_i = 0.80 \quad \text{[NDS Table 4.3.8]}$$

The adjusted shear stress is

$$\begin{aligned} F'_v &= F_v C_D C_M C_t C_i \\ &= \left(175 \; \frac{\text{lbf}}{\text{in}^2}\right)(1.6)(0.97)(0.50)(0.80) \\ &= 109 \; \text{lbf/in}^2 \end{aligned}$$

Flat Use Factor, C_{fu}

As specified in NDS Sec. 4.3.7, the reference bending design value for sawn lumber members, loaded on their wide face, may be multiplied by the flat use factor, C_{fu}. The adjustment factors are given in NDS Supp. Tables 4A and 4B for visually graded dimension lumber, and in NDS Supp. Table 4C for machine graded dimension lumber. Design values for visually graded decking in NDS Supp. Table 4E already incorporate the applicable flat use factor.

As specified in NDS Sec. 5.3.7, the reference bending design value for glued laminated members loaded parallel to the wide faces of the laminations may be multiplied by the flat use factor, C_{fu}. The adjustment factors for glued laminated members are given in NDS Supp. Tables 5A–D.

Example 5.5

A glued laminated beam of combination 24F-V10 western species with 1.5 in thick laminations has a width of $6^3/_4$ in, and a depth of 30 in. The beam has a moisture content exceeding 16%, and is subjected to sustained temperatures between 100°F and 125°F. The governing load combination is dead load plus live load plus wind load. Determine the allowable design value for the modulus of elasticity about the x-x axis.

Solution

The reference design value for the modulus of elasticity about the x-x axis is tabulated in NDS Supp. Table 5A, Expanded Combinations, as

$$E_x = 1.8 \times 10^6 \ \frac{\text{lbf}}{\text{in}^2}$$

The applicable adjustment factors for the modulus of elasticity are C_M, for wet service conditions, and C_t, for elevated temperature in wet conditions.

$$C_M = 0.833 \quad \text{[NDS Supp. Table 5A]}$$
$$C_t = 0.9 \quad \text{[NDS Table 2.3.3]}$$

The adjusted modulus of elasticity is

$$\begin{aligned} E'_x &= E_x C_M C_t \\ &= \left(1.8 \times 10^6 \ \frac{\text{lbf}}{\text{in}^2}\right)(0.833)(0.90) \\ &= 1.3 \times 10^6 \ \text{lbf/in}^2 \end{aligned}$$

2. DESIGN FOR FLEXURE

Nomenclature

b	breadth of rectangular bending member	in
C_L	beam stability factor	–
d	beam depth	in
E	reference modulus of elasticity	lbf/in^2
E'	adusted modulus of elasticity	lbf/in^2
$E_{\min}$	reference modulus of elasticity for beam stability	lbf/in^2
$E'_{\min}$	adjusted modulus of elasticity for beam stability	lbf/in^2
F	ratio of F_{bE} to F_b^*	–
F_b	reference bending design value	lbf/in^2
F_b^*	reference bending design value multiplied by all applicable adjustment factors except C_{fu}, C_V, and C_L	lbf/in^2
F'_b	adusted bending design value	lbf/in^2
F_{bE}	critical buckling design value for bending members	lbf/in^2
l_e	effective span length of bending member	ft or in
l_u	laterally unsupported length of beam	ft or in
L	span length of bending member	ft or in
R_B	slenderness ratio of bending member	–

Beam Stability Factor, C_L

The beam stability factor, C_L, is applicable to the reference bending design value for sawn lumber and glued laminated members. In accordance with NDS Sec. 3.3.3, $C_L = 1.0$ when the depth of a bending member does not exceed its breadth, or when the compression edge of a bending member is provided with continuous lateral restraint, and the ends are restrained against rotation. Similarly, $C_L = 1.0$ when sawn lumber beams are supported in accordance with NDS Sec. 4.4.1, and any of the following conditions are met.

- $d/b \leq 2$
- $2 < d/b \leq 4$, and full depth bracing is provided at the ends of the member
- $4 < d/b \leq 5$, and the compression edge is continuously restrained
- $5 < d/b \leq 6$, and the compression edge is continuously restrained with full depth bracing provided at intervals not exceeding 8 ft
- $6 < d/b \leq 7$, and both edges are continuously restrained

For glued laminated members, C_L is not applied simultaneously with the volume factor, C_V, and the lesser of these two factors is applicable. The beam stability factor is given by NDS Sec. 3.3.3 as

$$C_L = \frac{1.0 + F}{1.9} - \sqrt{\left(\frac{1.0 + F}{1.9}\right)^2 - \frac{F}{0.95}} \quad \text{[NDS 3.3-6]}$$

$$F_b^* = F_b C_D C_M C_t C_F C_r C_c$$

$$F_{bE} = \frac{1.20 E'_{\min}}{R_B^2}$$

The adjusted modulus of elasticity for stability calculations is

$$E'_{\min} = E_{\min} C_M C_t$$

The slenderness ratio is

$$R_B = \sqrt{\frac{l_e d}{b^2}} \quad \text{[NDS 3.3-5]}$$
$$\leq 50$$

The effective span length, l_e, is determined in accordance with NDS Table 3.3.3. The value of l_e depends on the load configuration and the distance between lateral restraints, l_u. Typical values for l_e are given in Fig. 5.1.

Flexural Design of Sawn Lumber

When designing sawn lumber members for bending, both the size factor, C_F, and the stability factor, C_L, must be applied. The design span is defined in NDS Sec. 3.2.1 as clear span plus half the bearing length at each end.

Figure 5.1 Typical Values of Effective Length, l_e

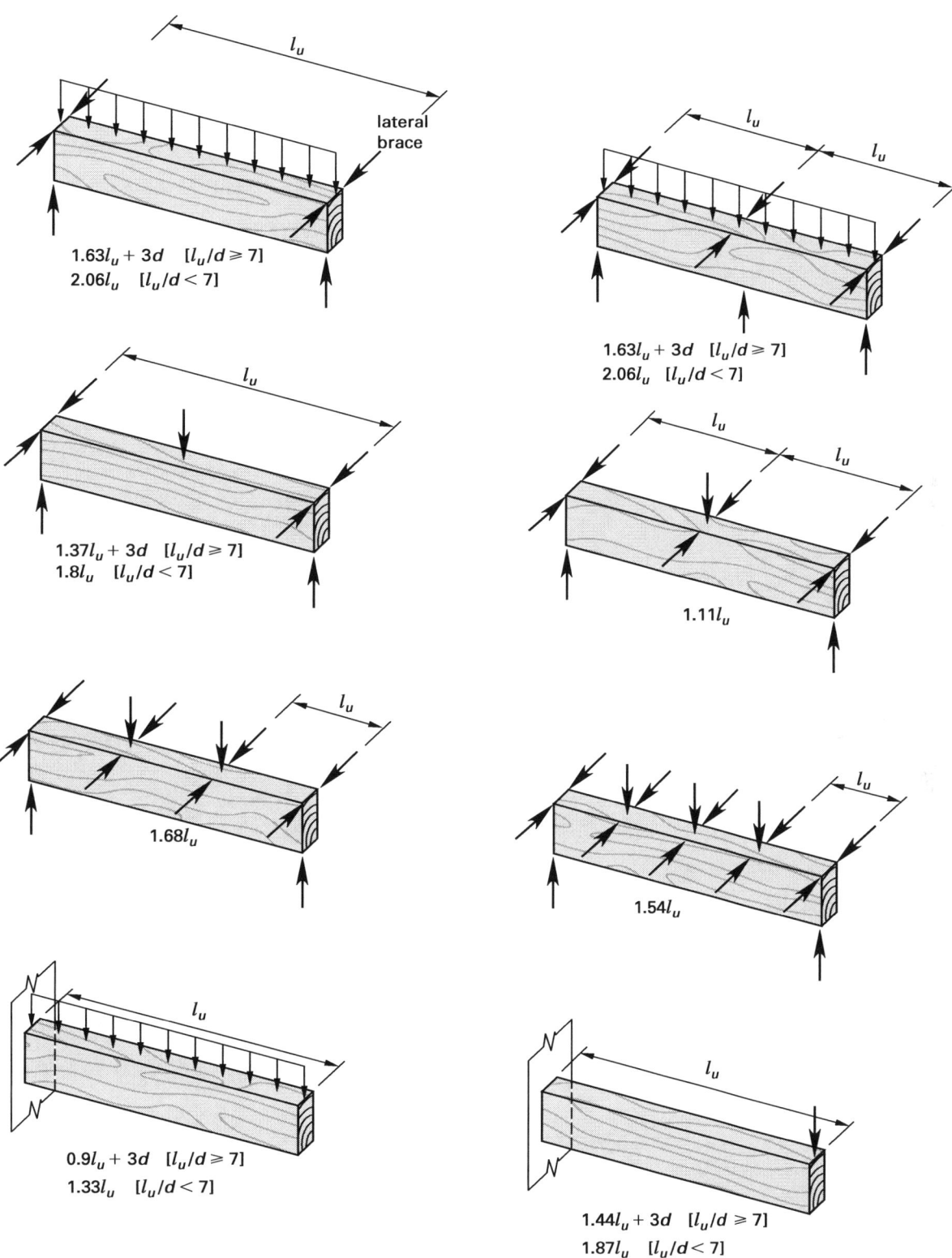

Example 5.6

A visually graded, no. 1 and btr., Douglas Fir-Larch 4×12 joist is simply supported over a span of 20 ft. The governing load combination is a uniformly distributed dead load plus live load plus snow load, and the joist is laterally braced at midspan. Determine the allowable design value in bending.

Solution

The reference design values for bending and the modulus of elasticity for beam stability calculations, tabulated in NDS Supp. Table 4A, are

$$F_b = 1200 \ \frac{\text{lbf}}{\text{in}^2}$$

$$E_{\min} = 0.66 \times 10^6 \ \frac{\text{lbf}}{\text{in}^2}$$

The applicable adjustment factors for modulus of elasticity are the following. The wet service factor is

$$C_M = 1.0 \quad [\text{NDS Supp. Table 4A}]$$

The temperature factor is

$$C_t = 1.0 \quad [\text{NDS Table 2.3.3}]$$

The incising factor is

$$C_i = 1.0 \quad [\text{NDS Table 4.3.8}]$$

The adjusted modulus of elasticity for stability calculations is

$$E'_{\min} = E_{\min} C_M C_t C_i$$
$$= \left(0.66 \times 10^6 \ \frac{\text{lbf}}{\text{in}^2}\right)(1.0)(1.0)(1.0)$$
$$= 0.66 \times 10^6 \ \text{lbf/in}^2$$

The distance between lateral restraints is

$$l_u = \frac{20 \ \text{ft}}{2}$$
$$= 10 \ \text{ft}$$

$$\frac{l_u}{d} = \frac{(10 \ \text{ft})\left(12 \ \frac{\text{in}}{\text{ft}}\right)}{11.25 \ \text{in}}$$
$$= 10.7$$
$$> 7$$

For a uniformly distributed load and an l_u/d ratio greater than 7, the effective length is obtained from Fig. 5.1 as

$$l_e = 1.63 l_u + 3d$$
$$= (1.63)(10 \ \text{ft})\left(12 \ \frac{\text{in}}{\text{ft}}\right) + (3)(11.25 \ \text{in})$$
$$= 229.4 \ \text{in}$$

The slenderness ratio is given by NDS Sec. 3.3.3 as

$$R_B = \sqrt{\frac{l_e d}{b^2}} \quad [\text{NDS 3.3-5}]$$
$$= \sqrt{\frac{(229.4 \ \text{in})(11.25 \ \text{in})}{(3.5 \ \text{in})^2}}$$
$$= 14.51$$
$$< 50 \quad [\text{satisfies NDS Sec. 3.3.3}]$$

The critical buckling design value is

$$F_{bE} = \frac{1.20 E'_{\min}}{R_B^2}$$
$$= \frac{(1.20)\left(0.66 \times 10^6 \ \frac{\text{lbf}}{\text{in}^2}\right)}{(14.51)^2}$$
$$= 3762 \ \text{lbf/in}^2$$

The applicable adjustment factors for flexure are C_F, the size factor for a 4×12 member, and C_D, the load duration factor for snow load.

$$C_F = 1.1 \quad [\text{NDS Supp. Table 4A}]$$
$$C_D = 1.15 \quad [\text{NDS Table 2.3.2}]$$

The reference flexural design value, multiplied by all applicable adjustment factors except C_L, is

$$F_b^* = F_b C_F C_D$$
$$= \left(1200 \ \frac{\text{lbf}}{\text{in}^2}\right)(1.1)(1.15)$$
$$= 1518 \ \text{lbf/in}^2$$

The ratio of F_{bE} to F_b^* is

$$F = \frac{F_{bE}}{F_b^*}$$
$$= \frac{3762 \ \frac{\text{lbf}}{\text{in}^2}}{1518 \ \frac{\text{lbf}}{\text{in}^2}}$$
$$= 2.48$$

The beam stability factor is given by NDS Sec. 3.3.3 as

$$C_L = \frac{1.0 + F}{1.9} - \sqrt{\left(\frac{1.0 + F}{1.9}\right)^2 - \frac{F}{0.95}}$$
$$= \frac{1.0 + 2.48}{1.9} - \sqrt{\left(\frac{1.0 + 2.48}{1.9}\right)^2 - \frac{2.48}{0.95}}$$
$$= 0.969$$

The allowable flexural design value is

$$F_b' = C_L C_F C_D F_b$$
$$= C_L F_b^*$$
$$= (0.969)\left(1518 \ \frac{\text{lbf}}{\text{in}^2}\right)$$
$$= 1500 \ \text{lbf/in}^2$$

Flexural Design of Glued Laminated Timber Members

When designing glued laminated timber members for bending, both the volume factor, C_V, and the stability factor, C_L, must be determined. However, only the lesser of these two factors is applied in determining the allowable design value in bending.

Example 5.7

A glued laminated beam of combination 24F-V10 western species with 1.5 in thick laminations has a width of $6^3/_4$ in, and a depth of 30 in. Tension laminations are provided as specified in NDS Table 5A, Expanded Combinations, Footnote 2. The beam is simply supported over a span of 40 ft, and is laterally braced at midspan. The governing load combination is a uniformly distributed dead load plus live load plus snow load. Determine the allowable design value in bending.

Solution

The reference design values for bending and the modulus of elasticity, tabulated in NDS Supp. Table 5A, Expanded Combinations, are

$$F_b = 2400 \ \text{lbf/in}^2$$
$$E_{x,\min} = 0.93 \times 10^6 \ \text{lbf/in}^2$$
$$E_{y,\min} = 0.78 \times 10^6 \ \text{lbf/in}^2$$

The applicable adjustment factors for the modulus of elasticity are C_M, the wet service factor, and C_t, the temperature factor.

$$C_M = 1.0 \quad \text{[NDS Supp. Table 5A]}$$
$$C_t = 1.0 \quad \text{[NDS Table 2.3.3]}$$

The adjusted modulus of elasticity is

$$E_{\min}' = E_{y,\min}'$$
$$= E_{y,\min} C_M C_t$$
$$= \left(0.78 \times 10^6 \ \frac{\text{lbf}}{\text{in}^2}\right)(1.0)(1.0)$$
$$= 0.78 \times 10^6 \ \text{lbf/in}^2$$

The distance between lateral restraints is

$$l_u = \frac{40 \ \text{ft}}{2}$$
$$= 20 \ \text{ft}$$
$$\frac{l_u}{d} = \frac{(20 \ \text{ft})\left(12 \ \frac{\text{in}}{\text{ft}}\right)}{30 \ \text{in}}$$
$$= 8.0$$
$$> 7$$

For a uniformly distributed load and an l_u/d ratio greater than 7, the effective length is obtained from Fig. 5.1 as

$$l_e = 1.63 l_u + 3d$$
$$= (1.63)(20 \ \text{ft})\left(12 \ \frac{\text{in}}{\text{ft}}\right) + (3)(30 \ \text{in})$$
$$= 481 \ \text{in}$$

The slenderness ratio is given by NDS Sec. 3.3.3 as

$$R_B = \sqrt{\frac{l_e d}{b^2}}$$
$$= \sqrt{\frac{(481 \ \text{in})(30 \ \text{in})}{(6.75 \ \text{in})^2}}$$
$$= 17.80$$
$$< 50 \quad \text{[satisfies NDS Sec. 3.3.3]}$$

The critical buckling design value is

$$F_{bE} = \frac{1.20 E_{\min}'}{R_B^2}$$
$$= \frac{(1.20)\left(0.78 \times 10^6 \ \frac{\text{lbf}}{\text{in}^2}\right)}{(17.80)^2}$$
$$= 2954 \ \text{lbf/in}^2$$

The reference flexural design value, multiplied by all applicable adjustment factors except C_L and C_V, is

$$F_b^* = F_b C_D$$
$$= \left(2400 \ \frac{\text{lbf}}{\text{in}^2}\right)(1.15)$$
$$= 2760 \ \text{lbf/in}^2$$

The ratio of F_{bE} to F_b^* is

$$F = \frac{F_{bE}}{F_b^*}$$
$$= \frac{2954 \ \frac{\text{lbf}}{\text{in}^2}}{2760 \ \frac{\text{lbf}}{\text{in}^2}}$$
$$= 1.07$$

The beam stability factor is given by NDS Sec. 3.3.3 as

$$C_L = \frac{1.0 + F}{1.9} - \sqrt{\left(\frac{1.0 + F}{1.9}\right)^2 - \frac{F}{0.95}}$$
$$= \frac{1.0 + 1.07}{1.9} - \sqrt{\left(\frac{1.0 + 1.07}{1.9}\right)^2 - \frac{1.07}{0.95}}$$
$$= 0.843$$

From Ex. 5.2, the volume factor is

$$C_V = 0.832$$
$$< C_L$$

The volume factor governs.

The allowable flexural design value is

$$F'_b = C_V C_D F_b$$
$$= (0.832)(1.15)\left(2400 \ \frac{\text{lbf}}{\text{in}^2}\right)$$
$$= 2300 \ \text{lbf/in}^2$$

3. DESIGN FOR COMPRESSION

Nomenclature

A	area of cross section	in^2
c	column parameter	–
C_{m1}	moment magnification factor for biaxial bending and axial compression, $1.0 - f_c/F_{cE1}$	–
C_{m2}	moment magnification factor for biaxial bending and axial compression, $1.0 - f_c/F_{cE2} - (f_{b1}/F_{bE})^2$	–
C_{m3}	moment magnification factor for axial compression and flexure with load applied to narrow face, $1.0 - f_c/F_{cE1}$	–
C_{m4}	moment magnification factor for axial compression and flexure with load applied to wide face, $1.0 - f_c/F_{cE2}$	–
C_{m5}	moment magnification factor for biaxial bending, $1.0 - (f_{b1}/F_{bE})^2$	–
C_P	column stability factor	–
d	least dimension of rectangular compression member	in
d_1	dimension of wide face	in
d_2	dimension of narrow face	in
E	tabulated modulus of elasticity	lbf/in^2
E'	allowable modulus of elasticity	lbf/in^2
F'	ratio of F_{cE} to F_c^*	–
F_{bE}	critical buckling design value for bending member, $1.20 E'_{min}/R_B^2$	lbf/in^2
F'_{b1}	allowable bending design value for load applied to the narrow face including adjustment for the slenderness ratio	lbf/in^2
F'_{b2}	allowable bending design value for load applied to the wide face including adjustment for the slenderness ratio	lbf/in^2
F_c^*	reference compressive design value multiplied by all applicable adjustment factors except C_P	lbf/in^2
F'_c	allowable compression design value including adjustment for the largest slenderness ratio	lbf/in^2
F_{cE}	critical buckling design value	lbf/in^2
F_{cE1}	critical buckling design value in plane of bending for load applied to the narrow face, $0.822 E'_{min}/(l_{e1}/d_1)^2$	lbf/in^2
F_{cE2}	critical buckling design value in plane of bending for load applied to the wide face, $0.822 E'_{min}/(l_{e2}/d_2)^2$	lbf/in^2
f_{b1}	actual edgewise bending stress for load applied to the narrow face	lbf/in^2
f_{b2}	actual flatwise bending stress for load applied to the wide face	lbf/in^2
f_c	actual compression stress parallel to grain	lbf/in^2
K_e	buckling length coefficient for compression members	–
L	span	ft or in
l_1	distance between points of lateral support restraining buckling about the strong axis of compression member	ft or in
l_2	distance between points of lateral support restraining buckling about the weak axis of compression member	ft or in
l_e	effective length of compression member	ft or in
l_{e1}	effective length between supports restraining buckling in plane of bending from load applied to narrow face of compression member, $K_e l_1$	ft or in
l_{e2}	effective length between supports restraining buckling in plane of bending from load applied to wide face of compression member, $K_e l_2$	ft or in
l_{e1}/d_1	slenderness ratio about the strong axis of compression member	–
l_{e2}/d_2	slenderness ratio about the weak axis of compression member	–
P	total concentrated load or total axial load	lbf or kips
S	section modulus	in^3
W	applied load	lbf

Column Stability Factor, C_P

The column stability factor is applicable to the reference compression design values parallel to the grain for sawn lumber and glued laminated members. In accordance with NDS Sec. 3.7.1, $C_L = 1.0$ when a compression member is provided with continuous lateral restraint in all directions. For other conditions, the column stability factor is specified by NDS Sec. 3.7.1 as

$$C_p = \frac{1.0 + F'}{2c} - \sqrt{\left(\frac{1.0 + F'}{2c}\right)^2 - \frac{F'}{c}} \quad \text{[NDS 3.7-1]}$$

The reference compression design value, multiplied by all applicable adjustment factors except C_P, is

$$F_c^* = F_c C_D C_M C_t C_i C_F$$

The critical buckling design value is

$$F_{cE} = \frac{0.822 E'_{\min}}{\left(\frac{l_e}{d}\right)^2}$$

The adjusted modulus of elasticity for stability calculations is

$$E'_{\min} = E_{\min} C_M C_t C_i$$

The column parameter is

$$c = 0.8 \quad \text{[sawn lumber]}$$
$$= 0.9 \quad \text{[glue laminated timber]}$$

Axial Load Only

The effective length of a column is defined in NDS App. G as the length of the column that is assumed to buckle in the shape of a sine wave. The effective length is dependent on the lateral translation and fixity at the ends of the column. For various standard column types, the effective length may be determined by multiplying the distance between lateral supports by a buckling length coefficient. Values of the buckling length coefficient, K_e, for various restraint conditions are given in NDS App. G Table G1, and summarized in Fig. 5.2. The effective length is specified in NDS Sec. 3.7.1.2 as

$$l_e = K_e l$$

The slenderness ratio is defined in NDS Sec. 3.7.1.3 as

$$\frac{l_e}{d} \leq 50$$

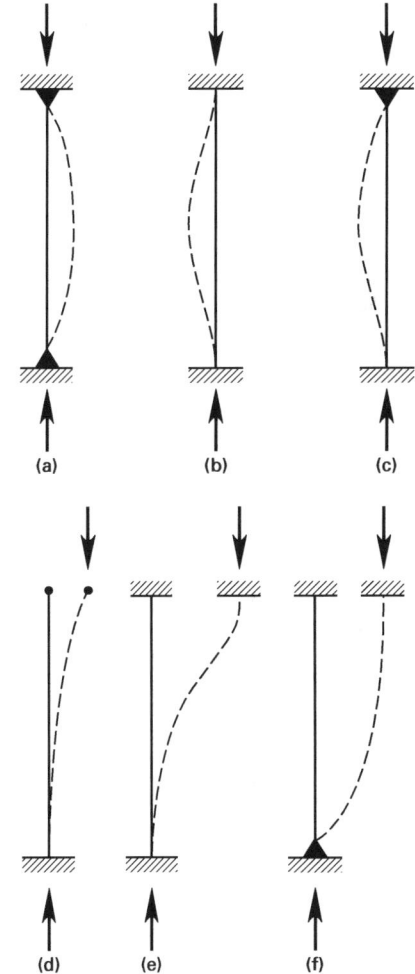

Figure 5.2 Buckling Length Coefficients

illus.	end conditions	K_e theoretical	design
(a)	both ends pinned	1	1.00
(b)	both ends built in	0.5	0.65
(c)	one end pinned, one end built in	0.7	0.8
(d)	one end built in, one end free	2	2.10
(e)	one end built in, one end fixed against rotation but free to translate	1	1.20
(f)	one end pinned, one end fixed against rotation but free to translate	2	2.40

As shown in Fig. 5.3, where the distances between lateral supports about the x-x axis and the y-y axis are different, two values of the slenderness ratio are obtained, l_{e1}/d_1 and l_{e2}/d_2, and the larger of these governs.

The applicable allowable compression design value, F'_c, is governed by the maximum slenderness ratio of the column. The maximum allowable axial load on a column is

$$P_{al} = A F'_c$$

Figure 5.3 Slenderness Ratios for a Rectangular Column

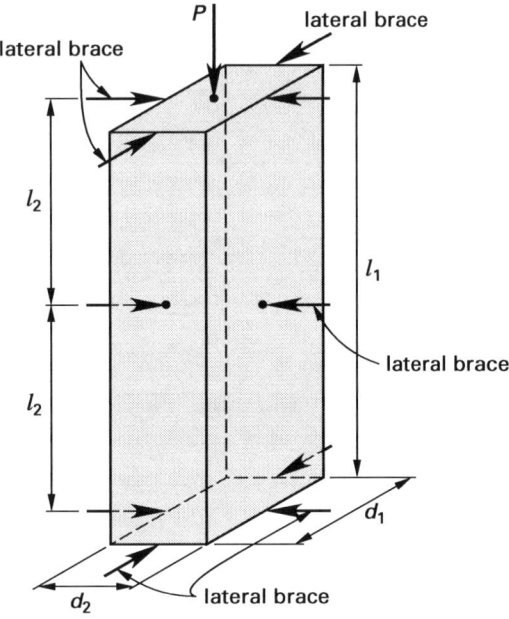

Example 5.8

A Douglas Fir-Larch, select structural, 6 × 14 visually graded timber post supports an axial load of 30 kips, as shown in the illustration. The post is 16 ft long, and is pinned at each end. The governing load combination consists of dead load plus live load plus snow load, and the moisture content exceeds 19%. The column is laterally braced at midheight about the weak axis, and the self-weight of the column and bracing members may be neglected. Determine whether the member is adequate.

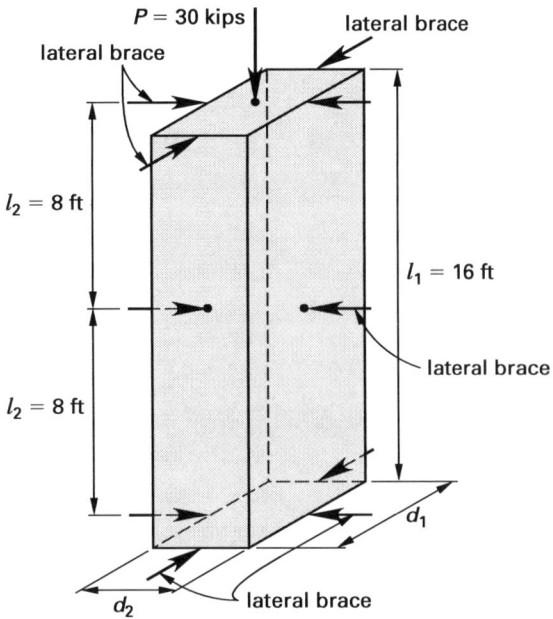

Solution

The reference design values for compression parallel to the grain are obtained from NDS Supp. Table 4D as

$$F_c = 1150 \ \frac{\text{lbf}}{\text{in}^2}$$

$$E_{\min} = 0.58 \times 10^6 \ \text{lbf/in}^2$$

The size factor is obtained from NDS Supp. Table 4D as

$$C_F = 1.00 \quad [\text{for } F_c]$$
$$= 1.00 \quad [\text{for } E]$$

The load duration factor is obtained from NDS Table 2.3.2 as

$$C_D = 1.15 \quad [\text{snow load}]$$

The wet service factor is obtained from NDS Supp. Table 4D as

$$C_M = 0.91 \quad [\text{for } F_c]$$
$$= 1.00 \quad [\text{for } E]$$

For pinned ended support conditions, the buckling length coefficient is obtained from NDS App. G Table G1 as

$$K_e = 1.0$$

The slenderness ratio about the strong axis is

$$\frac{K_e l_1}{d_1} = \frac{(1.0)(16 \ \text{ft})\left(12 \ \frac{\text{in}}{\text{ft}}\right)}{13.5 \ \text{in}}$$
$$= 14.22$$

The slenderness ratio about the weak axis is

$$\frac{K_e l_2}{d_2} = \frac{(1.0)(8 \ \text{ft})\left(12 \ \frac{\text{in}}{\text{ft}}\right)}{5.5 \ \text{in}}$$
$$= 17.45 \quad [\text{governs}]$$

The adjusted modulus of elasticity for stability calculations is

$$E'_{\min} = E_{\min} C_M C_F$$
$$= \left(0.58 \times 10^6 \ \frac{\text{lbf}}{\text{in}^2}\right)(1.0)(1.0)$$
$$= 0.58 \times 10^6 \ \text{lbf/in}^2$$

The reference compression design value, multiplied by all applicable adjustment factors except C_P, is

$$F_c^* = F_c C_M C_F C_D$$
$$= \left(1150 \, \frac{\text{lbf}}{\text{in}^2}\right)(0.91)(1.00)(1.15)$$
$$= 1203 \, \text{lbf/in}^2$$

The critical buckling design value is

$$F_{cE2} = \frac{0.822 E'_{\min}}{\left(\frac{l_{e2}}{d_2}\right)^2}$$
$$= \frac{(0.822)\left(0.58 \times 10^6 \, \frac{\text{lbf}}{\text{in}^2}\right)}{(17.45)^2}$$
$$= 1566 \, \text{lbf/in}^2$$

The ratio of F_{cE2} to F_c^* is

$$F' = \frac{F_{cE2}}{F_c^*}$$
$$= \frac{1566 \, \frac{\text{lbf}}{\text{in}^2}}{1203 \, \frac{\text{lbf}}{\text{in}^2}}$$
$$= 1.30$$

The column parameter is obtained from NDS Sec. 3.7.1.5 as

$$c = 0.8 \quad [\text{sawn lumber}]$$

The column stability factor is specified by NDS Sec. 3.7.1.5 as

$$C_P = \frac{1.0 + F'}{2c} - \sqrt{\left(\frac{1.0 + F'}{2c}\right)^2 - \frac{F'}{c}}$$
$$= \frac{1.0 + 1.30}{(2)(0.8)} - \sqrt{\left(\frac{1.0 + 1.30}{(2)(0.8)}\right)^2 - \frac{1.30}{0.8}}$$
$$= 0.77$$

The allowable compression design value parallel to the grain is

$$F_c' = F_c C_M C_F C_D C_P$$
$$= F_c^* C_P$$
$$= \left(1203 \, \frac{\text{lbf}}{\text{in}^2}\right)(0.77)$$
$$= 926 \, \text{lbf/in}^2$$

The actual compressive stress on the column is

$$f_c = \frac{P}{A}$$
$$= \frac{30{,}000 \, \text{lbf}}{74.25 \, \text{in}^2}$$
$$= 404 \, \text{lbf/in}^2$$
$$< F_c'$$

The column is adequate.

Combined Axial Compression and Flexure

Members subjected to combined compression and flexural stresses from axial and transverse loading must satisfy the interaction equation given in NDS Sec. 3.9.2 as

$$\left(\frac{f_c}{F_c'}\right)^2 + \frac{f_{b1}}{F_{b1}' C_{m1}} + \frac{f_{b2}}{F_{b2}' C_{m2}} \leq 1.00 \quad [\text{NDS 3.9-3}]$$

For bending load applied to the narrow face of the member and concentric axial compression load, the interaction equation reduces to

$$\left(\frac{f_c}{F_c'}\right)^2 + \frac{f_{b1}}{F_{b1}' C_{m3}} \leq 1.00$$

For bending load applied to the wide face of the member and concentric axial compression load, the equation reduces to

$$\left(\frac{f_c}{F_c'}\right)^2 + \frac{f_{b2}}{F_{b2}' C_{m4}} \leq 1.00$$

For bending loads applied to the narrow and wide faces of the member and no concentric axial load, the equation reduces to

$$\frac{f_{b1}}{F_{b1}'} + \frac{f_{b2}}{F_{b2}' C_{m5}} \leq 1.00$$

Example 5.9

A Douglas Fir-Larch, select structural, 6×14 visually graded timber post supports an axial load of 30 kips. A lateral load of 2 kips is applied to the narrow face of the column, as shown in the illustration. The post is 16 ft long, pinned at each end. The governing load combination consists of dead load plus live load plus snow load, and the moisture content exceeds 19%. The column is laterally braced at midheight about the weak axis, and the self-weight of the column and bracing members may be neglected. Determine whether the member is adequate.

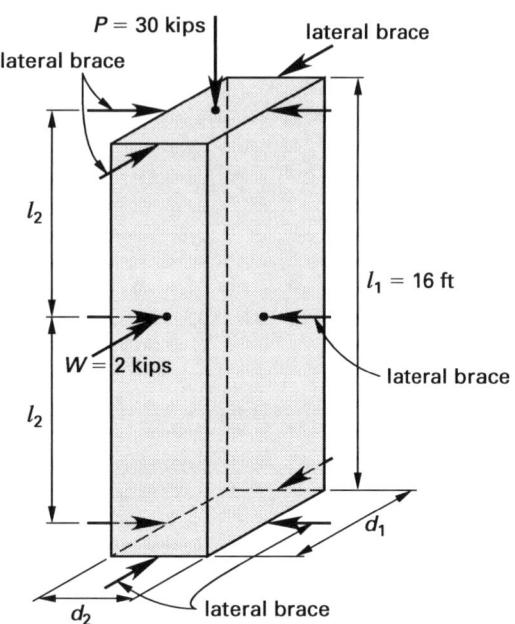

Solution

The relevant details from Ex. 5.8 are the following.

$$F_c = 1150 \ \frac{\text{lbf}}{\text{in}^2}$$
$$E_{\min} = 0.58 \times 10^6 \ \text{lbf/in}^2$$

The size factor for axial load and modulus of elasticity is

$$C_F = 1.00$$

The load duration factor is

$$C_D = 1.15 \quad [\text{snow load}]$$

The wet service factor is

$$C_M = 0.91 \quad [\text{for } F_c]$$
$$= 1.00 \quad [\text{for } E]$$

The buckling length coefficient is

$$K_e = 1.0$$

The slenderness ratio about the strong axis is

$$\frac{K_e l_1}{d_1} = 14.22$$

The adjusted modulus of elasticity for stability calculations is

$$E'_{\min} = 0.58 \times 10^6 \ \frac{\text{lbf}}{\text{in}^2}$$
$$F'_c = 926 \ \frac{\text{lbf}}{\text{in}^2}$$
$$f_c = 404 \ \frac{\text{lbf}}{\text{in}^2}$$

The reference design value for bending is obtained from NDS Supp. Table 4D as

$$F_b = 1500 \ \frac{\text{lbf}}{\text{in}^2}$$

The distance between lateral restraints for bending instability is

$$l_u = \frac{16 \ \text{ft}}{2}$$
$$= 8 \ \text{ft}$$

For a concentrated load at midspan and with lateral restraint at midspan, the effective length is obtained from Fig. 5.1 as

$$l_e = 1.11 l_u$$
$$= (1.11)(8 \ \text{ft})\left(12 \ \frac{\text{in}}{\text{ft}}\right)$$
$$= 106.56 \ \text{in}$$

The slenderness ratio for bending is given by NDS Sec. 3.3.3 as

$$R_B = \sqrt{\frac{l_e d_1}{d_2^2}}$$
$$= \sqrt{\frac{(106.56 \ \text{in})(13.5 \ \text{in})}{(5.5 \ \text{in})^2}}$$
$$= 6.90$$
$$< 50 \quad [\text{satisfies NDS Sec. 3.3.3}]$$

The critical buckling design value is

$$F_{bE} = \frac{1.20 E'_{\min}}{R_B^2}$$
$$= \frac{(1.20)\left(0.58 \times 10^6 \ \frac{\text{lbf}}{\text{in}^2}\right)}{(6.90)^2}$$
$$= 14{,}619 \ \text{lbf/in}^2$$

The applicable wet service factor for flexure, obtained from NDS Supp. Table 4D, is

$$C_M = 1.00$$

The size factor for flexure is obtained from NDS Supp. Table 4D as

$$C_F = \left(\frac{12}{d}\right)^{1/9}$$
$$= \left(\frac{12}{13.5 \ \text{in}}\right)^{1/9}$$
$$= 0.99$$

The basic flexural design value, multiplied by all applicable adjustment factors except C_L, is

$$F_b^* = F_b C_M C_F C_D$$
$$= \left(1500 \ \frac{\text{lbf}}{\text{in}^2}\right)(1.00)(0.99)(1.15)$$
$$= 1708 \ \text{lbf/in}^2$$

The ratio of F_{bE} to F_b^* is

$$F = \frac{F_{bE}}{F_b^*}$$
$$= \frac{14{,}619 \ \frac{\text{lbf}}{\text{in}^2}}{1708 \ \frac{\text{lbf}}{\text{in}^2}}$$
$$= 8.56$$

The beam stability factor is given by NDS Sec. 3.3.3 as

$$C_L = \frac{1.0 + F}{1.9} - \sqrt{\left(\frac{1.0 + F}{1.9}\right)^2 - \frac{F}{0.95}}$$
$$= \frac{1.0 + 8.56}{1.9} - \sqrt{\left(\frac{1.0 + 8.56}{1.9}\right)^2 - \frac{8.56}{0.95}}$$
$$= 0.99$$

The allowable flexural design value for the load applied to the narrow face is

$$F'_{b1} = F_b C_M C_F C_D C_L$$
$$= \left(1708 \ \frac{\text{lbf}}{\text{in}^2}\right)(0.99)$$
$$= 1690 \ \text{lbf/in}^2$$

The actual edgewise bending stress is

$$f_{b1} = \frac{WL}{4S}$$
$$= \frac{(2000 \ \text{lbf})(16 \ \text{ft})\left(12 \ \frac{\text{in}}{\text{ft}}\right)}{(4)(167 \ \text{in}^3)}$$
$$= 575 \ \text{lbf/in}^2$$

The critical buckling design value in the plane of bending for load applied to the narrow face is

$$F_{cE1} = \frac{0.822 E'_{\min}}{\left(\frac{l_{e1}}{d_1}\right)^2}$$
$$= \frac{(0.822)\left(0.58 \times 10^6 \ \frac{\text{lbf}}{\text{in}^2}\right)}{(14.22)^2}$$
$$= 2358 \ \text{lbf/in}^2$$

The moment magnification factor for axial compression and flexure with load applied to the narrow face is

$$C_{m3} = 1.0 - \frac{f_c}{F_{cE1}}$$
$$= 1.0 - \frac{404 \ \frac{\text{lbf}}{\text{in}^2}}{2358 \ \frac{\text{lbf}}{\text{in}^2}}$$
$$= 0.83$$

The interaction equation for bending load applied to the narrow face of the member and concentric axial compression load is given in NDS Sec. 3.9.2 as

$$\left(\frac{f_c}{F'_c}\right)^2 + \frac{f_{b1}}{F'_{b1} C_{m3}} \leq 1.0$$

The left-hand side of the expression is

$$\left(\frac{404 \ \frac{\text{lbf}}{\text{in}^2}}{926 \ \frac{\text{lbf}}{\text{in}^2}}\right)^2 + \frac{575 \ \frac{\text{lbf}}{\text{in}^2}}{\left(1690 \ \frac{\text{lbf}}{\text{in}^2}\right)(0.83)} = 0.190 + 0.410$$
$$= 0.60$$
$$< 1.0 \quad [\text{satisfactory}]$$

The post is adequate.

4. DESIGN FOR TENSION

Nomenclature

A	area of cross section	–
C_D	load duration factor	–
C_F	size factor for sawn lumber	–
C_i	incising factor	–
C_m	wet service factor	–
C_t	temperature factor	–
E	reference modulus of elasticity	lbf/in^2
E'	allowable modulus of elasticity	lbf/in^2
f_t	actual tension stress parallel to grain	lbf/in^2
F	ratio of F_{bE} to F_b^*	–
F_b^*	reference bending design value multiplied by all applicable adjustment factors except C_L	lbf/in^2
F_b^{**}	reference bending design value multiplied by all applicable adjustment factors except C_V	lbf/in^2
F_{bE}	critical buckling design value for bending members	–
F_t	reference tension design value parallel to grain	lbf/in^2
F'_t	allowable tension design value parallel to grain	lbf/in^2
l_e	effective length of compression member	ft
l_u	laterally unsupported length of beam	ft
R_B	slenderness ratio of bending member	–
T	tensile force on member	lbf

Axial Tension

The reference design values for tension parallel to the grain are tabulated in the NDS Supplements. Allowable design values are obtained by multiplying basic values by the applicable adjustment factors. As specified in NDS Sec. 3.8.2, tension perpendicular to the grain is to be avoided.

Example 5.10

The select structural 2×8 Douglas Fir-Larch bottom chord of a truss is axially loaded in tension as shown in the illustration. The governing load combination consists of dead load plus live load plus snow load, and the moisture content exceeds 19%. The self-weight of the chord may be neglected. At the end connections, the net area is $A_{\text{net}} = 9.3 \text{ in}^2$. Determine whether the member is adequate.

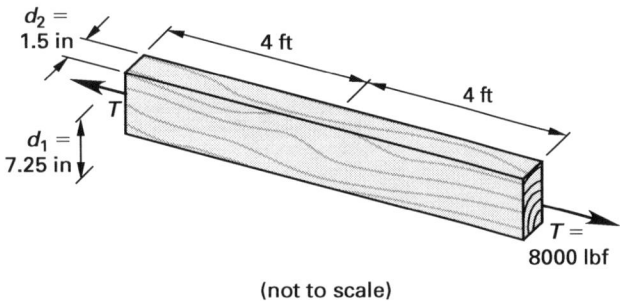

(not to scale)

Solution

The reference design value for tension, tabulated in NDS Supp. Table 4A, is

$$F_t = 1000 \, \frac{\text{lbf}}{\text{in}^2}$$

The size factor for tensile load is obtained from NDS Supp. Table 4A as

$$C_F = 1.20$$

The load duration factor is obtained from NDS Table 2.3.2 as

$$C_D = 1.5 \quad \text{[snow load]}$$

The wet service factor for tensile load is obtained from NDS Supp. Table 4A as

$$C_M = 1.00$$

The allowable tension design value parallel to grain is

$$\begin{aligned} F'_t &= F_t C_M C_F C_D \\ &= \left(1000 \, \frac{\text{lbf}}{\text{in}^2}\right)(1.0)(1.2)(1.15) \\ &= 1380 \, \text{lbf/in}^2 \end{aligned}$$

The actual tension stress on the net area of the chord is

$$\begin{aligned} f_{t,\text{net}} &= \frac{T}{A_{\text{net}}} \\ &= \frac{8000 \, \text{lbf}}{9.3 \, \text{in}^2} \\ &= 860 \, \frac{\text{lbf}}{\text{in}^2} \\ &< F'_t \quad \text{[satisfactory]} \end{aligned}$$

The member is adequate.

Combined Axial Tension and Flexure

Members subjected to combined tension and flexural stresses caused by axial and transverse loading must satisfy the two expressions given in NDS Sec. 3.9.1 as

$$\frac{f_t}{F'_t} + \frac{f_b}{F^*_b} \leq 1.0 \quad \text{[NDS 3.9-1]}$$

$$\frac{f_b - f_t}{F^{**}_b} \leq 1.0 \quad \text{[NDS 3.9-2]}$$

Example 5.11

The select structural 2×8 Douglas Fir-Larch bottom chord of a truss is loaded as shown in the illustration. The governing load combination consists of dead load plus live load plus snow load, and the moisture content exceeds 19%. The chord is laterally braced at midlength about the weak axis, and the self-weight of the chord and bracing members may be neglected. Determine whether the member is adequate.

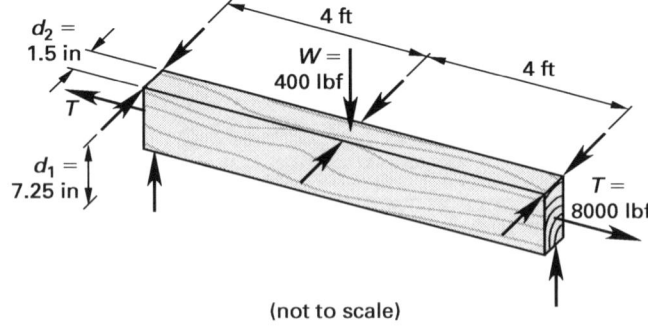

(not to scale)

Solution

The relevant details from Ex. 5.10 follow. The allowable tension design value parallel to the grain is

$$F'_t = 1380 \, \frac{\text{lbf}}{\text{in}^2}$$

The actual tension stress on the gross area of the chord is

$$f_t = (f_{t,\text{net}})\left(\frac{A_{\text{net}}}{A_{\text{gross}}}\right)$$
$$= \left(860\ \frac{\text{lbf}}{\text{in}^2}\right)\left(\frac{9.3\ \text{in}^2}{10.88\ \text{in}^2}\right)$$
$$= 735\ \text{lbf/in}^2$$

The reference design values for bending and the modulus of elasticity for stability calculations, tabulated in NDS Supp. Table 4A, are

$$F_b = 1500\ \frac{\text{lbf}}{\text{in}^2}$$
$$E_{\text{min}} = 0.69 \times 10^6\ \frac{\text{lbf}}{\text{in}^2}$$

The applicable adjustment factors for the modulus of elasticity are the following. The wet service factor is

$$C_M = 0.9 \quad [\text{NDS Supp. Table 4A}]$$

The temperature factor is

$$C_t = 1.0 \quad [\text{NDS Table 2.3.3}]$$

The incising factor is

$$C_i = 1.0 \quad [\text{NDS Table 4.3.8}]$$

The adjusted modulus of elasticity for stability calculations is

$$E'_{\text{min}} = E_{\text{min}} C_M C_t C_i$$
$$= \left(0.69 \times 10^6\ \frac{\text{lbf}}{\text{in}^2}\right)(0.9)(1.0)(1.0)$$
$$= 0.62 \times 10^6\ \text{lbf/in}^2$$

For a concentrated load with lateral restraint, both at midspan, the effective length is obtained from Fig. 5.1 as

$$l_e = 1.11 l_u$$
$$= (1.11)(4\ \text{ft})\left(12\ \frac{\text{in}}{\text{ft}}\right)$$
$$= 53.28\ \text{in}$$

The slenderness ratio is given by NDS Sec. 3.3.3 as

$$R_B = \sqrt{\frac{l_e d_1}{d_2^2}}$$
$$= \sqrt{\frac{(53.28\ \text{in})(7.25\ \text{in})}{(1.5\ \text{in})^2}}$$
$$= 13.10$$
$$< 50 \quad [\text{satisfies criteria of NDS Sec. 3.3.3}]$$

The critical buckling design value is

$$F_{bE} = \frac{1.20 E'_{\text{min}}}{R_B^2}$$
$$= \frac{(1.20)\left(0.62 \times 10^6\ \frac{\text{lbf}}{\text{in}^2}\right)}{(13.10)^2}$$
$$= 4335\ \text{lbf/in}^2$$

The applicable adjustment factors for bending are the following. The wet service factor is

$$C_M = 0.85 \quad [\text{NDS Supp. Table 4A}]$$

The size factor is

$$C_F = 1.2 \quad [\text{NDS Supp. Table 4A}]$$

The load duration factor is

$$C_D = 1.15 \quad [\text{NDS Table 2.3.2}]$$

The reference flexural design value, multiplied by all applicable adjustment factors except C_L, is

$$F_b^* = F_b C_M C_F C_D$$
$$= \left(1500\ \frac{\text{lbf}}{\text{in}^2}\right)(0.85)(1.2)(1.15)$$
$$= 1760\ \text{lbf/in}^2$$

The ratio of F_{bE} to F_b^* is

$$F = \frac{F_{bE}}{F_b^*}$$
$$= \frac{4335\ \frac{\text{lbf}}{\text{in}^2}}{1760\ \frac{\text{lbf}}{\text{in}^2}}$$
$$= 2.46$$

The beam stability factor is given by NDS Sec. 3.3.3 as

$$C_L = \frac{1.0 + F}{1.9} - \sqrt{\left(\frac{1.0 + F}{1.9}\right)^2 - \frac{F}{0.95}}$$
$$= \frac{1 + 2.46}{1.9} - \sqrt{\left(\frac{1 + 2.46}{1.9}\right)^2 - \frac{2.46}{0.95}}$$
$$= 0.97$$

The reference flexural design value, multiplied by all applicable adjustment factors except C_V, is

$$F_b^{**} = F_b C_M C_F C_D C_L$$
$$= F_b^* C_L$$
$$= \left(1760\ \frac{\text{lbf}}{\text{in}^2}\right)(0.97)$$
$$= 1707\ \text{lbf/in}^2$$

The actual edgewise bending stress is

$$f_{b1} = \frac{WL}{4S}$$

$$= \frac{(400 \text{ lbf})(8 \text{ ft})\left(12 \frac{\text{in}}{\text{ft}}\right)}{(4)(13.14 \text{ in}^3)}$$

$$= 731 \text{ lbf/in}^2$$

Substituting in the two expressions given in NDS Sec. 3.9.1 gives

$$\frac{f_t}{F'_t} + \frac{f_{b1}}{F_b^*} = \frac{735 \frac{\text{lbf}}{\text{in}^2}}{1380 \frac{\text{lbf}}{\text{in}^2}} + \frac{731 \frac{\text{lbf}}{\text{in}^2}}{1760 \frac{\text{lbf}}{\text{in}^2}}$$

$$= 0.533 + 0.415$$

$$= 0.948$$

$$< 1.0 \quad \text{[satisfactory]}$$

$$\frac{f_{b1} - f_t}{F_b^{**}} = \frac{731 \frac{\text{lbf}}{\text{in}^2} - 735 \frac{\text{lbf}}{\text{in}^2}}{1707 \frac{\text{lbf}}{\text{in}^2}}$$

$$= -0.002$$

$$< 1.0 \quad \text{[satisfactory]}$$

The chord is adequate.

5. DESIGN FOR SHEAR

Nomenclature

C_D	load duration factor	–
C_M	wet service factor	–
C_t	temperature factor	–
d	depth of unnotched bending member	in
d_e	depth of member, less the distance from the unloaded edge of the member to the nearest edge of the nearest split ring or shear plate connector	in
d_e	depth of member, less the distance from the unloaded edge of the member to the center of the nearest bolt or lag screw	in
d_n	depth of member remaining at a notch	in
e	distance a notch extends past the inner edge of a support	in
f_v	actual shear stress parallel to grain	lbf/in²
F_v	tabulated shear design value parallel to grain	lbf/in²
F'_v	allowable shear design value parallel to grain	lbf/in²
I	moment of inertia	in⁴
l_n	length of notch	in
Q	statical moment of an area about the neutral axis	in³
V	shear force	lbf or kips
V'_r	allowable design shear	lbf or kips
W	concentrated load	kips
W'	equivalent concentrated load	kips
x	distance from face of support to the load	in

General Requirements Applicable to Sawn Lumber and Glued Laminated Members

The shear stress parallel to the grain in a flexural member is given in NDS Sec. 3.4.2 as

$$f_v = \frac{VQ}{Ib} \quad \text{[NDS 3.4-1]}$$

For a rectangular beam, this reduces to

$$f_v = \frac{3V}{2bd} \quad \text{[NDS 3.4-2]}$$

$$= \frac{1.5V}{bd}$$

For a beam with distributed loads, the manner of calculating the shear force is specified in NDS Sec. 3.4.3.1, and illustrated in Fig. 5.4. In determining the shear force on a beam, uniformly distributed loads applied to the top of the beam within a distance from either support equal to the depth of the beam, are ignored.

Figure 5.4 Shear Caused by Distributed Loads

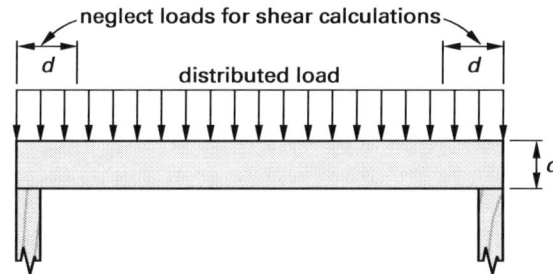

For a beam with concentrated loads, the manner of calculating the shear force is specified in NDS Sec. 3.4.3.1, and illustrated in Fig. 5.5. Concentrated loads within a distance from either support equal to the depth of the beam are multiplied by x/d, where x is the distance from the support to the load, to give an equivalent shear force.

Figure 5.5 Shear Caused by Concentrated Loads

To facilitate the selection of glued laminated beam sections, tables are available[3] that provide shear and bending capacities of sections. For lumber joists, tables are available[4] that assist in the selection of a joist size for various span and live load combinations.

Example 5.12

A glued laminated beam of combination 24F-V10 western species with a width of $6^3/_4$ in, and a depth of 30 in, is loaded as shown in the illustration. The load consists of dead load plus live load, and the distributed load of 1.0 kip/ft includes the self-weight of the beam. The moisture content exceeds 16%. Determine whether the beam is adequate.

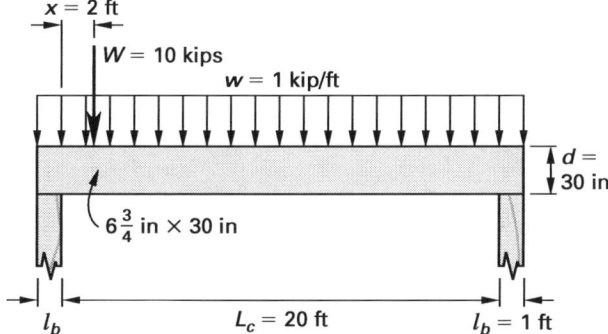

Solution

The reference design value for shear is tabulated in NDS Supp. Table 5A, Expanded Combinations, as

$$F_v = 215 \; \frac{\text{lbf}}{\text{in}^2}$$

The applicable adjustment factor for shear stress is the wet service factor.

$$C_M = 0.875 \quad \text{[NDS Supp. Table 5A]}$$

The adjusted shear stress is

$$F'_v = F_v C_M = \left(215 \; \frac{\text{lbf}}{\text{in}^2}\right)(0.875)$$
$$= 188 \; \text{lbf/in}^2$$

The clear span is obtained from the illustration as

$$L_c = 20 \; \text{ft}$$

The bearing length is obtained from the illustration as

$$l_b = 1.0 \; \text{ft}$$

---
[3]American Plywood Association, 2007 (See References and Codes)
[4]Western Wood Products Association, 1992 (See References and Codes)

The design span is defined in NDS Sec. 3.2.1 as

$$L = L_c + l_b = 20 \; \text{ft} + 1.0 \; \text{ft}$$
$$= 21 \; \text{ft}$$

The shear at the left support caused by the distributed load is given by NDS Sec. 3.4.3.1(a) as

$$V_D = \frac{w(L_c - 2x)}{2} = \frac{\left(1.0 \; \frac{\text{kips}}{\text{ft}}\right)(16 \; \text{ft})}{2}$$
$$= 8 \; \text{kips}$$

The concentrated load, W, is less than a distance, d, from the face of the left support. In accordance with NDS Sec. 3.4.3.1, this is equivalent to a load of

$$W' = \frac{Wx}{d} = \frac{(10 \; \text{kips})(24 \; \text{in})}{30 \; \text{in}}$$
$$= 8 \; \text{kips}$$

The shear at the left support caused by the equivalent load is

$$V_C = \frac{W'\left(L - x - \frac{l_b}{2}\right)}{L}$$
$$= \frac{(8 \; \text{kips})\left(21 \; \text{ft} - 2 \; \text{ft} - \frac{1 \; \text{ft}}{2}\right)}{21 \; \text{ft}}$$
$$= 7.05 \; \text{kips}$$

The total shear at the left support is

$$V = V_D + V_C$$
$$= 8 \; \text{kips} + 7.05 \; \text{kips}$$
$$= 15.05 \; \text{kips}$$

The shear stress parallel to the grain is given by NDS Sec. 3.4.2 as

$$f_v = \frac{1.5V}{bd} \quad \text{[NDS 3.4-2]}$$
$$= \frac{(1.5)(15{,}050 \; \text{lbf})}{(6.75 \; \text{in})(30 \; \text{in})}$$
$$= 111 \; \text{lbf/in}^2$$
$$< F'_v \quad \text{[satisfactory]}$$

The beam is adequate.

Notched Beams

Notches reduce the shear capacity of a beam. NDS Sec. 4.4.3 imposes restrictions on their size and location in sawn lumber members, as shown in Fig. 5.6. For glued laminated members, similar restrictions are specified in NDS Sec. 5.4.4, shown in Fig. 5.7.

Figure 5.6 Notches in Sawn Lumber Beams

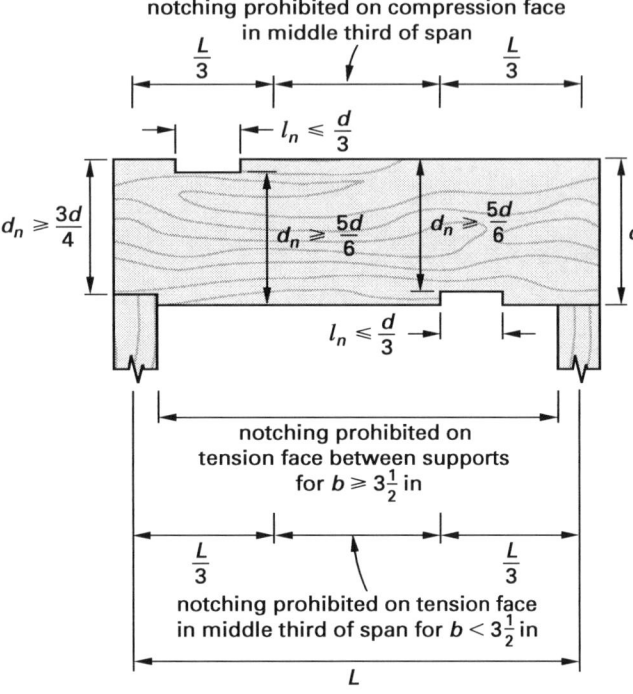

Figure 5.7 Notches in Glued Laminated Beams

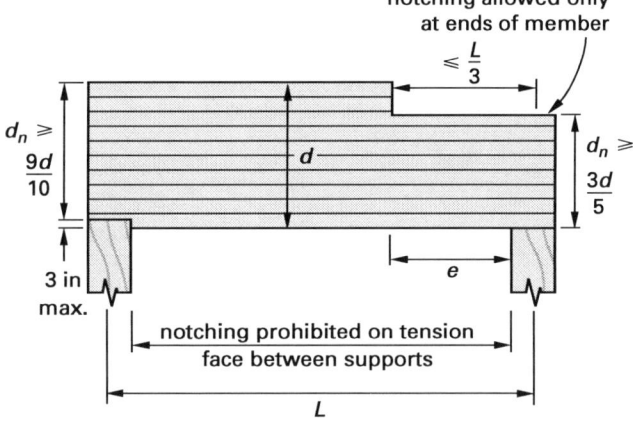

The allowable design shear at a notch on the tension side of a beam is given by NDS Sec. 3.4.3.2(a) as

$$V'_r = \left(\frac{F'_v b d_n}{1.5}\right)\left(\frac{d_n}{d}\right)^2 \quad \text{[NDS 3.4-3]}$$

When e, the distance the notch extends past the inner edge of the support, is less than or equal to d_n, the shear stress at the notch on the compression side of the beam is given by NDS Sec. 3.4.3.2(e) as

$$V'_r = \left(\frac{F'_v b}{1.5}\right)\left(d - \left(\frac{d - d_n}{d_n}\right)e\right) \quad \text{[NDS 3.4-5]}$$

When e is greater than d_n,

$$V'_r = \frac{F'_v b d_n}{1.5}$$

Example 5.13

A glued laminated beam of combination 24F-V10 western species, with a width of $6^{3}/_{4}$ in and a depth of 30 in, is notched as shown in the illustration. The beam has a moisture content exceeding 16%, and is subjected to sustained temperatures between 100°F and 125°F. The governing load combination is dead load plus live load plus snow load. Determine the maximum allowable shear force at each support.

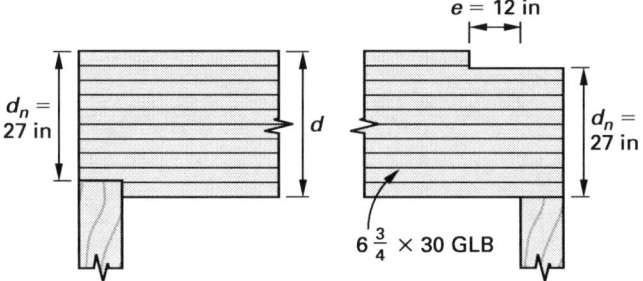

Solution

The reference design value for shear is tabulated in NDS Supp. Table 5A, Expanded Combinations, as

$$F_v = 215 \ \frac{\text{lbf}}{\text{in}^2}$$

The applicable adjustment factors for shear are the following. The wet service factor is

$$C_M = 0.875 \quad \text{[NDS Supp. Table 5A]}$$

The temperature factor for wet conditions is

$$C_t = 0.7 \quad \text{[NDS Table 2.3.3]}$$

The load duration factor is

$$C_D = 1.15$$

After allowing for the notch, as required by Table 5A, Expanded Combinations, Footnote 3, the adjusted shear stress is

$$\begin{aligned} F'_v &= 0.72 F_v C_M C_t C_D \\ &= (0.72)\left(215 \ \frac{\text{lbf}}{\text{in}^2}\right)(0.875)(0.70)(1.15) \\ &= 109 \ \text{lbf/in}^2 \end{aligned}$$

At the left support, the allowable shear force is

$$V'_r = \left(\frac{F'_v b d_n}{1.5}\right)\left(\frac{d_n}{d}\right)^2 \quad \text{[NDS 3.4-3]}$$

$$= \left(\frac{\left(109 \frac{\text{lbf}}{\text{in}^2}\right)(6.75 \text{ in})(27 \text{ in})}{1.5}\right)\left(\frac{27 \text{ in}}{30 \text{ in}}\right)^2$$

$$= 10{,}727 \text{ lbf}$$

At the right support, e is less than d_n, and the allowable shear force is

$$V'_r = \left(\frac{F'_v b}{1.5}\right)\left(d - \left(\frac{d - d_n}{d_n}\right)e\right) \quad \text{[NDS 3.4-5]}$$

$$= \left(\frac{\left(109 \frac{\text{lbf}}{\text{in}^2}\right)(6.75 \text{ in})}{1.5}\right)$$

$$\times \left(30 \text{ in} - \left(\frac{30 \text{ in} - 27 \text{ in}}{27 \text{ in}}\right)(12 \text{ in})\right)$$

$$= 14{,}061 \text{ lbf}$$

Shear at Connections

For a connection less than five times the depth of the member from its end, shown in Fig. 5.8, the allowable design shear is given by NDS Sec. 3.4.3.3 as

$$V'_r = \left(\frac{F'_v b d_e}{1.5}\right)\left(\frac{d_e}{d}\right)^2 \quad \text{[NDS 3.4-6]}$$

Figure 5.8 Bolted Connections

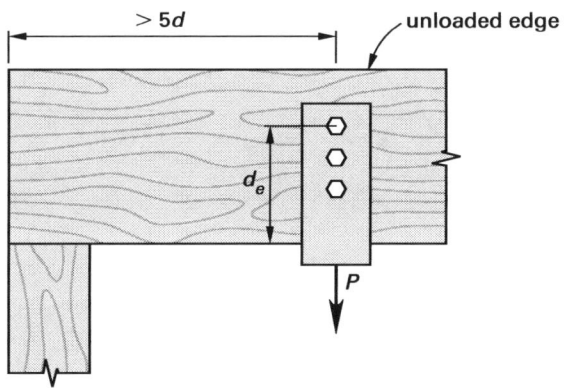

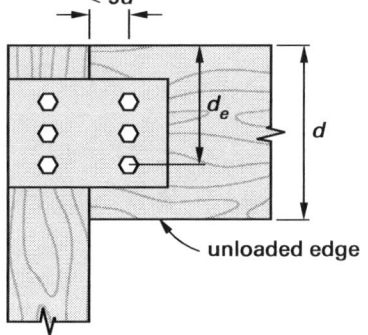

When the connection is at least five times the depth of the member from its end, the allowable design shear is

$$V'_r = \frac{F'_v b d_e}{1.5} \quad \text{[NDS 3.4-7]}$$

Example 5.14

A glued laminated beam of combination 24F-V10 (24F-1.7E) western species with a width of $6^3/_4$ in and a depth of 30 in is loaded with a bolted hanger connection, as shown in the illustration. The beam has a moisture content exceeding 16%. The governing load combination is dead load plus live load. Determine the maximum allowable shear force at the hanger connection.

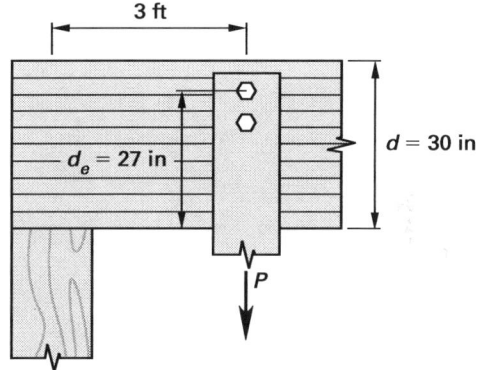

Solution

The reference design value for shear is tabulated in NDS Supp. Table 5A as

$$F_v = 215 \frac{\text{lbf}}{\text{in}^2}$$

The applicable adjustment factor for shear is the wet service factor.

$$C_M = 0.875 \quad \text{[NDS Supp. Table 5A]}$$

The adjusted shear stress is

$$F'_v = F_v C_M$$

$$= \left(215 \frac{\text{lbf}}{\text{in}^2}\right)(0.875)$$

$$= 188 \text{ lbf/in}^2$$

The hanger connection is less than $5d$ from the end of the beam, and from NDS Eq. (3.4-6) the allowable shear force is

$$F'_r = \left(\frac{b d_e F'_v}{1.5}\right)\left(\frac{d_e}{d}\right)^2$$

$$= \left(\frac{(6.75 \text{ in})(27 \text{ in})\left(188 \frac{\text{lbf}}{\text{in}^2}\right)}{1.5}\right)\left(\frac{27 \text{ in}}{30 \text{ in}}\right)^2$$

$$= 18{,}502 \text{ lbf}$$

6. DESIGN OF CONNECTIONS

Nomenclature

a	center to center spacing between adjacent rows of fasteners	in
a_e	minimum edge distance with load parallel to grain	in
a_p	minimum end distance with load parallel to grain	in
a_q	minimum end distance with load perpendicular to grain	in
A	area of cross section	in^2
A_m	gross cross-sectional area of main wood member(s)	in^2
A_n	net area of member	in^2
A_s	sum of gross cross-sectional areas of side member(s)	in^2
C_d	penetration depth factor for connections	–
C_D	load duration factor	–
C_{di}	diaphragm factor for nailed connections	–
C_{eg}	end grain factor for connections	–
C_g	group action factor for connections	–
C_M	wet service factor	–
C_{st}	metal side plate factor for 4 in shear plate connections	–
C_{tn}	toe-nail factor for nailed connections	–
C_Δ	geometry factor for connections	–
d	pennyweight of nail or spike	–
d_e	effective depth of member at a connection	in
D	diameter	in
e_p	minimum edge distance, unloaded edge	in
e_q	minimum edge distance, loaded edge	in
g	gage of screw	–
l_m	length of bolt in wood main member	in
l_s	total length of bolt in wood side member(s)	in
L	length of nail	in
n	number of fasteners in a row	–
N	reference lateral design value at an angle, θ, to the grain for a single split ring connector unit or shear plate connector unit	lbf
N'	allowable lateral design value at an angle, θ, to the grain for a single split ring connector unit or shear plate connector unit	lbf
p	depth of fastener penetration into wood member	in
P	reference lateral design value parallel to grain for a single split ring connector unit or shear plate connector unit	lbf
P'	allowable lateral design value parallel to grain for a single split ring connector unit or shear plate connector unit	lbf
Q	reference lateral design value perpendicular to grain for a single split ring connector unit or shear plate connector unit	lbf
Q'	allowable lateral design value perpendicular to grain for a single split ring connector or shear plate connector unit	lbf
s	center to center spacing between adjacent fasteners in a row	in
s_p	minimum spacing	in
t_m	thickness of main member	in
t_s	thickness of side member	in
T	tensile force on member	lbf
W	reference withdrawal design value for fastener	lbf/in of penetration
W'	allowable withdrawal design valve for fastener	lbf/in of penetration
Z	reference lateral design value for a single fastener connection	lbf
Z'	allowable lateral design value for a single fastener connection	lbf
$Z_\parallel$	reference lateral design value for a single bolt or lag screw connection with all wood members loaded parallel to grain	lbf
$Z_\perp$	reference lateral design value for a single bolt or lag screw, wood-to-wood, wood-to-metal or wood-to-concrete connection with all wood members loaded perpendicular to grain	lbf
Z'_α	allowable design value for lag screw with load applied at an angle, α, to the wood surface	lbf
$Z_{m\perp}$	reference lateral design value for a single bolt or lag screw wood-to-wood connection with main member loaded perpendicular to grain and side member loaded parallel to grain	lbf
$Z_{s\perp}$	reference lateral design value for a single bolt or lag screw wood-to-wood connection with main member loaded parallel to grain and side member loaded perpendicular to grain	lbf

Symbols

α	angle between wood surface and direction of applied load	deg
γ	load/slip modulus for a connection	lbf/in
θ	angle between direction of load and direction of grain (longitudinal axis of member)	deg

Adjustment of Design Values

The allowable design values for wood fasteners depend on the type of fastener and on the service conditions. The reference design values, tabulated in NDS Secs. 10–13, are applicable to normal conditions of use and normal load duration. To determine the relevant design values for other conditions of service, the reference design values are multiplied by adjustment factors specified in NDS Sec. 10.3. A summary of adjustment factors follows. The applicability of each to the basic design values is given in NDS Table 10.3.1, and summarized in Table 5.3.

Load Duration Factor, C_D

Normal load duration is equivalent to applying the maximum allowable load to a member for a period of 10 years. For loads of shorter duration, a member has the capacity to sustain higher loads, and the basic design values are multiplied by the load duration factor, C_D. With the exception of the impact load duration factor, which does not apply to fasteners, values of the load duration factor given in Table 5.2 are also applicable to connections.

Wet Service Factor, C_M

Basic design values apply to fasteners in wood seasoned to a moisture content of 19% or less. When the moisture content of the member exceeds 19%, the adjustment factors given in NDS Table 10.3.3 are applicable.

Temperature Factor, C_t

The temperature factor is applicable to all fasteners exposed to sustained temperatures up to 150°F. It is specified by NDS Table 10.3.4.

Group Action Factor, C_g

The group action factors for various connection geometries and fastener types are given in NDS Tables 10.3.6A–D. This factor is dependent on the ratio of the area of the side members in a connection to the area of the main member, A_s/A_m. A_m and A_s are calculated using gross areas without deduction for holes. When adjacent rows of fasteners are staggered, as shown in Fig. 5.9, the adjacent rows are considered as a single row.

Figure 5.9 Staggered Fasteners

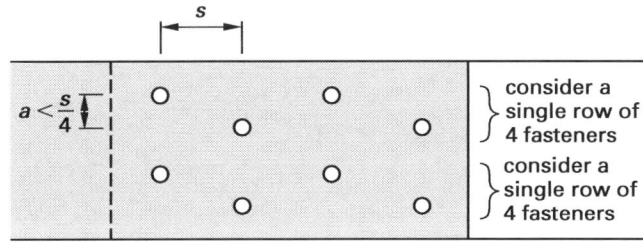

Geometry Factor, C_Δ

The geometry factor applies to bolts, lag screws, split rings, and shear plates. The factor is applied, in accordance with NDS Sec. 11.5.1, to dowel-type fasteners when the end distance or spacing is less than the minimum required for full design value. The smallest geometry factor for any fastener in a group shall apply to the whole group. NDS Tables 11.5.1B and 11.5.1C stipulate the minimum end distance and spacing requirements for dowel-type fasteners.

The factor is applied, in accordance with NDS Sec. 12.3.2, to split ring and shear plate connectors when the end distance, edge distance, or spacing is less than the minimum required for full design value. The smallest geometry factor for any connector in a group shall apply to the whole group. NDS Table 12.3 specifies the minimum end distance, edge distance, spacing requirements, and geometry factors for split ring and shear plate connectors.

Table 5.3 Adjustment Factors for Fasteners

adjustment factor	bolts	lag screws		split rings and shear plates		screws		nails	
design value	Z^a	W^b	Z^a	P^c	Q^d	W^b	Z^a	W^b	Z^a
C_D load duration factor	√	√	√	√	√	√	√	√	√
C_M wet service factor	√	√	√	√	√	√	√	√	√
C_t temperature factor	√	√	√	√	√	√	√	√	√
C_g group action factor	√	–	√	√	√	–	–	–	–
C_Δ geometry factor	√	–	√	√	√	–	–	–	√
C_d penetration depth factor	–	–	√	√	√	–	√	–	√
C_{eg} end grain factor	–	√	√	–	–	–	√	–	√
C_{st} metal side plate factor	–	–	–	√	–	–	–	–	–
C_{di} diaphragm factor	–	–	–	–	–	–	–	–	√
C_{tn} toe-nail factor	–	–	–	–	–	–	–	√	√

[a] lateral design value
[b] withdrawal design value
[c] parallel to grain design value
[d] perpendicular to grain design value

Penetration Depth Factor, C_d

The penetration depth factor applies to lag screws, split rings, shear plates, screws, and nails. The factor is applied in accordance with NDS Tables 11J–R and 12.2.3 when the penetration is less than the minimum specified for full design values.

End Grain Factor, C_{eg}

The end grain factor applies to lag screws, screws, and nails. The factor is applied in accordance with NDS Sec. 11.5.2 when the fastener is inserted in the end grain of a member.

Metal Side Plate Factor, C_{st}

The metal side plate factor is applicable to split rings and shear plates. The factor is applied in accordance with NDS Sec. 12.2.4 when metal side plates are used instead of wood side members.

Diaphragm Factor, C_{di}

The diaphragm factor applies to nails and spikes in accordance with NDS Sec. 11.5.3. When the fasteners are used in diaphragm construction, reference design values are multiplied by the diaphragm factor

$$C_{di} = 1.1$$

Toe-Nail Factor, C_{tn}

The toe-nail factor applies to nails and spikes in accordance with NDS Sec. 11.5.4. The reference withdrawal values for toe-nailed connections are multiplied by the toe-nail factor

$$C_{tn} = 0.67$$

The reference lateral design values for toe-nailed connections are multiplied by the toe-nail factor

$$C_{tn} = 0.83$$

Example 5.15

A select structural 2×8 Douglas Fir-Larch tie is connected to a 4 in $\times$ ¼ in steel gusset plate by a single row of eight ⅞ in diameter bolts in single shear. The governing load combination consists of dead load plus live load plus snow load, and the in-service moisture content exceeds 19%. The bolt spacing and end distance are 4 in. Determine the capacity of the connection.

Solution

From Ex. 5.10, the allowable tension design value for the tie, parallel to the grain, is

$$F'_t = F_t C_M C_F C_D$$
$$= \left(1000 \ \frac{\text{lbf}}{\text{in}^2}\right)(1.0)(1.2)(1.15)$$
$$= 1380 \ \text{lbf/in}^2$$

The allowable tension capacity of the Douglas Fir-Larch tie is

$$T = F'_t A_n$$
$$= F'_t\left(A - \left(D + \frac{1}{16}\right)b\right)$$
$$= \left(1380 \ \frac{\text{lbf}}{\text{in}^2}\right)\left(10.88 \ \text{in}^2 - \left(0.875 \ \text{in} + \frac{1}{16}\right)(1.5 \ \text{in})\right)$$
$$= 13{,}074 \ \text{lbf}$$

The reference ⅞ in diameter bolt design value for a 1½ in thick member loaded parallel to the grain, with a steel side plate in single shear, is tabulated in NDS Table 11B as

$$Z_{\parallel} = 1020 \ \text{lbf}$$

The gross cross-sectional area of the main wood member is

$$A_m = 10.88 \ \text{in}^2$$

The gross cross-sectional area of the steel gusset plate is

$$A_s = (4 \ \text{in})(0.25 \ \text{in})$$
$$= 1.0 \ \text{in}^2$$
$$\frac{A_m}{A_s} = \frac{10.88 \ \text{in}^2}{1.0 \ \text{in}^2}$$
$$= 10.88$$

The specified minimum spacing between bolts for the full bolt design value is specified in NDS Table 11.5.1C as

$$s_{p,\text{full}} = 4D$$
$$= (4)(0.875 \ \text{in})$$
$$= 3.5 \ \text{in}$$

The actual spacing is

$$s_p = 4 \ \text{in}$$
$$> s_{p,\text{full}}$$

The geometry factor for bolt spacing is given by NDS Sec. 11.5.1 as

$$C_\Delta = 1.0$$

The minimum end distance for the full bolt design value, for loading parallel to the grain, is specified in NDS Table 11.5.1B as

$$a_{p,\text{full}} = 7D$$
$$= (7)(0.875 \ \text{in})$$
$$= 6.125 \ \text{in}$$

The minimum end distance for the reduced bolt design value, for loading parallel to the grain, is specified in NDS Table 11.5.1B as

$$a_{p,\text{red}} = 3.5D$$
$$= (3.5)(0.875 \text{ in})$$
$$= 3.063 \text{ in}$$

The actual end distance is

$$a_p = 4 \text{ in}$$
$$< a_{p,\text{full}}$$
$$> a_{p,\text{red}}$$

The geometry factor for bolt end spacing is given by NDS Sec. 11.5.1 as

$$C_\Delta = \frac{a_p}{a_{p,\text{full}}}$$
$$= \frac{4 \text{ in}}{6.125 \text{ in}}$$
$$= 0.653 \quad [\text{governs}]$$

The additional applicable adjustment factors for the bolts are the following. The wet service factor from NDS Table 10.3.3 is

$$C_M = 0.7$$

The group action factor from NDS Table 10.3.6C for $A_m/A_s = 10.88$ and $A_m = 10.88 \text{ in}^2$ is

$$C_g = 0.58$$

The load duration factor from NDS Table 2.3.2 is

$$C_D = 1.15$$

The allowable lateral design value for 8 bolts is

$$T = nZ_\parallel C_M C_g C_\Delta C_D$$
$$= (8)(1020 \text{ lbf})(0.70)(0.58)(0.653)(1.15)$$
$$= 2500 \text{ lbf} \quad [\text{governs}]$$

Bolted Connections

Bolt holes shall be a minimum of $^1\!/_{32}$ in to a maximum of $^1\!/_{16}$ in larger than the bolt diameter, as specified in NDS. Sec. 11.1.2. A metal washer or plate is required between the wood and the nut, and between the wood and the bolt head. To ensure that the full design values of bolts are developed, spacing, edge, and end distances are specified in NDS Sec. 11.5. These distances are illustrated in Fig. 5.10.

Reference design values for single shear and symmetric double shear connections are specified in NDS Sec. 11.3 and tabulated in NDS Table 11A for two sawn lumber members of identical species; in NDS Table 11B for a sawn lumber member with a steel side plate; in NDS Table 11C for a glued laminated member with sawn lumber side member; in NDS Table 11D for a glued laminated member with a steel side plate; and in NDS Table 11E for connections to concrete.

Design values for double shear connections are specified in NDS Sec. 11.3 and tabulated in NDS Table 11F for three sawn lumber members of identical species; in NDS Table 11G for a sawn lumber member with steel side plates; in NDS Table 11H for a glued laminated member with sawn lumber side members; and in NDS Table 11I for a glued laminated member with steel side plates.

Example 5.16

A 3×8 select structural Douglas Fir-Larch ledger is attached to a concrete wall with $^3\!/_4$ in bolts at 4 ft centers, as shown in the illustration. What is the maximum dead load plus live load that the ledger can support?

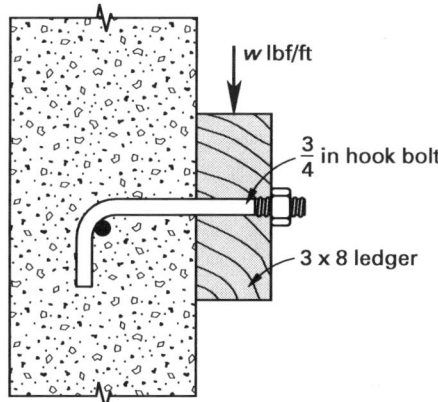

Solution

The nominal $^3\!/_4$ in diameter bolt design value for a $2^1\!/_2$ in thick member loaded perpendicular to the grain and attached to a concrete wall is tabulated in NDS Table 11E as

$$Z_\perp = 800 \text{ lbf}$$

The minimum edge distance for the full bolt design value, for loading perpendicular to grain, is specified in NDS Table 11.5.1A as

$$e_{q,\text{full}} = 4D$$
$$= (4)(0.75 \text{ in})$$
$$= 3.0 \text{ in}$$

The actual edge distance is

$$e_q = \frac{7.25 \text{ in}}{2}$$
$$= 3.625 \text{ in}$$
$$> e_{q,\text{full}} \quad [\text{satisfactory}]$$

The geometry factor is given by NDS Sec. 11.5.1 as

$$C_\Delta = 1.0$$

The maximum load that the ledger can support is

$$w = \frac{Z_\perp}{4}$$
$$= \frac{800 \text{ lbf}}{4 \text{ ft}}$$
$$= 200 \text{ lbf/ft}$$

Lag Screw Connections

In accordance with NDS Sec. 11.1.3, a clearance hole equal in diameter to the diameter of the shank, must be bored in the member for the full length of the unthreaded shank. A lead hole at least equal in length to the threaded portion of the screw must be provided. For wood of specific gravity in excess of 0.6, the lead hole diameter must equal 65–85% of the shank diameter. For wood of specific gravity of 0.5 or less, the lead hole diameter must equal 40–70% of the shank diameter. For wood of intermediate specific gravity, the lead hole diameter must equal 60–75% of the shank diameter. The lag screw must be inserted into the lead hole by turning with a wrench. For lag screws of $3/8$ in diameter or smaller, loaded primarily in withdrawal in wood with specific gravity of 0.5 or less, lead holes and clearance holes are not required.

Lateral Design Values in Side Grain

Minimum edge distances, end distances, spacing, and geometry factors are given in NDS Tables 11.5.1A–E, and are identical with those for bolts with a diameter equal to the shank diameter of the lag screw. The minimum allowable penetration of the lag screw into the member, not including the length of the tapered tip, is specified in NDS Sec. 11.1.3.6 as

$$p = 4D$$

As specified in NDS Table 11J, for full design values to be applicable, the depth of penetration shall not be less than

$$p = 8D$$

Figure 5.10 Bolt Spacing Requirements for Full Design Values

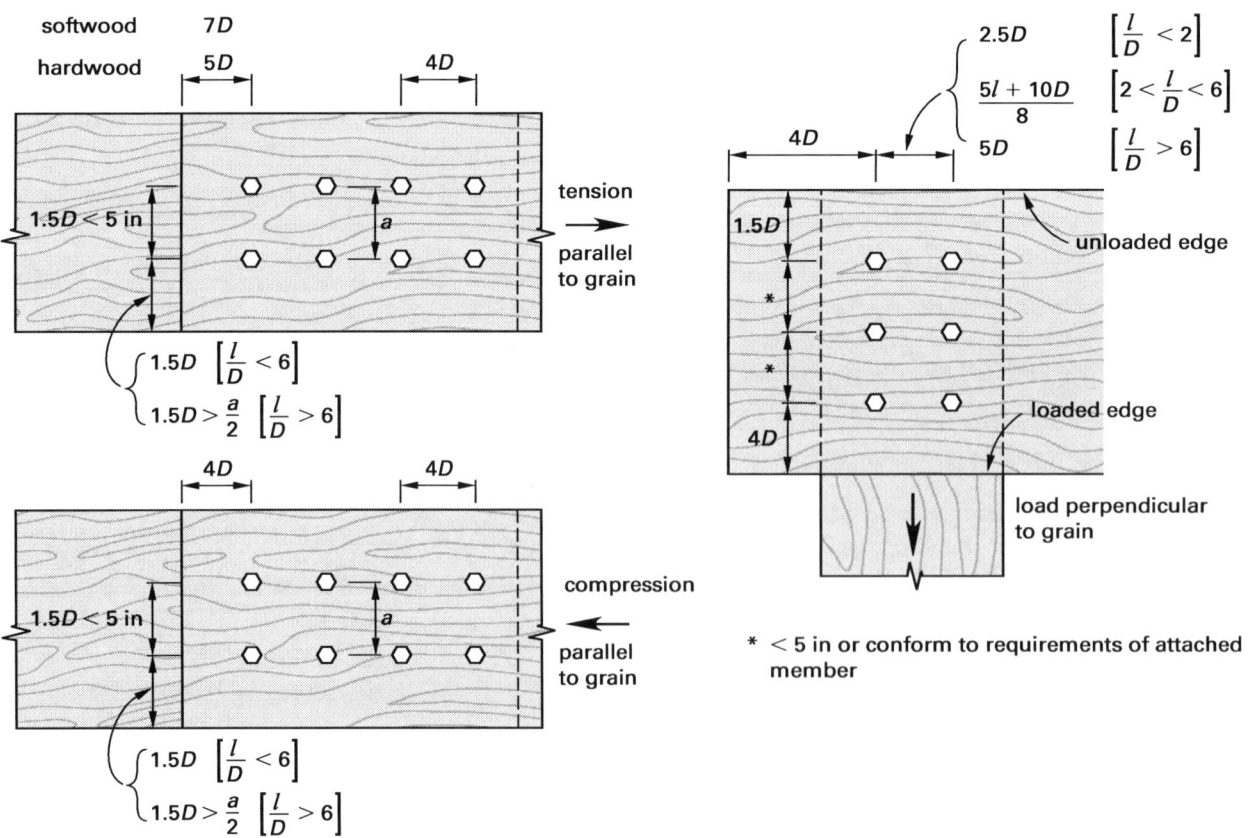

l is the lesser of length of bolt in main member or total length of bolt in side member(s)
D is the diameter of bolt

When the penetration is between $4D$ and $8D$, the nominal design value is multiplied by the penetration factor, which is defined in NDS Table 11J as

$$C_d = \frac{p}{8D}$$

Reference design values for single shear connections are specified in NDS Sec. 11.3. They are tabulated in NDS Table 11J for connections with a wood side member, and in NDS Table 11K for connections with a steel side plate.

Withdrawal Design Values in Side Grain Without Lateral Load

Minimum edge distance, end distance, and spacing, specified in NDS Table 11.5.1E, are

$$\text{edge distance} = 1.5D$$
$$\text{end distance} = 4D$$
$$\text{spacing} = 4D$$

Withdrawal design values in pounds per inch of thread penetration, not including the length of tapered tip, are given in NDS Table 11.2A.

Combined Lateral and Withdrawal Loads

When the load applied to a lag screw is at an angle, α, to the wood surface, as shown in Fig. 5.11, the lag screw is subjected to combined lateral and withdrawal loading. In accordance with NDS Sec. 11.4.1, the design value is now determined by Hankinson's formula,

$$Z'_\alpha = \frac{W'pZ'}{W'p\cos^2\alpha + Z'\sin^2\alpha} \quad \text{[NDS 11.4-1]}$$

Figure 5.11 Combined Lateral and Withdrawal Load

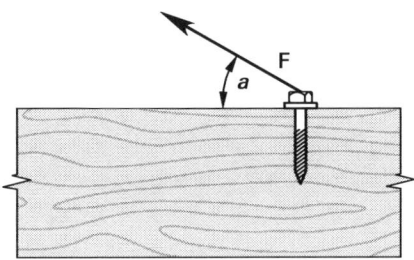

Example 5.17

A $\frac{1}{4}$ in thick steel plate is attached to a Douglas Fir-Larch member, as shown in the illustration, using two 3 in long $\times$ $\frac{3}{8}$ in diameter lag screws. The governing load combination consists of dead load plus live load, and the in-service moisture content exceeds 19%. The bolt spacing and end distance are 3 in. What is the maximum allowable load that can be applied to the plate?

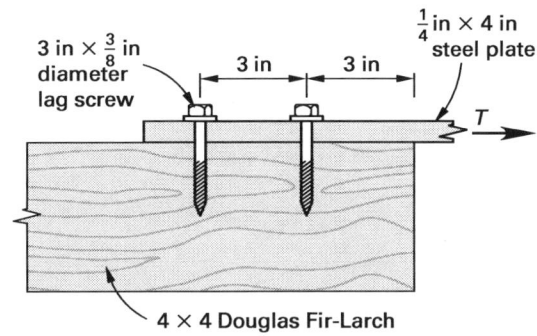

Solution

From NDS Table 11K, the reference lateral design value for load applied parallel to the grain is

$$Z_\| = 280 \text{ lbf}$$

The gross cross-sectional area of the 4×4 main wood member is

$$A_m = 12.25 \text{ in}^2$$

The gross cross-sectional area of the steel plate is

$$A_s = (4 \text{ in})(0.25 \text{ in})$$
$$= 1.0 \text{ in}^2$$
$$\frac{A_m}{A_s} = \frac{12.25 \text{ in}^2}{1.0 \text{ in}^2}$$
$$= 12.25$$

The specified minimum spacing between lag screws for the full design value is specified in NDS Table 11.5.1C as

$$s_{p,\text{full}} = 4D$$
$$= (4)(0.375 \text{ in})$$
$$= 1.50 \text{ in}$$

The actual spacing is

$$s_p = 3 \text{ in}$$
$$> s_{p,\text{full}}$$

The geometry factor for lag screw spacing is given by NDS Sec. 11.5.1 as

$$C_\Delta = 1.0$$

The minimum end distance for the full lag screw design value, for loading parallel to grain, is specified in NDS Table 11.5.1B as

$$a_{p,\text{full}} = 7D$$
$$= (7)(0.375 \text{ in})$$
$$= 2.625 \text{ in}$$

The actual end distance is

$$a_p = 3 \text{ in}$$
$$> a_{p,\text{full}}$$

The geometry factor for lag screw end spacing is given by NDS Sec. 11.5.1 as

$$C_\Delta = 1.0$$

From NDS App. L, the penetration into the main member by the screw shank, plus the threaded length, less the length of the tapered tip is

$$p = S + T - E - t_s$$
$$= 1.0 \text{ in} + 1.781 \text{ in} - 0.25 \text{ in}$$
$$= 2.531 \text{ in}$$

From NDS Table 11K, the penetration factor is

$$C_d = \frac{p}{8D}$$
$$= \frac{2.531 \text{ in}}{(8)(0.375 \text{ in})}$$
$$= 0.84$$

The additional applicable adjustment factors for the lag screws are the following. The wet service factor from NDS Table 10.3.3 is

$$C_M = 0.7$$

The group action factor from NDS Table 10.3.6C for $A_m/A_s = 12.25$ and $A_m = 12.25 \text{ in}^2$ is

$$C_g = 0.985$$

The load duration factor from NDS Table 2.3.2 is

$$C_D = 1.0$$

The allowable lateral design value for two lag screws is

$$T = nZ_\| C_M C_g C_\Delta C_D C_d$$
$$= (2)(280 \text{ lbf})(0.70)(0.985)(1.0)(1.0)(0.84)$$
$$= 320 \text{ lbf}$$

Split Ring and Shear Plate Connections

Dimensions for split ring and shear plate connectors are provided in NDS App. K. Edge and end distances, spacing, and geometry factors, C_Δ, for various sizes of split ring and shear plate connectors are specified in NDS Table 12.3. When lag screws are used instead of bolts, reference design values must, where appropriate, be multiplied by the penetration depth factors specified in NDS Table 12.2.3.

NDS Table 12.2.4 provides metal side plate factors, C_{st}, for 4 in shear plate connectors, loaded parallel to the grain, when metal side plates are substituted for wood side members. Group action factors, C_g, for 4 in split ring or shear plate connectors with wood side members are tabulated in NDS Table 10.3.6B. Group action factors, C_g, for 4 in shear plate connectors with steel side plates are tabulated in NDS Table 10.3.6D.

Reference design values for split ring connectors are provided in NDS Table 12.2A and, for shear plate connectors, in NDS Table 12.2B. When a load acts in the plane of the wood surface at an angle, θ, to the grain, the allowable design value is given by NDS Sec. 12.2.5 as

$$N' = \frac{P'Q'}{P'\sin^2\theta + Q'\cos^2\theta} \quad \text{[NDS 12.2-1]}$$

Example 5.18

The Douglas Fir-Larch select structural members shown in the illustration are connected with $2^5/_8$ in shear plate connectors. The governing load combination consists of dead load plus live load, and the in-service moisture content exceeds 19%. The connector spacing and end distances are as shown. Determine the capacity of the connection.

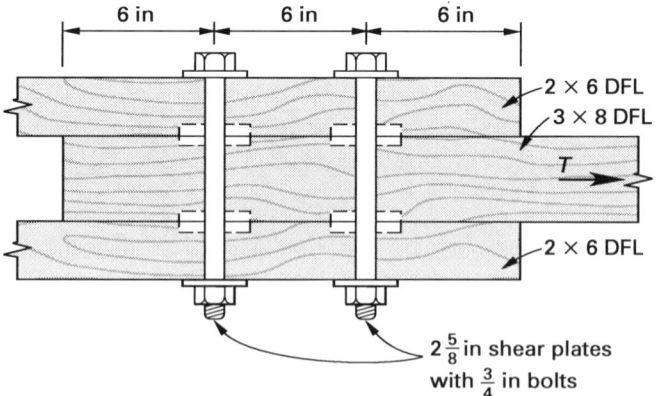

Solution

The reference $2^5/_8$ in shear plate design value for the $2^1/_2$ in thick main member of Group B species, with a connector on two faces, is tabulated in NDS Table 12.2B as

$$P_{\text{main}} = 2860 \text{ lbf}$$

The reference $2\frac{5}{8}$ in shear plate design value for a $1\frac{1}{2}$ in thick side member of Group B species, with a connector on one face, is tabulated in NDS Table 12.2B as

$$P_{\text{side}} = 2670 \text{ lbf} \quad [\text{governs}]$$

The gross cross-sectional area of the 3×8 main wood member is

$$A_m = 18.13 \text{ in}^2$$

The gross cross-sectional area of the two 2×6 side members is

$$A_s = (2)(8.25 \text{ in}^2)$$
$$= 16.50 \text{ in}^2$$
$$\frac{A_s}{A_m} = \frac{16.50 \text{ in}^2}{18.13 \text{ in}^2}$$
$$= 0.91$$

The group action factor from NDS Table 10.3.6B for $A_s/A_m = 0.91$ and $A_s = 16.50 \text{ in}^2$ is

$$C_g = 0.98$$

The minimum end distance for the full shear plate design value, for loading parallel to the grain, is specified in NDS Table 12.3 as

$$a_{p,\text{full}} = 5.5 \text{ in}$$

The actual edge distance is

$$a_p = 6.0 \text{ in}$$
$$> a_{p,\text{full}}$$

The geometry factor is given by NDS Sec. 12.3.4 as

$$C_\Delta = 1.0$$

The specified minimum spacing for the full shear plate design value, for loading parallel to the grain, is given in NDS Table 12.3 as

$$s_{\text{full}} = 6.75 \text{ in}$$

The specified minimum spacing for the reduced shear plate design value, for loading parallel to the grain, is specified in NDS Table 12.3 as

$$s_{\text{red}} = 3.5 \text{ in}$$

The actual spacing is

$$s = 6 \text{ in}$$
$$< s_{\text{full}}$$
$$> s_{\text{red}}$$

The geometry factor for shear plate spacing is given by interpolation, as specified in NDS Sec. 12.3.5, as

$$C_\Delta = 0.5 + \frac{(0.5)(6 \text{ in} - 3.5 \text{ in})}{6.75 \text{ in} - 3.5 \text{ in}}$$
$$= 0.885 \quad [\text{governs}]$$

The wet service factor is obtained from NDS Table 10.3.3 as

$$C_M = 0.7$$

The allowable design value for four shear plates is

$$T = nP_{\text{side}} C_g C_\Delta C_M$$
$$= (4)(2670 \text{ lbf})(0.98)(0.885)(0.7)$$
$$= 6500 \text{ lbf}$$

Wood Screw Connections

In accordance with NDS Sec. 11.1.4.3, for screws loaded laterally in wood of specific gravity in excess of 0.6, a clearance hole approximately equal in diameter to the diameter of the shank must be bored in the member for the full length of the unthreaded shank. The lead hole receiving the threaded portion of the screw must have a diameter approximately equal to the wood screw root diameter. For wood of specific gravity not exceeding 0.6, a clearance hole approximately equal in diameter to $\frac{7}{8}$ the diameter of the shank must be bored in the member for the full length of the unthreaded shank. The lead hole receiving the threaded portion of the screw must have a diameter approximately equal to $\frac{7}{8}$ the diameter of the wood screw root diameter.

In accordance with NDS Sec. 11.1.4.2, for screws loaded in withdrawal in wood of specific gravity in excess of 0.6, a clearance hole approximately equal in diameter to 90% of the wood screw root diameter must be provided for the full length of the screw. For wood with specific gravity of 0.5 or less, lead holes are not required. For wood of intermediate specific gravity, the lead hole diameter must approximately equal 70% of the wood screw root diameter. The screw must be inserted into the lead hole by turning with a screw driver, not by driving with a hammer.

Lateral Design Values in Side Grain

Recommended edge distances, end distances, and spacing are tabulated in NDS Commentary Table C11.1.4.7 for wood and steel side plates with and without prebored holes. Wood screws are not subject to the group action factor, C_g.

As specified in NDS Table 11L, for full design values to be applicable, the depth of penetration must not be less than

$$p_{\text{full}} = 10D$$

The minimum allowable penetration is given by NDS Sec. 11.1.4.6 as

$$p_{min} = 6D$$

When the actual penetration, p, is between $6D$ and $10D$, the nominal design value is multiplied by the penetration factor, which is defined in NDS Table 11L as

$$C_d = \frac{p}{p_{full}}$$
$$\leq 1.0$$

Reference design values for single shear connections are specified in NDS Sec. 11.3, tabulated in NDS Table 11L for connections with a wood side member, and tabulated in NDS Table 11M for connections with a steel side plate.

Withdrawal Design Values in Side Grain

Withdrawal design values in pounds per inch of thread penetration are tabulated in NDS Table 11.2B. The length of thread is specified in the NDS App. L as $2/3$ the total screw length or four times the screw diameter, whichever is the greater.

Wood screws shall not be loaded in withdrawal from end grain of wood.

Combined Lateral and Withdrawal Loads

When the load applied to a wood screw is at an angle, α, to the wood surface, as shown in Fig. 5.11, the wood screw is subjected to combined lateral and withdrawal loading. The design value is determined by the Hankinson's formula, given in NDS Sec. 11.4.1 as

$$Z'_\alpha = \frac{W'pZ'}{W'p\cos^2\alpha + Z'\sin^2\alpha} \quad \text{[NDS 11.4-1]}$$

Example 5.19

A 2×4 select structural Douglas Fir-Larch collector is secured to the Douglas Fir-Larch top plate of a shear wall with a single row of eight $14 \text{ g} \times 3^{1}/_2$ in wood screws. Edge distances, end distances, and spacing are sufficient to prevent splitting of the wood. Determine the maximum tensile force, T, due to seismic load, which can be resisted by the wood screws.

Solution

The diameter of a 14 g wood screw is obtained from NDS App. L as

$$D = 0.242 \text{ in}$$

The reference design value for a 14 g wood screw in a $1^{1}/_2$ in Douglas Fir-Larch side member in single shear is tabulated in NDS Table 11L as

$$Z = 163 \text{ lbf}$$

As specified in NDS Table 11L, for full design values to be applicable, the depth of penetration shall not be less than

$$p_{full} = 10D$$
$$= (10)(0.242 \text{ in})$$
$$= 2.42 \text{ in}$$

The minimum allowable penetration is given by NDS Sec. 11.1.4.6 as

$$p_{min} = 6D$$
$$= (6)(0.242 \text{ in})$$
$$= 1.45 \text{ in}$$

The actual penetration of the wood screw into the top plate of the shear wall is

$$p = 3.5 \text{ in} - 1.5 \text{ in}$$
$$= 2.0 \text{ in}$$
$$< p_{full}$$
$$> p_{min}$$

The penetration depth factor is

$$C_d = \frac{p}{p_{full}}$$
$$= \frac{2 \text{ in}}{2.42 \text{ in}}$$
$$= 0.826$$

The load duration factor for seismic load is obtained from NDS Table 2.3.2 as

$$C_D = 1.60$$

The allowable lateral design value for eight screws is

$$T = nZC_DC_d$$
$$= (8)(163 \text{ lbf})(1.60)(0.826)$$
$$= 1723 \text{ lbf}$$

Connections with Nails and Spikes

The NDS specifications apply to common nails and spikes, box nails, sinker nails, and threaded hardened-steel nails. The tabulated reference design values apply to nailed connections either with or without prebored holes. A prebored hole may be used to prevent splitting of the wood. For wood of specific gravity in excess of 0.6, the hole diameter must not exceed 90% of diameter of the nail. For wood of specific gravity not exceeding 0.6, the hole diameter must not exceed 75% of the diameter of the nail.

As shown in Fig. 5.12 and specified in NDS Sec. 11.1.5, toe-nails are driven at an angle of approximately 30° with the face of the member, with the point of penetration approximately $1/3$ the length of the nail from the member end. In accordance with NDS Commentary Sec. C11.5.4, the side member thickness, t_s, is taken as equal to this end distance, and

$$t_s = \frac{L}{3}$$

Figure 5.12 Toe-Nail Connection

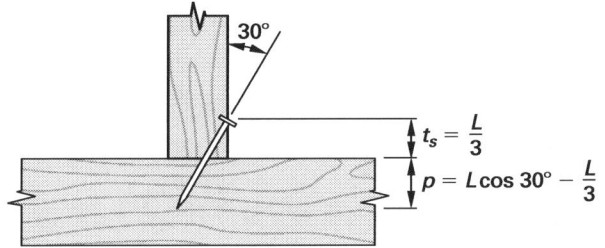

Lateral Design Values in Side Grain

Recommended edge distances, end distances, and spacing are tabulated in NDS Commentary Table C11.1.5.6 for wood and steel side plates with and without prebored holes. Nails and spikes are not subject to the group action factor, C_g.

As specified in NDS Table 11N, for full design values to be applicable, the depth of penetration must not be less than

$$p_{\text{full}} = 10D$$

The minimum allowable penetration is given by NDS Table 11N as

$$p_{\min} = 6D$$

When the actual penetration, p, is between $6D$ and $10D$ the reference design value is multiplied by the penetration factor, which is defined in NDS Table 11N as

$$C_d = \frac{p}{p_{\text{full}}} \leq 1.0$$

In accordance with NDS Sec. 11.5.3, reference design values for nails and spikes used in diaphragm construction must be multiplied by the diaphragm factor, which is

$$C_{di} = 1.1$$

In accordance with NDS Sec. 11.5.4, reference lateral design values for nails and spikes used in toe-nailed connections must be multiplied by the toe-nail factor, which is

$$C_{tn} = 0.83$$

Reference design values for single shear connections for two sawn lumber members of identical species are tabulated in NDS Table 11N.

Reference design values for single shear connections for a sawn lumber member with a steel side plate are tabulated in NDS Table 11P.

In accordance with NDS Sec. 11.3.7. the nominal double shear value for a three-member sawn lumber connection is twice the lesser of the nominal design values for each shear plane. The minimum allowable penetration into the side members is given by NDS Sec. 11.1.5.5 as

$$p_{\min} = 6D$$

An exception is permitted when the side member is at least $3/8$ in thick, and $12d$ or smaller nails extend at least three diameters beyond the side member and are clinched.

Withdrawal Design Values in Side Grain

Reference withdrawal design values in pounds per inch of penetration are tabulated in NDS Table 11.2C. In accordance with NDS Sec. 11.5.4, reference withdrawal design values for nails and spikes used in toe-nailed connections shall be multiplied by the toe-nail factor, which is

$$C_{tn} = 0.67$$

As specified in NDS Sec. 11.2.3.2, nails and spikes must not be loaded in withdrawal from end grain of wood.

Combined Lateral and Withdrawal Loads

When the load applied to a wood screw is at an angle, α, to the wood surface, the wood screw is subjected to combined lateral and withdrawal loading. In accordance with NDS Sec. 11.4.2, the design value is determined by the interaction equation

$$Z'_\alpha = \frac{W'pZ'}{W'p\cos\alpha + Z'\sin\alpha} \quad \text{[NDS 11.4-2]}$$

Example 5.20

A 4×8 select structural Douglas Fir-Larch collector is secured to the Douglas Fir-Larch top plate of a shear wall with a 12 gage steel strap, as shown in the illustration. Fourteen $16d$ common nails, $2^{1}/_{2}$ in long, are provided on each side of the strap. Edge distances, end distances, and spacing are sufficient to prevent splitting of the wood. Determine the maximum tensile force, T, due to seismic load, which can be resisted by the nails.

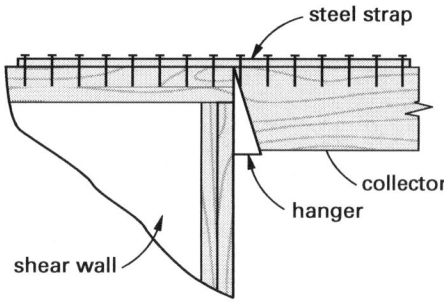

Solution

The diameter of a $16d$ common nail is obtained from NDS Table 11P as

$$D = 0.162 \text{ in}$$

The reference single shear design value for a $16d$ common nail in a Douglas Fir-Larch member with a 12 gauge side plate is tabulated in NDS Table 11P as

$$Z = 149 \text{ lbf}$$

As specified in NDS Table 11N, for full design values to be applicable, the depth of penetration shall not be less than

$$\begin{aligned} p_{\text{full}} &= 10D \\ &= (10)(0.162 \text{ in}) \\ &= 1.62 \text{ in} \end{aligned}$$

The actual penetration of the nails is

$$\begin{aligned} p &= 2.5 \text{ in} - 0.105 \text{ in} \\ &= 2.395 \text{ in} \\ &> p_{\text{full}} \end{aligned}$$

The penetration depth factor is

$$C_d = 1.0$$

The load duration factor for seismic load is obtained from NDS Table 2.3.2 as

$$C_D = 1.60$$

The allowable lateral design value for fourteen nails is

$$\begin{aligned} T &= nZC_DC_d \\ &= (14)(149 \text{ lbf})(1.60)(1.0) \\ &= 3340 \text{ lbf} \end{aligned}$$

PRACTICE PROBLEMS

Problems 1–3 refer to a glued laminated beam of combination 24F-V4 (24F-1.8E) western species that has a width of $6^3/_4$ in, and a depth of 24 in. The beam is simply supported over an effective span of 24 ft, and is laterally braced at 8 ft on center. The governing load combination is a uniformly distributed dead load plus live load, and the moisture content exceeds 16%.

1. What is most nearly the volume factor for the beam?

(A) 0.75

(B) 0.80

(C) 0.85

(D) 0.90

2. What is most nearly the stability factor?

(A) 0.91

(B) 0.95

(C) 0.99

(D) 1.0

3. What is most nearly the design value in bending?

(A) 1728 lbf/in^2

(B) 1824 lbf/in^2

(C) 1901 lbf/in^2

(D) 1920 lbf/in^2

Problems 4–6 refer to a 6×6 Douglas Fir-Larch no. 1 visually graded timber post that is 12 ft long and pinned at each end. The governing load combination consists of dead load plus live load, and the column is not braced laterally.

4. What is most nearly the stability factor of the post?

(A) 0.52

(B) 0.56

(C) 0.64

(D) 0.75

5. Neglecting the self-weight of the column, what is most nearly the maximum load that the post can support?

(A) 15,125 lbf

(B) 16,940 lbf

(C) 19,360 lbf

(D) 22,687 lbf

6. Determine whether the post is adequate to support an axial load of 5 kips and a horizontal wind load of 500 lbf applied on one face at midheight.

(A) No, the interaction equation for bending and concentric axial compression is less than 1.0.

(B) No, the interaction equation for bending and concentric axial compression is greater than 1.0.

(C) Yes, the interaction equation for bending and concentric axial compression is less than 1.0.

(D) Yes, the interaction equation for bending and concentric axial compression is greater than 1.0.

SOLUTIONS

1. The volume factor is given by NDS Eq. (5.3-1) as

$$C_V = \left(\frac{1291.5}{bdL}\right)^{1/x}$$
$$= \left(\frac{1291.5 \text{ in}^2\text{-ft}}{(6.75 \text{ in})(24 \text{ in})(24 \text{ ft})}\right)^{1/10}$$
$$= 0.90$$

The answer is (D).

2. The reference design values for bending and the modulus of elasticity, tabulated in NDS Supp. Table 5A, are

$$F_b = 2400 \; \frac{\text{lbf}}{\text{in}^2}$$

$$E_{y,\min} = 0.83 \times 10^6 \; \frac{\text{lbf}}{\text{in}^2}$$

The applicable adjustment factor for the modulus of elasticity is the wet service factor.

$$C_M = 0.833 \quad \text{[NDS Supp. Table 5A]}$$

The adjusted modulus of elasticity for stability calculations is

$$E'_{\min} = E'_{y,\min}$$
$$= E_{y,\min} C_M$$
$$= \left(0.83 \times 10^6 \; \frac{\text{lbf}}{\text{in}^2}\right)(0.833)$$
$$= 0.69 \times 10^6 \text{ lbf/in}^2$$

The distance between lateral restraints is

$$l_u = 8 \text{ ft}$$
$$\frac{l_u}{d} = \frac{(8 \text{ ft})\left(12 \; \frac{\text{in}}{\text{ft}}\right)}{24 \text{ in}}$$
$$= 4.0$$
$$< 7$$

For a uniformly distributed load and an l_u/d ratio less than 7, the effective length is obtained from NDS Table 3.3.3 as

$$l_e = 2.06 l_u$$
$$= (2.06)(8 \text{ ft})\left(12 \; \frac{\text{in}}{\text{ft}}\right)$$
$$= 198 \text{ in}$$

The slenderness ratio is given by NDS Sec. 3.3.3 as

$$R_B = \sqrt{\frac{l_e d}{b^2}}$$

$$= \sqrt{\frac{(198 \text{ in})(24 \text{ in})}{(6.75 \text{ in})^2}}$$

$$= 10.21$$

$$< 50 \quad \text{[satisfies criteria of NDS Sec. 3.3.3]}$$

The critical buckling design value is

$$F_{bE} = \frac{1.20 E'_{\min}}{R_B^2}$$

$$= \frac{(1.20)\left(0.69 \times 10^6 \,\frac{\text{lbf}}{\text{in}^2}\right)}{(10.21)^2}$$

$$= 7943 \text{ lbf/in}^2$$

The reference flexural design value, multiplied by all applicable adjustment factors except C_V and C_L, is

$$F_b^* = F_b C_M$$

$$= \left(2400 \,\frac{\text{lbf}}{\text{in}^2}\right)(0.80)$$

$$= 1920 \text{ lbf/in}^2$$

The ratio of F_{bE} to F_b^* is

$$F = \frac{F_{bE}}{F_b^*}$$

$$= \frac{7943 \,\frac{\text{lbf}}{\text{in}^2}}{1920 \,\frac{\text{lbf}}{\text{in}^2}}$$

$$= 4.14$$

The beam stability factor is given by NDS Sec. 3.3.3 as

$$C_L = \frac{1.0 + F}{1.9} - \sqrt{\left(\frac{1.0 + F}{1.9}\right)^2 - \frac{F}{0.95}}$$

$$= \frac{1.0 + 4.14}{1.9} - \sqrt{\left(\frac{1.0 + 4.14}{1.9}\right)^2 - \frac{4.14}{0.95}}$$

$$= 0.98$$

The answer is (C).

3. From Prob. 2, the beam stability factor is

$$C_L = 0.98$$

From Prob. 1, the volume factor is

$$C_V = 0.90$$

$$< C_L$$

The volume factor governs.

The allowable flexural design value is

$$F_b' = C_V F_b C_M$$

$$= C_V F_b^*$$

$$= (0.90)\left(1920 \,\frac{\text{lbf}}{\text{in}^2}\right)$$

$$= 1728 \text{ lbf/in}^2$$

The answer is (A).

4. The basic design values for compression parallel to the grain are obtained from NDS Supp. Table 4D as

$$F_c = 1000 \,\frac{\text{lbf}}{\text{in}^2}$$

$$E_{\min} = 0.58 \times 10^6 \,\frac{\text{lbf}}{\text{in}^2}$$

The size factor for compression from NDS Supp. Table 4D is

$$C_F = 1.00$$

For pinned-ended support conditions, the buckling length coefficient from NDS App. Table G1 is

$$K_e = 1.0$$

The slenderness ratio about both axes is

$$\frac{K_e l}{d} = \frac{(1.0)(12 \text{ ft})\left(12 \,\frac{\text{in}}{\text{ft}}\right)}{5.5 \text{ in}}$$

$$= 26.18$$

The adjusted modulus of elasticity for compression for stability calculations is

$$E'_{\min} = E_{\min} C_F$$

$$= \left(0.58 \times 10^6 \,\frac{\text{lbf}}{\text{in}^2}\right)(1.00)$$

$$= 0.58 \times 10^6 \text{ lbf/in}^2$$

The tabulated compression design value, multiplied by all applicable adjustment factors except C_P, is

$$F_c^* = F_c C_F$$

$$= \left(1000 \,\frac{\text{lbf}}{\text{in}^2}\right)(1.00)$$

$$= 1000 \text{ lbf/in}^2$$

The critical buckling design value is

$$F_{cE} = \frac{0.822 E'_{min}}{\left(\frac{l_e}{d}\right)^2}$$

$$= \frac{(0.822)\left(0.58 \times 10^6 \frac{\text{lbf}}{\text{in}^2}\right)}{(26.18)^2}$$

$$= 696 \text{ lbf/in}^2$$

The ratio of F_{cE} to F_c^* is

$$F' = \frac{F_{cE}}{F_c^*}$$

$$= \frac{696 \frac{\text{lbf}}{\text{in}^2}}{1000 \frac{\text{lbf}}{\text{in}^2}}$$

$$= 0.696$$

The column parameter is obtained from NDS Sec. 3.7.1.5 as

$$c = 0.8 \quad [\text{sawn lumber}]$$

The column stability factor specified by NDS Sec. 3.7.1 is

$$C_p = \frac{1.0 + F'}{2c} - \sqrt{\left(\frac{1.0 + F'}{2c}\right)^2 - \frac{F'}{c}}$$

$$= \frac{1.0 + 0.696}{(2)(0.8)} - \sqrt{\left(\frac{1.0 + 0.696}{(2)(0.8)}\right)^2 - \frac{0.696}{0.8}}$$

$$= 0.56$$

The answer is (B).

5. From Prob. 4, the column stability factor is

$$C_P = 0.56$$

The allowable compression design value parallel to the grain is

$$F'_c = F_c C_F C_P$$

$$= \left(1000 \frac{\text{lbf}}{\text{in}^2}\right)(0.56)$$

$$= 560 \text{ lbf/in}^2$$

The allowable axial load on the column is

$$P_{al} = A F'_c$$

$$= (30.25 \text{ in}^2)\left(560 \frac{\text{lbf}}{\text{in}^2}\right)$$

$$= 16{,}940 \text{ lbf}$$

The answer is (B).

6. The reference design value for bending from NDS Supp. Table 4D is

$$F_b = 1200 \frac{\text{lbf}}{\text{in}^2}$$

The applicable load duration factor for flexure from NDS Table 2.3.2 is

$$C_D = 1.6$$

The size factor for flexure from NDS Supp. Table 4D is

$$C_F = 0.74$$

The beam stability factor for a square section, given by NDS Sec. 3.3.3.1, is

$$C_L = 1.0$$

The allowable flexural design value is

$$F'_b = F_b C_F C_D C_L$$

$$= \left(1200 \frac{\text{lbf}}{\text{in}^2}\right)(0.74)(1.6)(1.0)$$

$$= 1421 \text{ lbf/in}^2$$

The actual edgewise bending stress is

$$f_{b1} = \frac{WL}{4S}$$

$$= \frac{(500 \text{ lbf})(12 \text{ ft})\left(12 \frac{\text{in}}{\text{ft}}\right)}{(4)(27.73 \text{ in}^3)}$$

$$= 649 \text{ lbf/in}^2$$

$$< F'_b \quad [\text{satisfactory}]$$

The allowable compression design value parallel to the grain, obtained from Prob. 5, is

$$F'_c = 560 \frac{\text{lbf}}{\text{in}^2}$$

The actual compressive stress on the column is

$$f_c = \frac{P}{A}$$

$$= \frac{5000 \text{ lbf}}{30.25 \text{ in}^2}$$

$$= 165 \frac{\text{lbf}}{\text{in}^2}$$

$$< F_c \quad [\text{satisfactory}]$$

The distance between lateral restraints for bending instability is

$$l_u = 12 \text{ ft}$$

$$\frac{l_u}{d} = \frac{(12 \text{ ft})\left(12 \frac{\text{in}}{\text{ft}}\right)}{5.5 \text{ in}}$$

$$= 26$$

$$> 7$$

For a concentrated load at midheight and an l_u/d ratio greater than 7, the effective length from NDS Table 3.3.3 is

$$l_e = 1.37 l_u + 3d$$

$$= (1.37)(12 \text{ ft})\left(12 \frac{\text{in}}{\text{ft}}\right) + (3)(5.5 \text{ in})$$

$$= 214 \text{ in}$$

The adjusted modulus of elasticity for flexure is

$$E'_{min} = E_{min} C_F$$

$$= \left(0.58 \times 10^6 \frac{\text{lbf}}{\text{in}^2}\right)(0.90)$$

$$= 0.52 \times 10^6 \text{ lbf/in}^2$$

The critical buckling design value in the plane of bending is

$$F_{cE1} = \frac{0.822 E'_{min}}{\left(\frac{l_{e1}}{d_1}\right)^2} = \frac{(0.822)\left(0.52 \times 10^6 \frac{\text{lbf}}{\text{in}^2}\right)}{\left(\frac{214 \text{ in}}{5.5 \text{ in}}\right)^2}$$

$$= 282 \text{ lbf/in}^2$$

$$> f_c \quad \text{[satisfactory]}$$

The moment magnification factor for axial compression and flexure is

$$C_{m3} = 1.0 - \frac{f_c}{F_{cE1}} = 1.0 - \frac{165 \frac{\text{lbf}}{\text{in}^2}}{282 \frac{\text{lbf}}{\text{in}^2}}$$

$$= 0.41$$

The interaction equation for bending and concentric axial compression is given in NDS Sec. 3.9.2 as

$$\left(\frac{f_c}{F'_c}\right)^2 + \frac{f_{b1}}{F'_{b1} C_{m3}} \leq 1.0$$

The left-hand side of the expression is

$$\left(\frac{165 \frac{\text{lbf}}{\text{in}^2}}{560 \frac{\text{lbf}}{\text{in}^2}}\right)^2 + \frac{649 \frac{\text{lbf}}{\text{in}^2}}{\left(1421 \frac{\text{lbf}}{\text{in}^2}\right)(0.41)} = 0.087 + 1.114$$

$$= 1.201$$

$$> 1.0$$

[unsatisfactory]

The post is inadequate.

The answer is (B).

6 Design of Reinforced Masonry

1. Design Principles 6-1
2. Design for Flexure 6-4
3. Design for Shear 6-10
4. Design of Masonry Columns 6-12
5. Design of Masonry Shear Walls 6-14
6. Wall Design for Out-of-Plane Loads 6-18
7. Design of Anchor Bolts 6-24
8. Design of Prestressed Masonry 6-26
   Practice Problems 6-33
   Solutions 6-34

1. DESIGN PRINCIPLES

Nomenclature

b_s	width of support	in
d	effective depth, distance from extreme compression fiber to centroid of tension reinforcement	in
D	dead load	kips or kips/ft
E	earthquake load	kips or kips/ft
F	load caused by fluids	kips or kips/ft
h	overall depth of beam	in
H	load caused by lateral earth pressure	kips or kips/ft
l	clear span length of beam between supports	ft
l_e	effective span length of beam	ft
l_s	distance between centers of supports	ft
L	live load	kips or kips/ft
M_D	service level bending moment caused by applied loads	ft-kips
M_s	service level bending moment caused by member self-weight	ft-kips
Q	required strength	kips or ft-kips
R	rain load	kips or kips/ft
S	snow load	kips or kips/ft
T	self-straining force	kips or kips/ft
V	unfactored shear force at section under consideration	kips
V_D	shear force caused by dead loads	kips
V_L	shear force caused by live loads	kips
V_s	service level shear force caused by member self-weight	kips
w	distributed load	lbf/ft
W	wind load	kips or kips/ft

Symbols

ϕ	strength reduction factor	–

General Design Requirements

Allowable stress design is the method of proportioning structural members so that the allowable stress of the material is not exceeded under the action of design loads calculated using ASD load combinations.

The design of masonry structures using allowable stress design is permitted by IBC[1] Sec. 2107.1. Design must comply with MSJC[2] Chs. 1 and 2, with the exception of Secs. 2.1.2, 2.1.9.7, and 2.1.9.7.1.1. Hence, the load combinations in MSJC Ch. 2 are not adopted.

Load Combinations

Where the legally adopted building code does not provide load combinations, the combinations specified in ASCE[3] Sec. 2.4.1 must be used. The governing load condition for dead load and fluid pressure is given by ASCE Sec. 2.4.1 load combination 1 as

$$Q = D + F$$

The governing load condition when the structure is subjected to maximum values of occupancy live loads is given by load combination 2 as

$$Q = D + F + T + L + H$$

The governing load condition when the structure is subjected to maximum values of roof live load, rainwater, or snow load is given by load combination 3 as

$$Q = D + H + F + (L_r \text{ or } S \text{ or } R)$$

The governing load condition when the structure is subjected to simultaneous values of occupancy live loads, roof live load, rainwater, or snow load is given by load combination 4 as

$$Q = D + H + F + 0.75(L + T) + 0.75(L_r \text{ or } S \text{ or } R)$$

The governing load condition when the structure is subjected to simultaneous values of occupancy live loads, roof live load, rainwater, or snow load plus wind

---
[1]International Code Council, 2009 (See References and Codes)
[2]American Concrete Institute, 2008 (See References and Codes)
[3]American Society of Civil Engineers, 2005 (See References and Codes)

load or seismic load increasing the effects of dead load is given by load combination 6 as

$$Q = D + H + F + 0.75(W \text{ or } 0.75E)$$
$$+ 0.75L + 0.75(L_r \text{ or } S \text{ or } R)$$

The governing load condition when the structure is subjected to maximum values of seismic load or wind load, which increase the effects of dead load, is given by load combination 5 as

$$Q = D + H + F + (W + 0.7E)$$

The governing load condition when the structure is subjected to maximum values of wind load, which oppose the effects of dead load, is given by load combination 7 as

$$Q = 0.6D + W + H$$

The governing load condition when the structure is subjected to maximum values of seismic load, which oppose the effects of dead load, is given by load combination 8 as

$$Q = 0.6D + 0.7E + H$$

No increase in allowable stresses is permitted with these load combinations.

Effective Span Length of Beams

In accordance with MSJC Sec. 1.13, the effective span length of a simply supported beam is defined as the clear span plus the depth of the member, but it is not to exceed the distance between the centers of the supports. As shown in Fig. 6.1(b) and (c),

$$l_e = l + h$$
$$\leq l_s$$

For a continuous beam, the effective span length is defined as the distance between the centers of the supports. As shown in Fig. 6.1(d),

$$l_e = l_s$$

The effective span length of beams built integrally with supports is customarily taken as being equal to the clear span. As shown in Fig. 6.1(a),

$$l_e = l$$

Figure 6.1 Effective Span Length

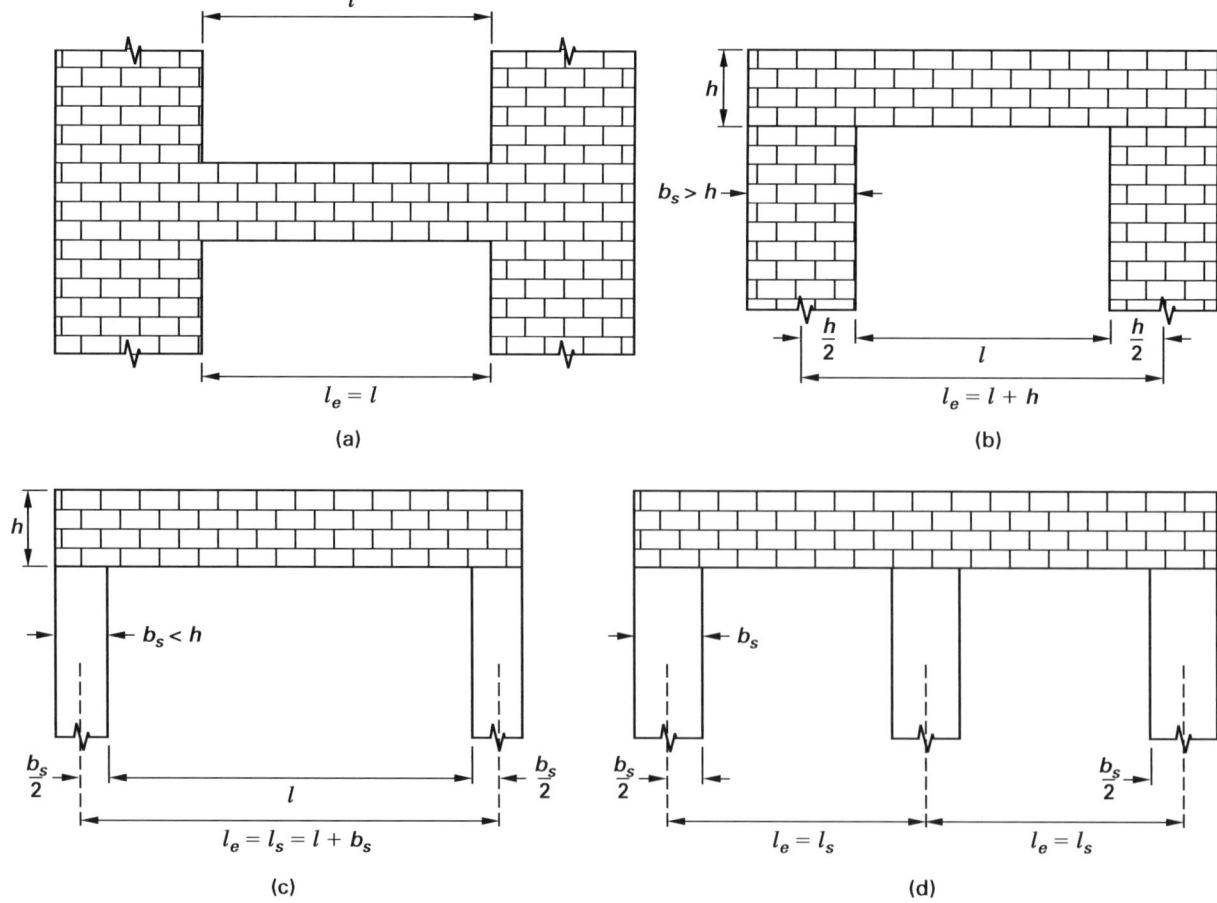

Example 6.1

The nominal 8 in, solid grouted concrete block masonry beam shown in the following illustration is simply supported over a clear span of 14 ft. The overall depth of the beam is 48 in, and its effective depth, d, is 45 in. The weight may be assumed to be 69 lbf/ft². Determine the maximum bending moment on the beam. Determine the shear force at a distance of $d/2$ from the face of the support.

Illustration for Example 6.1

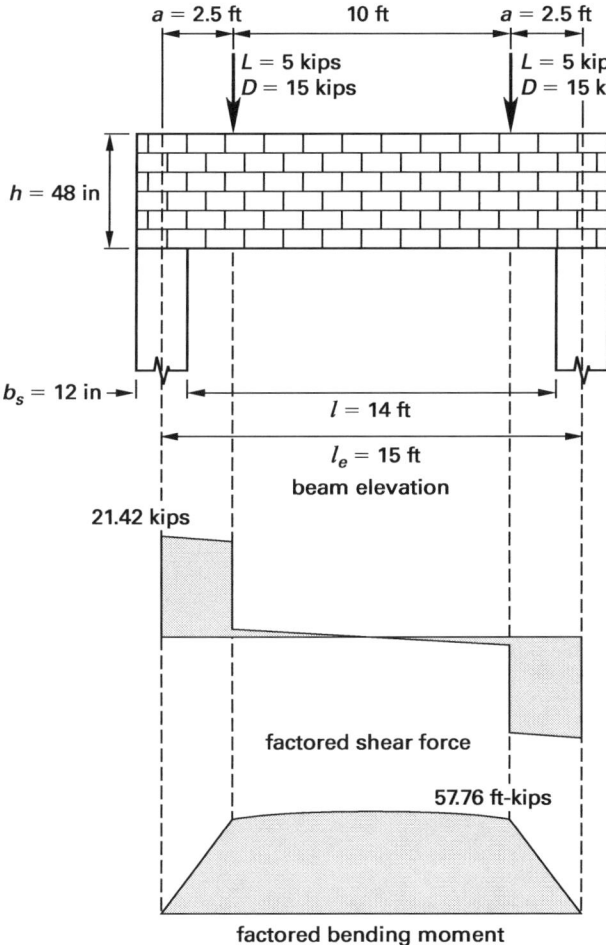

Solution

The effective span length is given by MSJC Sec. 1.13.1.2 as

$$l_e = l_s$$
$$= l + b_s$$
$$= 14 \text{ ft} + 1 \text{ ft}$$
$$= 15 \text{ ft}$$

The beam self-weight is

$$w = \left(69 \ \frac{\text{lbf}}{\text{ft}^2}\right)(4 \text{ ft})$$
$$= 276 \text{ lbf/ft}$$

The bending moment at midspan, produced by this self-weight, is

$$M_s = \frac{wl_e^2}{8}$$
$$= \frac{\left(276 \ \frac{\text{lbf}}{\text{ft}}\right)(15 \text{ ft})^2}{(8)\left(1000 \ \frac{\text{lbf}}{\text{kip}}\right)}$$
$$= 7.76 \text{ ft-kips}$$

The bending moment at midspan, produced by the concentrated dead loads, is

$$M_D = Da$$
$$= (15 \text{ kips})(2.5 \text{ ft})$$
$$= 37.5 \text{ ft-kips}$$

The bending moment at midspan, produced by the concentrated live loads, is

$$M_L = La$$
$$= (5 \text{ kips})(2.5 \text{ ft})$$
$$= 12.5 \text{ ft-kips}$$

The total moment at midspan is given by ASCE Sec. 2.4.1 as

$$M = M_s + M_D + M_L$$
$$= 7.76 \text{ ft-kips} + 37.5 \text{ ft-kips}$$
$$\quad + 12.5 \text{ ft-kips}$$
$$= 57.76 \text{ ft-kips}$$

The shear force at a distance of $d/2$ from the face of each support, produced by the beam self-weight, is

$$V_s = \frac{w(l_e - d - b_s)}{2}$$
$$= \frac{\left(0.276 \ \frac{\text{kips}}{\text{ft}}\right)(15 \text{ ft} - 3.75 \text{ ft} - 1 \text{ ft})}{2}$$
$$= 1.41 \text{ kips}$$

The shear force at a distance of $d/2$ from the face of each support, produced by the concentrated dead loads, is

$$V_D = D$$
$$= 15 \text{ kips}$$

The shear force at a distance of $d/2$ from each support, produced by the concentrated live loads, is

$$V_L = L = 5 \text{ kips}$$

The total shear force at a distance of $d/2$ from the face of each support is given by ASCE Sec. 2.4.1 as

$$\begin{aligned} V &= V_s + V_D + V_L \\ &= 1.41 \text{ kips} + 15 \text{ kips} + 5 \text{ kips} \\ &= 21.41 \text{ kips} \end{aligned}$$

The relevant values are shown in the illustration.

2. DESIGN FOR FLEXURE

Nomenclature

A_s	area of tension reinforcement	in^2
b_w	width of beam	in
d	effective depth, distance from extreme compression fiber to centroid of tension reinforcement	in
d_b	bar diameter	in
E_m	modulus of elasticity of masonry in compression, $900 f'_m$	lbf/in^2
E_s	modulus of elasticity of steel reinforcement, $29{,}000{,}000$	lbf/in^2
f_b	calculated compressive stress in masonry due to flexure	lbf/in^2
f'_m	specified masonry compressive strength	lbf/in^2
f_s	calculated tensile stress in reinforcement due to flexure	lbf/in^2
f_y	yield strength of reinforcement	lbf/in^2
F_b	allowable compressive stress in masonry due to flexure	lbf/in^2
F_s	allowable tensile stress in reinforcement due to flexure	lbf/in^2
h	overall dimension of member	in
j	lever-arm factor, ratio of distance between centroid of flexural compressive forces and centroid of tensile forces to effective depth, $1 - (k/3)$	–
k	neutral axis depth factor, $\sqrt{2\rho n + (\rho n)^2} - \rho n$	–
K	lesser of masonry cover, clear spacing of reinforcement, or 5 times the bar diameter	in
l_c	distance between locations of lateral support	in
l_d	development length or lap length of straight reinforcement	in
l_e	development length or lap length of a standard hook	in
l_e	effective span	in
M	moment on the member	in-kips
n	modular ratio, E_s/E_m	–
s_c	clear spacing of reinforcement	in

Symbols

γ	reinforcement size factor	–
ϵ_m	strain in masonry	–
ϵ_s	strain in tension reinforcement	–
ρ	ratio of tension reinforcement, $A_s/b_w d$	–

Dimensional Limitations of Beams

In accordance with MSJC Sec. 1.13.2, the maximum distance between lateral supports on the compression side of the beam must not exceed

$$l_c = 32 b_w$$

In accordance with MSJC Sec. 2.3.3.3, the length of bearing of a beam must not be less than

$$b_r = 4 \text{ in}$$

Beam Reinforcement Requirements

Limits are placed on the size of reinforcing bars, in order to control the bond stresses developed in the bars, to reduce congestion, and to aid grout consolidation. As specified in MSJC Sec. 1.15.2 and IBC Sec. 2107.5, the bar diameter, d_b, must not exceed the lesser of the following.

- one-eighth of the nominal member thickness
- one-quarter of the least clear dimension of the cell or course
- no. 11 bar

In accordance with MSJC Sec. 1.15.3.4, not more than 2 reinforcing bars may be bundled.

In accordance with MSJC Sec. 1.15.3.1, the clear distance between bars must not be less than the greater of d_b or 1 in. In addition, the thickness of grout between the reinforcement and the masonry unit must be a minimum of $^1/_4$ in for fine grout, and a minimum of $^1/_2$ in for coarse grout.

Example 6.2

For the nominal 8 in beam shown in the illustration, determine the maximum permissible size of reinforcing bar.

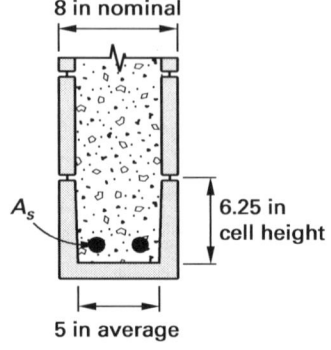

Solution

From MSJC Sec. 1.15.2 and IBC Sec. 2107.5, the maximum size of reinforcing bar allowed is

$$d_b = \text{no. 11 bar}$$
$$d_b = \frac{5 \text{ in}}{4}$$
$$= 1.25 \text{ in}$$
$$d_b = \frac{8 \text{ in}}{8}$$
$$= 1 \text{ in} \quad [\text{governs}]$$

To conform to the governing bar diameter of 1 in, two no. 8 bars are satisfactory. This provides a clear spacing between bars of 2 in, and a thickness of grout between the reinforcement and the masonry unit of $1/2$ in, which satisfies MSJC Secs. 1.15.3.1 and 1.15.5.5.

Development Length and Splice Length of Reinforcement

The required development length of compression and tension reinforcement is specified in MSJC Sec. 2.1.9.3 as

$$l_d = \frac{0.13 d_b^2 f_y \gamma}{K \sqrt{f'_m}} \quad [\text{MSJC 2-12}]$$
$$\geq 12 \text{ in}$$
$$\gamma = 1.0 \quad [\text{bar nos. 3, 4, and 5}]$$
$$\gamma = 1.3 \quad [\text{bar nos. 6 and 7}]$$
$$\gamma = 1.5 \quad [\text{bar nos. 8 through 11}]$$

The equivalent development length of a standard hook in tension is specified in MSJC Sec. 2.1.9.5.1 as

$$l_e = 11.25 d_b$$

The minimum length of lap splices for reinforcing bars in tension or compression is given by IBC Sec. 2107.3 as

$$l_d = 0.002 d_b f_s$$
$$\geq 12 \text{ in}$$
$$\geq 40 d_b$$

When the design tensile stress in the reinforcement is greater than 80% of the allowable tension stress, lap length must be increased 50%.

Where epoxy coated bars are used, lap length must be increased 50%.

Welded or mechanical splices are required to develop a minimum of $1.25 f_y$.

Example 6.3

For the beam described in Ex. 6.2, in which the masonry has a compressive strength of 1500 lbf/in², and the reinforcement consists of grade 60 bars, determine the required development length for straight bars, and for bars provided with a standard hook.

Solution

The development length parameter, K, is the lesser of masonry cover, clear spacing of reinforcement, or 5 times the bar diameter. For masonry cover,

$$K = \frac{b - s_c - 2d_b}{2}$$
$$= \frac{7.625 \text{ in} - 2.0 \text{ in} - (2)(1.0 \text{ in})}{2}$$
$$= 1.81 \text{ in}$$

For clear spacing of reinforcement

$$K = s_c$$
$$= 2 \text{ in} \quad [\text{from Ex. 6.2}]$$

For the bar diameter,

$$K = 5 d_b$$
$$= (5)(1.0 \text{ in})$$
$$= 5.0 \text{ in}$$

$K = 1.81$ in governs.

The reinforcement size factor for a no. 8 bar is given by MSJC Sec. 2.1.9.3 as

$$\gamma = 1.5$$

The required development length is given by MSJC Eq. (2-12) as

$$l_d = \frac{0.13 d_b^2 f_y \gamma}{K \sqrt{f'_m}}$$
$$= \frac{(0.13)(1.0 \text{ in})^2 \left(60{,}000 \ \frac{\text{lbf}}{\text{in}^2}\right)(1.5)}{(1.81 \text{ in})\sqrt{1500 \ \frac{\text{lbf}}{\text{in}^2}}}$$
$$= 167 \text{ in}$$

The equivalent development length provided by a standard hook is given by MSJC Sec. 2.1.9.5.1 as

$$l_e = 11.25 d_b$$
$$= (11.25)(1.0 \text{ in})$$
$$= 11.25 \text{ in}$$

The required development length of the no. 8 bars, for bars provided with a standard hook, is

$$l_d = 167 \text{ in} - 11.25 \text{ in}$$
$$= 155.75 \text{ in}$$

Design of Beams with Tension Reinforcement Only

The allowable stress design method, illustrated in Fig. 6.2 for a beam reinforced in tension, is used to calculate the stresses in a masonry beam under the action of the applied service loads, in order to ensure that these stresses do not exceed allowable values.

In the allowable stress design method, it is assumed that the strain distribution over the depth of the member is linear and that stresses in the concrete and the reinforcement are proportional to the strain. Tensile stress in the concrete is neglected.

The allowable compressive stress in the masonry caused by flexure is given by MSJC Sec. 2.3.3.2.2 as

$$F_b = \frac{f'_m}{3}$$

For grade 60 reinforcement, the allowable tensile stress is given by MSJC Sec. 2.3.2.1 as

$$F_s = 24{,}000 \text{ lbf/in}^2$$

The assumptions adopted are shown in Fig. 6.2. The depth of the neutral axis is obtained by equating tensile and compressive forces acting on the section and is given by

$$kd = \frac{2A_s f_s}{b_w f_b}$$

The tension reinforcement ratio is

$$\rho = \frac{A_s}{b_w d}$$

The neutral axis depth factor is then

$$k = \frac{2\rho f_s}{f_b}$$

From the strain diagram,

$$\frac{\epsilon_s}{\epsilon_m} = \frac{d - kd}{kd}$$
$$= \frac{1 - k}{k}$$
$$= \frac{f_s E_m}{f_b E_s}$$
$$= \frac{f_s}{n f_b}$$

The modular ratio is given by

$$n = \frac{E_s}{E_m}$$
$$= \frac{29{,}000{,}000 \frac{\text{lbf}}{\text{in}^2}}{900 f'_m}$$

The ratio of steel stress to masonry stress is then

$$\frac{f_s}{f_b} = \frac{n(1 - k)}{k}$$

The neutral axis depth factor is given by

$$k = \frac{2\rho n(1 - k)}{k}$$
$$= \sqrt{2\rho n + (\rho n)^2} - \rho n$$

Figure 6.2 Allowable Stress Design of Concrete Masonry Beams

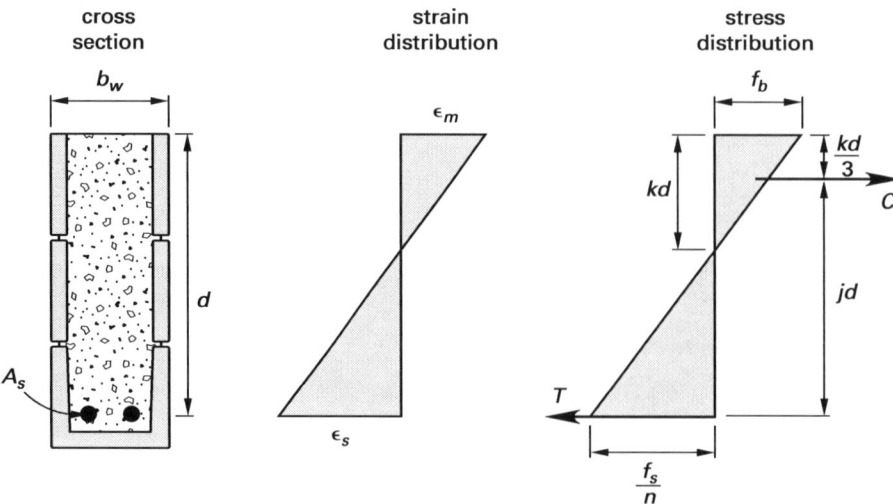

Table 6.1 Values of the Neutral Axis Factor k

ρn	0.000	0.001	0.002	0.003	0.004	0.005	0.006	0.007	0.008	0.009
0	0.0000	0.0437	0.0613	0.0745	0.0855	0.0951	0.1037	0.1115	0.1187	0.1255
0.01	0.1318	0.1377	0.1434	0.1488	0.1539	0.1589	0.1636	0.1682	0.1726	0.1769
0.02	0.1810	0.1850	0.1889	0.1927	0.1964	0.2000	0.2035	0.2069	0.2103	0.2136
0.03	0.2168	0.2199	0.2230	0.2260	0.2290	0.2319	0.2347	0.2375	0.2403	0.2430
0.04	0.2457	0.2483	0.2509	0.2534	0.2559	0.2584	0.2608	0.2632	0.2655	0.2679
0.05	0.2702	0.2724	0.2747	0.2769	0.2790	0.2812	0.2833	0.2854	0.2875	0.2895
0.06	0.2916	0.2936	0.2956	0.2975	0.2995	0.3014	0.3033	0.3051	0.3070	0.3088
0.07	0.3107	0.3125	0.3142	0.3160	0.3178	0.3195	0.3212	0.3229	0.3246	0.3263
0.08	0.3279	0.3296	0.3312	0.3328	0.3344	0.3360	0.3376	0.3391	0.3407	0.3422
0.09	0.3437	0.3452	0.3467	0.3482	0.3497	0.3511	0.3526	0.3540	0.3554	0.3569
0.10	0.3583	0.3597	0.3610	0.3624	0.3638	0.3651	0.3665	0.3678	0.3691	0.3705
0.11	0.3718	0.3731	0.3744	0.3756	0.3769	0.3782	0.3794	0.3807	0.3819	0.3832
0.12	0.3844	0.3856	0.3868	0.3880	0.3892	0.3904	0.3916	0.3927	0.3939	0.3951
0.13	0.3962	0.3974	0.3985	0.3996	0.4007	0.4019	0.4030	0.4041	0.4052	0.4063
0.14	0.4074	0.4084	0.4095	0.4106	0.4116	0.4127	0.4137	0.4148	0.4158	0.4169
0.15	0.4179	0.4189	0.4199	0.4209	0.4219	0.4229	0.4239	0.4249	0.4259	0.4269
0.16	0.4279	0.4288	0.4298	0.4308	0.4317	0.4327	0.4336	0.4346	0.4355	0.4364
0.17	0.4374	0.4383	0.4392	0.4401	0.4410	0.4419	0.4429	0.4437	0.4446	0.4455
0.18	0.4464	0.4473	0.4482	0.4491	0.4499	0.4508	0.4516	0.4525	0.4534	0.4542
0.19	0.4551	0.4559	0.4567	0.4576	0.4584	0.4592	0.4601	0.4609	0.4617	0.4625
0.20	0.4633	0.4641	0.4649	0.4657	0.4665	0.4673	0.4681	0.4689	0.4697	0.4705
0.21	0.4712	0.4720	0.4728	0.4736	0.4743	0.4751	0.4758	0.4766	0.4774	0.4781
0.22	0.4789	0.4796	0.4803	0.4811	0.4818	0.4825	0.4833	0.4840	0.4847	0.4855
0.23	0.4862	0.4869	0.4876	0.4883	0.4890	0.4897	0.4904	0.4911	0.4918	0.4925
0.24	0.4932	0.4939	0.4946	0.4953	0.4960	0.4966	0.4973	0.4980	0.4987	0.4993
0.25	0.5000	0.5007	0.5013	0.5020	0.5026	0.5033	0.5040	0.5046	0.5053	0.5059
0.26	0.5066	0.5072	0.5078	0.5085	0.5091	0.5097	0.5104	0.5110	0.5116	0.5123
0.27	0.5129	0.5135	0.5141	0.5147	0.5154	0.5160	0.5166	0.5172	0.5178	0.5184
0.28	0.5190	0.5196	0.5202	0.5208	0.5214	0.5220	0.5226	0.5232	0.5238	0.5243
0.29	0.5249	0.5255	0.5261	0.5267	0.5272	0.5278	0.5284	0.5290	0.5295	0.5301
0.30	0.5307	0.5312	0.5318	0.5323	0.5329	0.5335	0.5340	0.5346	0.5351	0.5357
0.31	0.5362	0.5368	0.5373	0.5379	0.5384	0.5389	0.5395	0.5400	0.5406	0.5411
0.32	0.5416	0.5422	0.5427	0.5432	0.5437	0.5443	0.5448	0.5453	0.5458	0.5464
0.33	0.5469	0.5474	0.5479	0.5484	0.5489	0.5494	0.5499	0.5505	0.5510	0.5515
0.34	0.5520	0.5525	0.5530	0.5535	0.5540	0.5545	0.5550	0.5554	0.5559	0.5564
0.35	0.5569	0.5574	0.5579	0.5584	0.5589	0.5593	0.5598	0.5603	0.5608	0.5613
0.36	0.5617	0.5622	0.5627	0.5632	0.5636	0.5641	0.5646	0.5650	0.5655	0.5660
0.37	0.5664	0.5669	0.5674	0.5678	0.5683	0.5687	0.5692	0.5696	0.5701	0.5705
0.38	0.5710	0.5714	0.5719	0.5723	0.5728	0.5732	0.5737	0.5741	0.5746	0.5750
0.39	0.5755	0.5759	0.5763	0.5768	0.5772	0.5776	0.5781	0.5785	0.5789	0.5794

Table 6.1[4] provides a design aid which tabulates ρn against k and facilitates the analysis of a given section. For a given section, the known values of reinforcement area, masonry strength, and section dimensions enable the determination of ρn and the corresponding value of k is obtained from Table 6.1.

The lever-arm of the internal resisting moment is obtained from Fig. 6.2 as

$$jd = d - \frac{kd}{3}$$

The lever-arm factor is

$$j = 1 - \frac{k}{3}$$

The resisting moment of the masonry is

$$M_m = Cjd$$
$$= \frac{f_b k j b_w d^2}{2}$$

---
[4]Williams, A., 2011 (See References and Codes)

For a given applied service moment M, the maximum masonry stress is given by

$$f_m = \frac{2M}{jkb_wd^2}$$

The resisting moment of the reinforcement is given by

$$M_s = Tjd$$
$$= f_s j\rho b_w d^2$$

For a given applied service moment M, the reinforcement stress is given by

$$f_s = \frac{M}{j\rho b_w d^2}$$

For a permissible masonry stress of F_b, and a permissible reinforcement stress of F_s, the service moment capacity of the section is the lesser of

$$M_m = \frac{F_b jkb_w d^2}{2}$$
$$M_s = F_s j\rho b_w d^2$$

Example 6.4

The 8 in, solid grouted concrete block masonry beam shown in the illustration is simply supported over a clear span of 14 ft. The overall depth of the beam is 48 in. Its effective depth, d, is 45 in. The unit weight may be assumed to be 69 lbf/ft^2. The masonry has a compressive strength of 1500 lbf/in^2 and the reinforcement consists of grade 60 bars. Determine the number of no. 6 grade 60 reinforcing bars required to resist the applied loads.

Solution

The total moment at midspan is derived in Ex. 6.1 as

$$M = 57.76 \text{ ft-kips}$$

Try 2 no. 6 bars with an area of

$$A_s = 0.88 \text{ in}^2$$

The relevant parameters of the beam are

$$b_w = 7.63 \text{ in}$$
$$d = 45 \text{ in}$$
$$f'_m = 1500 \text{ lbf/in}^2$$
$$f_y = 60{,}000 \text{ lbf/in}^2$$
$$l_e = 15 \text{ ft}$$
$$\frac{l_e}{b_w} = \frac{(15 \text{ ft})\left(12 \frac{\text{in}}{\text{ft}}\right)}{7.63 \text{ in}}$$
$$= 23.6$$
$$< 32 \quad [\text{satisfies MSJC Sec. 1.13.2}]$$

The allowable stresses are

$$F_b = \frac{f'_m}{3}$$
$$= \frac{1500 \frac{\text{lbf}}{\text{in}^2}}{3}$$
$$= 500 \frac{\text{lbf}}{\text{in}^2}$$
$$F_s = 24{,}000 \text{ lbf/in}^2$$

Illustration for Example 6.4

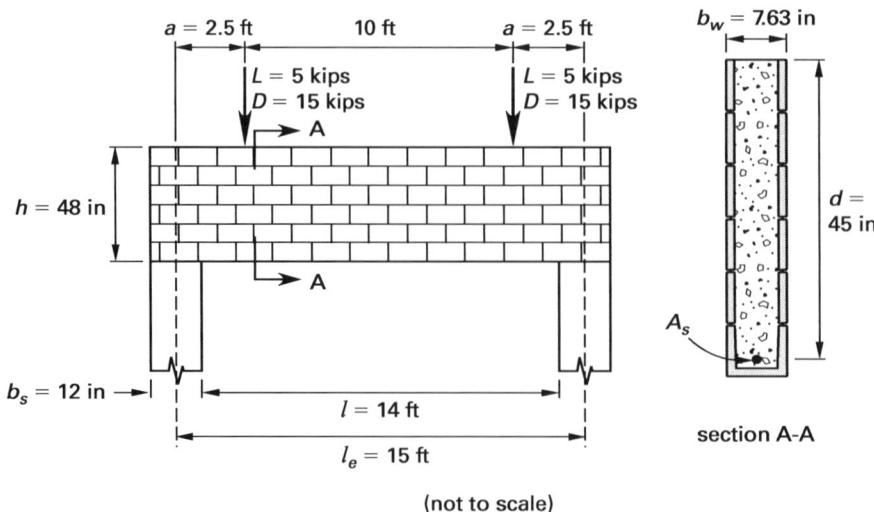

(not to scale)

The modular ratio is

$$n = \frac{E_s}{E_m}$$

$$= \frac{29{,}000{,}000 \ \frac{\text{lbf}}{\text{in}^2}}{900 f'_m}$$

$$= \frac{29{,}000{,}000 \ \frac{\text{lbf}}{\text{in}^2}}{(900)\left(1500 \ \frac{\text{lbf}}{\text{in}^2}\right)}$$

$$= 21.48$$

The tension reinforcement ratio is

$$\rho = \frac{A_s}{b_w d}$$

$$= \frac{0.88 \ \text{in}^2}{(7.63 \ \text{in})(45 \ \text{in})}$$

$$= 0.00256$$

$$\rho n = (0.00256)(21.48)$$

$$= 0.0550$$

Using this value of ρn, the neutral axis depth factor is obtained from Table 6.1 as

$$k = \sqrt{2\rho n + (\rho n)^2} - \rho n$$

$$= \sqrt{(2)(0.0550) + (0.0550)^2} - 0.0550$$

$$= 0.2811$$

The lever-arm factor is

$$j = 1 - \frac{k}{3}$$

$$= 1 - \frac{0.2811}{3}$$

$$= 0.906$$

The moment capacity of the section is the lesser of

$$M_m = \frac{F_b j k b_w d^2}{2}$$

$$= \frac{\left(500 \ \frac{\text{lbf}}{\text{in}^2}\right)(0.906)(0.2811)(7.63 \ \text{in})(45 \ \text{in})^2}{(2)\left(12 \ \frac{\text{in}}{\text{ft}}\right)\left(1000 \ \frac{\text{lbf}}{\text{kip}}\right)}$$

$$= 81.98 \ \text{ft-kips}$$

$$M_s = F_s j \rho b_w d^2$$

$$= \frac{\left(24{,}000 \ \frac{\text{lbf}}{\text{in}^2}\right)(0.906)(0.00256)(7.63 \ \text{in})(45 \ \text{in})^2}{\left(12 \ \frac{\text{in}}{\text{ft}}\right)\left(1000 \ \frac{\text{lbf}}{\text{kip}}\right)}$$

$$= 71.67 \ \text{ft-kips} \quad [\text{governs}]$$

$$> M \quad [\text{satisfactory}]$$

Two no. 6 bars are adequate.

Analysis of Beams with Tension Reinforcement Only

The design strength of a masonry beam is readily determined from the given beam dimensions and the area of reinforcing steel (see Fig. 6.3). The neutral axis depth factor, obtained by equating tensile and compressive forces acting on the section, is

$$k = \sqrt{2\rho n + (\rho n)^2} - \rho n$$

Figure 6.3 Analysis of Concrete Masonry Beams

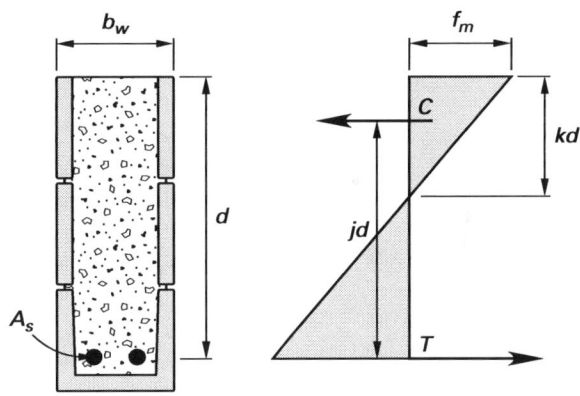

The lever arm of the compressive and tensile forces is

$$jd = d - \frac{kd}{3}$$

For a permissible masonry stress of F_b and a permissible reinforcement stress of F_s, the service moment capacity of the section is the lesser of

$$M_m = \frac{F_b j k b_w d^2}{2}$$

$$M_s = F_s j \rho b_w d^2$$

For a given applied service moment M, the maximum masonry stress is given by

$$f_m = \frac{2M}{jkb_wd^2}$$

For a given applied service moment M, the reinforcement stress is given by

$$f_s = \frac{M}{j\rho b_wd^2}$$

Example 6.5

The nominal 8 in, solid grouted concrete block masonry beam shown in the illustration has an effective depth, d, of 45 in. The masonry has a compressive strength of 1500 lbf/in². The reinforcement consists of one no. 7 grade 60 bar. Determine the stresses in the beam if the applied moment is 48 ft-kips.

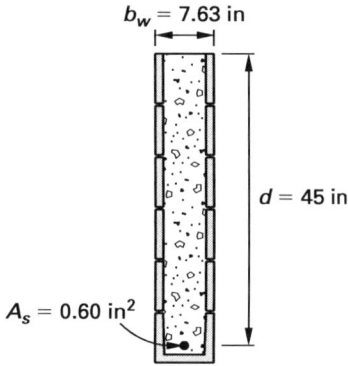

Solution

The tension reinforcement ratio is

$$\rho = \frac{A_s}{b_wd}$$
$$= \frac{0.60 \text{ in}^2}{(7.63 \text{ in})(45 \text{ in})}$$
$$= 0.00175$$
$$\rho n = (0.00175)(21.48)$$
$$= 0.0376$$

Using this value of ρn, the neutral axis depth factor is obtained from Table 6.1 as

$$k = \sqrt{2\rho n + (\rho n)^2} - \rho n$$
$$= \sqrt{(2)(0.0376) + (0.0376)^2} - 0.0376$$
$$= 0.239$$

The lever-arm factor is

$$j = 1 - \frac{k}{3}$$
$$= 1 - \frac{0.239}{3}$$
$$= 0.920$$

For an applied moment of 48 ft-kips, the maximum masonry stress is given by

$$f_m = \frac{2M}{jkb_wd^2}$$
$$= \frac{(2)(48{,}000 \text{ ft-lbf})\left(12 \frac{\text{in}}{\text{ft}}\right)}{(0.920)(0.239)(7.63 \text{ in})(45 \text{ in})^2}$$
$$= 339 \text{ lbf/in}^2$$
$$< 500 \text{ lbf/in}^2 \quad \text{[satisfactory]}$$

For an applied moment of 48 ft-kips, the reinforcement stress is given by

$$f_s = \frac{M}{j\rho b_wd^2}$$
$$= \frac{(48{,}000 \text{ ft-lbf})\left(12 \frac{\text{in}}{\text{ft}}\right)}{(0.920)(0.00175)(7.63 \text{ in})(45 \text{ in})^2}$$
$$= 23{,}155 \text{ lbf/in}^2$$
$$< 24{,}000 \text{ lbf/in}^2 \quad \text{[satisfactory]}$$

3. DESIGN FOR SHEAR

Nomenclature

A_n	net area of cross section	in²
A_v	area of shear reinforcement	in²
b_w	width of beam	in
f'_m	specified masonry compressive strength	lbf/in²
f_v	calculated shear stress in masonry	lbf/in²
F_v	allowable shear stress in masonry	lbf/in²
l_e	effective span length of beam	ft
s	spacing of shear reinforcement	in
V	shear force at the section under consideration	kips

Shear Reinforcement Requirements

When the shear stress on a section, f_v, exceeds the allowable shear stress of the masonry, F_v, MSJC Commentary Sec. 2.3.5 requires the provision of shear reinforcement to resist the total shear. As shown in Fig. 6.4, and specified by MSJC Sec. 3.3.4.2.3, shear reinforcement must comply with the following requirements.

Figure 6.4 Shear Reinforcement Requirements

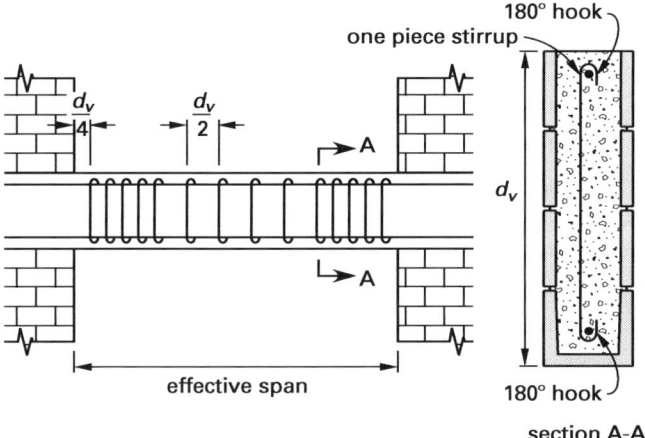

- To reduce congestion in the member, shear reinforcement must consist of a single bar with a standard 180° hook at each end.
- Shear reinforcement must be hooked around longitudinal reinforcement at each end to develop the shear reinforcement.
- To prevent brittle shear failure, the area of shear reinforcement must be not less than $0.0007 b_w d$.
- The first shear reinforcing bar must be located not more than one-fourth the depth of the beam from the end of the beam, in order to intersect any diagonal crack formed at the support.
- To ensure that every potential shear crack is crossed by at least one stirrup, the spacing of shear reinforcing bars must not exceed one-half the beam depth, nor 48 in.

Design of Beams for Shear

In accordance with MSJC Sec. 2.3.5.2.1, the shear stress in masonry is given by the expression

$$f_v = \frac{V}{b_w d} \quad \text{[MSJC 2-23]}$$

The allowable shear stress in a flexural member without shear reinforcement is given by MSJC Sec. 2.3.5.2.2 as

$$F_v = \sqrt{f'_m} \quad \text{[MSJC 2-24]}$$
$$\leq 50 \text{ lbf/in}^2$$

When this value of the shear stress is exceeded, shear reinforcement is provided to carry the full shear load without any contribution from the masonry. The area of shear reinforcement required is given by MSJC Sec. 2.3.5.3 as

$$\frac{A_v}{s} = \frac{V}{F_s d} \quad \text{[MSJC 2-30]}$$

The shear stress, with shear reinforcement designed to take the entire shear force, is limited by MSJC Sec. 2.3.5.2.3 to a maximum value of

$$F_v = 3\sqrt{f'_m} \quad \text{[MSJC 2-27]}$$
$$\leq 150 \text{ lbf/in}^2$$

When necessary, the dimensions of the masonry member must be increased to conform to this requirement.

As specified in MSJC Sec. 2.3.5.5, the maximum design shear for a beam may be calculated at a distance of $d/2$ from the face of the support, provided that no concentrated load occurs between the face of the support and a distance of $d/2$ from the face.

Example 6.6

The nominal 8 in, solid grouted concrete block masonry beam shown in the illustration has an effective depth, d, of 45 in and an overall depth of 48 in. The masonry has a compressive strength of 1500 lbf/in^2. The reinforcement consists of one no. 8 grade 60 bar. For the loading indicated, determine whether shear stirrups are required at the ends of the beam.

Solution

The critical section for shear for a beam, in accordance with MSJC Sec. 2.3.5.5, occurs at a distance of $d/2$ from the face of the support. The shear at the critical section is derived in Ex. 6.1 as

$$V = 21.42 \text{ kips}$$

The shear stress at the critical section is given by MSJC Eq. (2-23) as

$$f_v = \frac{V}{b_w d}$$
$$= \frac{(21.42 \text{ kips})\left(1000 \frac{\text{lbf}}{\text{kips}}\right)}{(7.63 \text{ in})(45 \text{ in})}$$
$$= 62.39 \text{ lbf/in}^2$$

The allowable shear stress without shear reinforcement is given by MSJC Eq. (2-24) as

$$F_v = \sqrt{f'_m}$$
$$= \sqrt{1500 \frac{\text{lbf}}{\text{in}^2}}$$
$$= 38.7 \text{ lbf/in}^2$$
$$< f_v \quad \text{[shear reinforcement is required]}$$

Illustration for Example 6.6

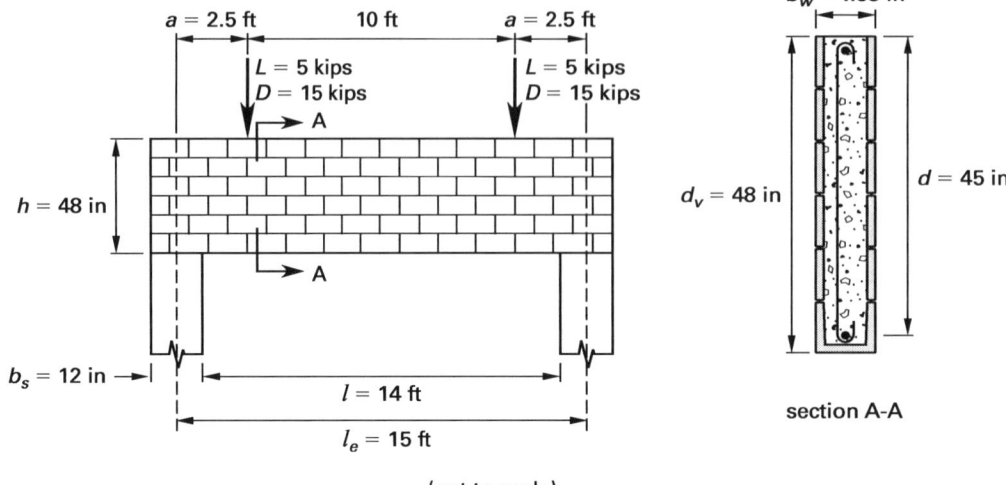

(not to scale)

The shear stress with shear reinforcement provided to carry the total shear force is limited by MSJC Eq. (2-27) to

$$F_v = 3\sqrt{f'_m}$$
$$= 3\sqrt{1500 \ \frac{\text{lbf}}{\text{in}^2}}$$
$$= 116 \ \text{lbf/in}^2$$
$$\leq 150 \ \text{lbf/in}^2 \quad \text{[satisfies MSJC Sec. 2.3.5.2.3]}$$
$$> f_v \quad \text{[satisfactory]}$$

The minimum area of shear reinforcement required is given by MSJC Eq. (2-30) as

$$\frac{A_v}{s} = \frac{V}{F_s d}$$
$$= \frac{(21.55 \ \text{kips})\left(12 \ \frac{\text{in}}{\text{ft}}\right)}{\left(24 \ \frac{\text{kips}}{\text{in}^2}\right)(45 \ \text{in})}$$
$$= 0.239 \ \text{in}^2/\text{ft}$$

Providing no. 4 grade 60 stirrups at 8 in centers supplies a value of

$$\frac{A_v}{s} = 0.300 \ \frac{\text{in}^2}{\text{ft}}$$
$$= 0.238 \ \text{in}^2/\text{ft} \quad \text{[satisfactory]}$$

4. DESIGN OF MASONRY COLUMNS

Nomenclature

a	distance between column reinforcement	in
A_n	net effective area of column	in^2
A_{st}	area of laterally tied reinforcement	in^2
b	width of section	in
d_{lat}	diameter of lateral ties	in
d_{long}	diameter of longitudinal reinforcement	in
E_m	modulus of elasticity of masonry in compression	lbf/in^2
E_s	modulus of elasticity of steel reinforcement	lbf/in^2
f'_m	specified masonry compressive strength	lbf/in^2
f_y	yield strength of reinforcement	lbf/in^2
h	effective height of column	in
P	unfactored axial load	kips
P_a	allowable axial strength	kips
r	radius of gyration	in
s	center-to-center spacing of items	in
t	nominal thickness of member	in

Symbols

ρ	reinforcement ratio	–

Dimensional Limitations of Columns

Limitations are imposed on column dimensions in MSJC Secs. 1.6, 1.14, and 2.1.6.1 to prevent lateral instability. As shown in Fig. 6.5, the minimum nominal column width is limited to 8 in. The ratio of the distance between lateral supports and least nominal width must not exceed 25. The minimum nominal column depth is limited to 8 in, and the maximum depth must not exceed three times the nominal width.

Column Reinforcement Requirements

Limitations are imposed on column longitudinal reinforcement in MSJC Sec. 1.14.1.2. As shown in Fig. 6.5, reinforcement must comply with the following requirements.

Figure 6.5 Column Details

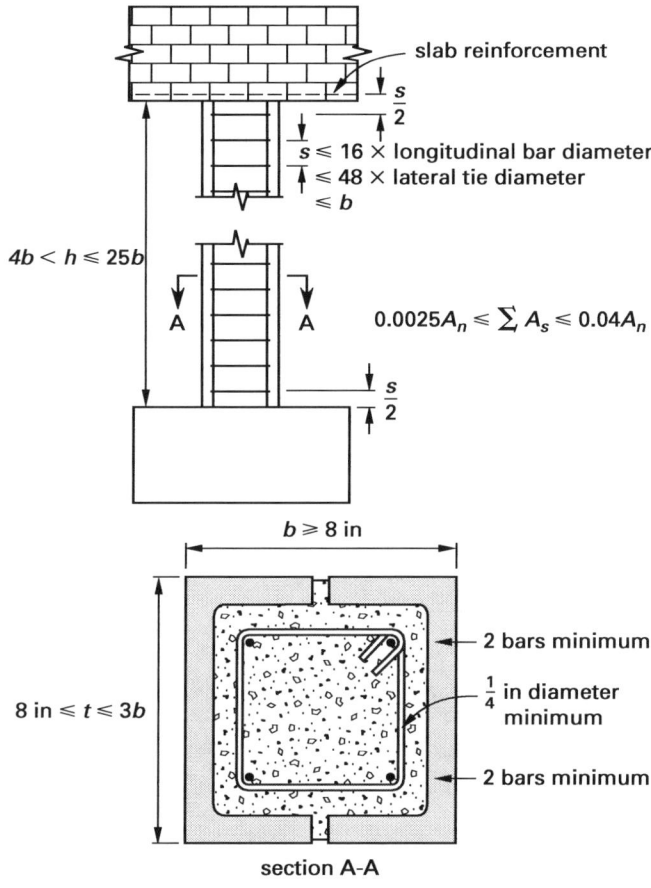

- Longitudinal reinforcement is limited to a maximum of 4% of the net column area of cross section to reduce congestion in the column.
- To prevent brittle failure, the area of longitudinal reinforcement must not be less than a minimum of 0.25% of the net column area of cross section.
- At least four longitudinal reinforcing bars must be provided, one in each corner of the column.

Lateral ties, for the confinement of longitudinal reinforcement in a column, are specified by MSJC Sec. 1.14.1.3. As shown in Fig. 6.5, lateral ties must comply with the following requirements.

- Lateral ties for the confinement of longitudinal reinforcement must not be less than $1/4$ in diameter.
- Lateral ties must be placed at a spacing not exceeding the lesser of 16 longitudinal bar diameters, 48 lateral tie diameters, or the least cross-sectional dimension of the column.
- Lateral ties must be arranged such that every corner and alternative longitudinal bar will have support provided by the corner of a lateral tie, and no bar will be farther than 6 in clear on each side from a supported bar.

- Lateral ties must be located not more than one-half lateral tie spacing above the top of the footing or slab in any story, and must be placed not more than one-half lateral tie spacing below the horizontal reinforcement in the beam or slab reinforcement above.
- Where beams or brackets frame into a column from four directions, lateral ties may be terminated not more than 3 in below the lowest reinforcement in the shallowest beam or bracket.

Axial Compression in Columns

The allowable axial compressive strength of an axially loaded reinforced masonry column, is given by MSJC Sec. 2.3.3.2.1. For columns having a ratio of effective height to radius of gyration not greater than 99, the allowable axial strength is

$$P_a = (0.25 f'_m A_n + 0.65 A_{st} F_s)\left(1 - \left(\frac{h}{140r}\right)^2\right)$$
[MSJC 2-20]

For columns having a ratio of effective height to radius of gyration greater than 99, the allowable axial load is

$$P_a = (0.25 f'_m A_n + 0.65 A_{st} F_s)\left(\frac{70r}{h}\right)^2$$
[MSJC 2-21]

Example 6.7

The nominal 16 in square, solid grouted concrete block masonry column shown in the illustration has a specified strength of 3000 lbf/in^2. It is reinforced with four no. 4 grade 60 bars. The column has a height of 15 ft, and may be considered pinned at each end. Determine the design axial strength of the column and the required size and spacing of lateral ties. Neglect the effects of accidental eccentricity.

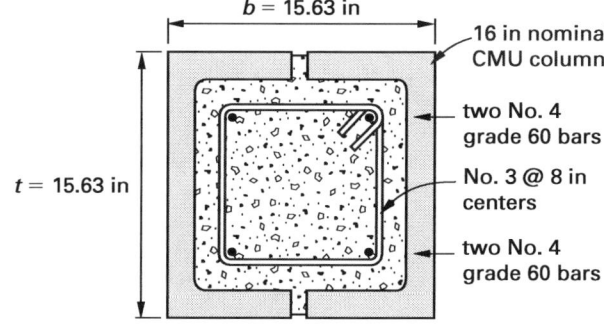

Solution

The relevant properties of the column are the following. The effective column width is

$$b = 15.63 \text{ in}$$

The effective column height is

$$h = 15 \text{ ft}$$

The reinforcement area is

$$A_s = 0.80 \text{ in}^2$$

The effective column area is

$$A_n = b^2$$
$$= (15.63 \text{ in})^2$$
$$= 244 \text{ in}^2$$
$$\rho = \frac{A_{st}}{A_n}$$
$$= \frac{0.80 \text{ in}^2}{244 \text{ in}^2}$$
$$= 0.0033$$
$$< 0.04$$
$$> 0.0025 \quad \text{[satisfies MSJC Sec. 1.14.1.2]}$$

The radius of gyration of the column is given by MSJC Commentary Sec. 1.9.3 as

$$r = \sqrt{\frac{I_n}{A_n}}$$
$$= 0.289b$$
$$= (0.289)(15.63 \text{ in})$$
$$= 4.52 \text{ in}$$

The slenderness ratio of the column is

$$\frac{h}{r} = \frac{(15 \text{ ft})\left(12 \frac{\text{in}}{\text{ft}}\right)}{4.52 \text{ in}}$$
$$= 39.82$$
$$< 99 \quad \text{[MSJC Eq. (2-20) is applicable]}$$

The allowable axial strength is

$$P_a = (0.25 f'_m A_n + 0.65 A_{st} F_s)\left(1 - \left(\frac{h}{140r}\right)^2\right)$$
$$\text{[MSJC 2-20]}$$
$$= \begin{pmatrix} (0.25)\left(3\, \frac{\text{kips}}{\text{in}^2}\right)(244 \text{ in}^2) \\ + (0.65)(0.80 \text{ in}^2)\left(24\, \frac{\text{kips}}{\text{in}^2}\right) \end{pmatrix}$$
$$\times \left(1.0 - \left(\frac{39.82}{140}\right)^2\right)$$
$$= 180 \text{ kips}$$

As specified by MSJC Sec. 1.14.1.3, lateral ties for the confinement of longitudinal reinforcement must not be less than $1/4$ in diameter. Using no. 3 bars for the ties, the spacing must not exceed the lesser of

$$s = 48 d_{\text{lat}}$$
$$= (48)(0.375 \text{ in})$$
$$= 18 \text{ in}$$
$$s = 16 \text{ in} \quad \text{[least cross-sectional column dimension]}$$
$$s = 16 d_{\text{long}}$$
$$= (16)(0.5 \text{ in})$$
$$= 8 \text{ in} \quad \text{[governs]}$$

5. DESIGN OF MASONRY SHEAR WALLS

Nomenclature

A_g	gross cross-sectional area of the wall using specified dimensions	in^2
A_s	area of reinforcement	in^2
A_{sh}	area of horizontal reinforcement	in^2
A_{sv}	area of vertical reinforcement	in^2
A_v	area of shear reinforcement	in^2
b	width of section	in
d	effective depth of tension reinforcement	in
d_v	depth of wall in direction of shear	in
f'_m	specified masonry compressive strength	lbf/in^2
f_v	calculated shear stress in masonry	lbf/in^2
F_s	allowable stress in reinforcement	lbf/in^2
F_v	allowable shear stress in masonry	lbf/in^2
h	effective height of shear wall	in
M	unfactored bending moment associated with V	in-kips
s	spacing of reinforcement	in
V	unfactored shear force	kips

Shear Wall Types

The following are several types of shear walls classified in MSJC Sec. 1.6. Determining which type of wall to adopt depends on the seismic design category of the structure.

- *Ordinary plain (unreinforced) masonry shear walls* are shear walls designed to resist lateral forces without reinforcement or where stresses in the reinforcement, if present, are neglected. This type of wall may be used only in seismic design categories A and B.

- *Detailed plain (unreinforced) masonry shear walls* are shear walls with specific minimum reinforcement and connection requirements that are designed to resist lateral forces with the stresses in the reinforcement neglected. This type of wall may be used only in seismic design categories A and B. The reinforcement requirements are specified in MSJC Sec. 1.17.3.2.3.1 as horizontal and vertical reinforcement of at least no. 4 bars at a maximum spacing of 120 in. Additional reinforcement is required at wall openings and corners.

- *Ordinary reinforced masonry shear walls* are shear walls with the minimum reinforcement specified in MSJC Sec. 1.17.3.2.3.1 that are designed to resist lateral forces while considering the stresses in the reinforcement. This type of wall may be used only in seismic design categories A, B, or C. The maximum permitted height in seismic design category C is 160 ft.

- *Intermediate reinforced masonry shear walls* are shear walls with the minimum reinforcement specified in MSJC Sec. 1.17.3.2.3.1 with the exception that the spacing of vertical reinforcement is limited to a maximum of 48 in. The walls are designed to resist lateral forces while considering the stresses in the reinforcement. This type of wall may be used only in seismic design category A, B, or C. There is no limitation on height in seismic design category C.

- *Special reinforced masonry shear walls* are shear walls with the minimum reinforcement specified in MSJC Sec. 1.17.3.2.6 that are designed to resist lateral forces while considering the stresses in the reinforcement. This type of wall must be used in seismic design categories D, E, and F. When used in bearing wall or building frame systems, the maximum permitted height in seismic design category D and E is 160 ft and in seismic design category F is 100 ft.

Special Reinforced Shear Wall Reinforcement Requirements

Reinforcement is provided in special reinforced masonry shear walls in order to provide ductile behavior in the walls under seismic loads. To ensure this in seismic design categories D, E, and F, MSJC Sec. 1.17.3.2.6 requires walls to be reinforced with uniformly distributed vertical and horizontal reinforcement. Shear reinforcement shall be anchored around vertical reinforcement with a standard 180° hook. The minimum required combined area of shear reinforcement and vertical reinforcement is

$$A_{sh} + A_{sv} = 0.002 A_g$$

The following requirements for horizontal shear reinforcement are shown in Fig. 6.6.

- The maximum spacing must not exceed the lesser of one-third the length of the wall, one-third the height of the wall, 48 in, or 24 in in stack bond masonry walls.

- For masonry laid in running bond, the minimum cross-sectional area must not be less than 0.0007 times the gross cross-sectional area of the wall, using specified dimensions.

- For masonry laid in other than running bond, the minimum cross-sectional area must not be less than 0.0015 times the gross cross-sectional area of the wall.

Figure 6.6 *Reinforcement Details for Special Reinforced Masonry Shear Walls Laid in Running Bond*

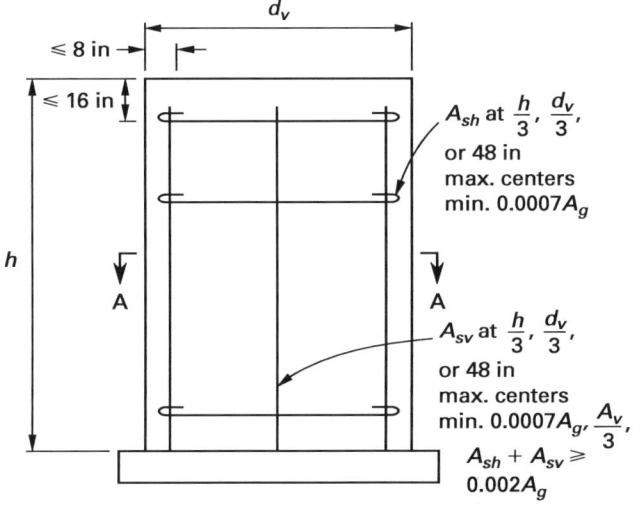
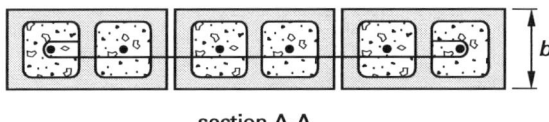

section A-A

The requirements for vertical reinforcement follow.

- The maximum spacing must not exceed the lesser of one-third the length of the wall, one-third the height of the wall, 48 in, or 24 in in stack bond masonry walls.

- The minimum cross-sectional area must not be less than the greater of 0.0007 times the gross cross-sectional area of the wall, using specified dimensions, or one-third the required shear reinforcement.

As specified by MSJC Sec. 1.17.3.2.6.1.2, special reinforced shear walls that are designed by the allowable stress method must be designed for 1.5 times the seismic force calculated by IBC Ch. 16.

Shear Capacity of a Shear Wall without Shear Reinforcement

The allowable shear stress depends on the ratio M/Vd, where M is the moment acting at the location where the applied shear force V is calculated. In a masonry wall without shear reinforcement and with $M/Vd < 1.0$, the allowable shear stress is given by MSJC Sec. 2.3.5.2.2(b) as

$$F_v = \left(\frac{1}{3}\right)\left(4 - \frac{M}{Vd}\right)\sqrt{f'_m} \quad \text{[MSJC 2-25]}$$
$$\leq \left(80 - \frac{45M}{Vd}\right) \quad \text{[lbf/in}^2\text{]}$$

In a masonry wall without shear reinforcement and with $M/Vd \geq 1.0$, the allowable shear stress is given by MSJC Sec. 2.3.5.2.2(b) as

$$F_v = \sqrt{f'_m} \quad \text{[MSJC 2-26]}$$
$$\leq 35 \text{ lbf/in}^2$$

The shear stress in the masonry is determined from MSJC Sec. 2.3.5.2.1 and MSJC Commentary Sec. 2.3.5.3 as

$$f_v = \frac{V}{bd} \quad \text{[MSJC 2-23]}$$

For shear walls without shear reinforcement and with shear parallel to the plane of the wall, MSJC Commentary Sec. 2.3.5.3 specifies the substitution of the overall depth of the wall d_v in place of the effective depth d. Similarly, for shear walls with horizontal shear reinforcement and with vertical reinforcement uniformly distributed along the depth of the wall, d_v may be substituted for d.

Example 6.8

The nominal 8 in, solid grouted concrete block masonry shear wall shown in the illustration has a specified strength of 3000 lbf/in^2.

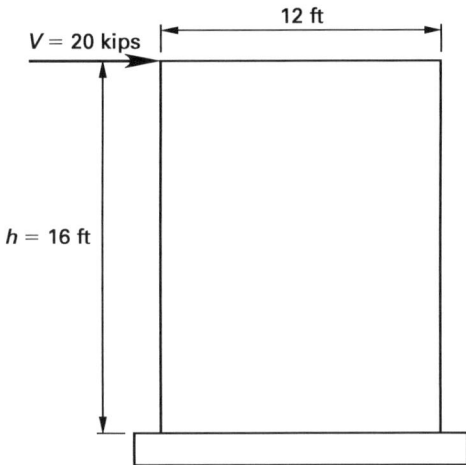

The wall has a height of 16 ft, and is assigned to seismic design category B. A wind load of 20 kips acts at the top of the wall and this is the governing shear load. Determine the shear reinforcement required.

Solution

The shear stress in the masonry wall is given by MSJC Eq. (2-23) and MSJC Commentary Sec. 2.3.5.3 as

$$f_v = \frac{V}{bd_v}$$
$$= \frac{20{,}000 \text{ lbf}}{(7.63 \text{ in})(144 \text{ in})}$$
$$= 18.20 \text{ lbf/in}^2$$

The allowable stress is obtained by applying MSJC Sec. 2.3.5.2.2. The ratio

$$\frac{M}{Vd} = \frac{Vh}{Vd_v}$$
$$= \frac{h}{d_v}$$
$$= \frac{(16 \text{ ft})\left(12 \frac{\text{in}}{\text{ft}}\right)}{144 \text{ in}}$$
$$= 1.33 \quad \text{[MSJC Eq. (2-26) is applicable]}$$

Applying MSJC Eq. (2-26), the allowable stress is the smaller value given by

$$F_v = \sqrt{f'_m}$$
$$= \sqrt{3000 \frac{\text{lbf}}{\text{in}^2}}$$
$$= 54.8 \text{ lbf/in}^2$$

Or,

$$F_v = 35 \frac{\text{lbf}}{\text{in}^2} \quad \text{[governs]}$$
$$> f_v \quad \text{[satisfactory]}$$

The masonry takes all the shear force, and nominal reinforcement is required as detailed in MSJC Sec. 1.17.3.2.3.1 for a structure assigned to seismic design category A or B.

Shear Capacity of a Shear Wall with Shear Reinforcement

When the allowable shear stress of the masonry is exceeded, shear reinforcement must be provided to carry the entire shear force without any contribution from the masonry. In a masonry wall with shear reinforcement designed to carry the entire shear force, and with $M/Vd < 1.0$, the shear stress is limited by MSJC Sec. 2.3.5.2.3(b) to

$$F_v = \left(\frac{1}{2}\right)\left(4 - \frac{M}{Vd}\right)\sqrt{f'_m} \quad \text{[MSJC 2-28]}$$
$$\leq 120 - \frac{45M}{Vd} \quad \text{[lbf/in}^2\text{]}$$

In a masonry wall, with shear reinforcement designed to carry the entire shear force and with $M/Vd \geq 1.0$, the shear stress is limited by MSJC Sec. 2.3.5.2.3(b) to

$$F_v = 1.5\sqrt{f'_m} \quad \text{[MSJC 2-29]}$$
$$\leq 75 \text{ lbf/in}^2$$

The area of shear reinforcement required when the allowable shear stress in the masonry is exceeded, is given by MSJC Sec. 2.3.5.3 as

$$\frac{A_v}{s} = \frac{V}{F_s d} \quad \text{[MSJC 2-30]}$$

The spacing of shear reinforcement must not exceed the lesser of $d/2$ or 48 in. Reinforcement must be provided perpendicular to the shear reinforcement and must be at least equal to $A_v/3$. This perpendicular reinforcement must be uniformly distributed and must not exceed a spacing of 8 ft.

Example 6.9

The nominal 8 in solid grouted, concrete block masonry shear wall shown in the illustration has a strength of 3000 lbf/in². An in-plane wind load of 50 kips acts at the top of the wall, and this is the governing shear load. The wall is located in a structure assigned to seismic design category D and is laid in running bond. Determine the shear reinforcement required in the wall.

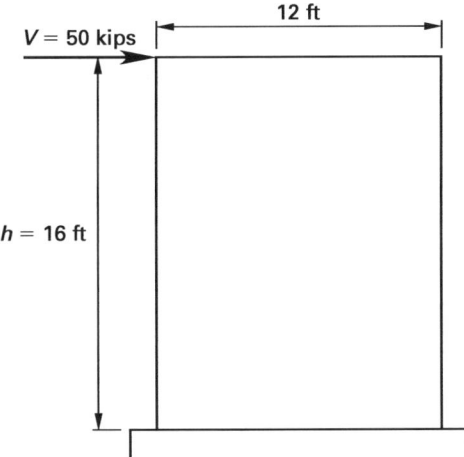

Solution

Since the structure is assigned to seismic design category D, the reinforcement details of a special reinforced shear wall must be provided. From Ex. 6.8, the allowable shear stress of the masonry section without shear reinforcement is

$$F_v = 35 \ \frac{\text{lbf}}{\text{in}^2}$$

The shear stress in the masonry wall is given by MSJC Eq. (2-23) and MSJC Commentary Sec. 2.3.5.3 as

$$f_v = \frac{V}{b d_v}$$

$$= \frac{50{,}000 \text{ lbf}}{(7.63 \text{ in})(144 \text{ in})}$$

$$= 45.51 \text{ lbf/in}^2$$

$$> F_v$$

Reinforcement is required to carry the entire shear force. In a masonry wall with shear reinforcement designed to carry the entire shear force, and with $M/Vd > 1.0$, the allowable stress is limited by MSJC Eq. (2-29) to the lesser of

$$F_v = 1.5\sqrt{f'_m}$$

$$= 1.5\sqrt{3000 \ \frac{\text{lbf}}{\text{in}^2}}$$

$$= 82.16 \text{ lbf/in}^2$$

Or,

$$F_v = 75 \ \frac{\text{lbf}}{\text{in}^2} \quad \text{[governs]}$$

$$> f_v \quad \text{[satisfies MSJC Sec. 2.3.5.2.3(b)]}$$

The minimum area of shear reinforcement required is given by MSJC Eq. (2-30) as

$$\frac{A_v}{s} = \frac{V}{F_s d}$$

$$= \frac{(50 \text{ kips})\left(12 \ \frac{\text{in}}{\text{ft}}\right)}{\left(24 \ \frac{\text{kips}}{\text{in}^2}\right)(144 \text{ in})}$$

$$= 0.174 \text{ in}^2/\text{ft}$$

Providing no. 5 horizontal bars at 16 in on center gives a reinforcement area of

$$A_{sh} = 0.233 \ \frac{\text{in}^2}{\text{ft}}$$

$$> 0.174 \ \frac{\text{in}^2}{\text{ft}} \quad \text{[satisfies MSJC Sec. 2.3.5.3]}$$

$$> 0.0007 A_g \quad \text{[satisfies MSJC Sec. 1.17.3.2.6(c)]}$$

To comply with MSJC Sec. 1.17.3.2.6(c), the vertical reinforcement must not be less than

$$A_v = \frac{A_v}{3}$$

$$= \frac{0.174 \ \frac{\text{in}^2}{\text{ft}}}{3}$$

$$= 0.058 \text{ in}^2/\text{ft}$$

Providing no. 5 vertical bars at 48 in on center gives a reinforcement area of

$$A_{sv} = 0.078 \ \frac{\text{in}^2}{\text{ft}}$$

$$\left[\begin{array}{l}\text{this will be augmented by flexural}\\ \text{reinforcement that is not considered here}\end{array}\right]$$

$$> 0.058 \ \frac{\text{in}^2}{\text{ft}} \quad \text{[satisfactory]}$$

$$> 0.0007 A_g \quad \text{[satisfies MSJC Sec. 1.17.3.2.6(c)]}$$

The sum of the horizontal and vertical reinforcement areas provided is

$$A_{sh} + A_{sv} = 0.233 \ \frac{\text{in}^2}{\text{ft}} + 0.078 \ \frac{\text{in}^2}{\text{ft}}$$

$$= 0.311 \ \text{in}^2/\text{ft} \quad \text{[satisfactory]}$$

The required sum, in accordance with MSJC Sec. 1.17.3.2.6(c), is

$$A_{sh} + A_{sv} = 0.002 A_g$$
$$= (0.002)(7.63 \ \text{in})(12 \ \text{in})$$
$$= 0.18 \ \text{in}^2/\text{ft}$$
$$< 0.311 \ \text{in}^2/\text{ft} \quad \text{[satisfactory]}$$

6. WALL DESIGN FOR OUT-OF-PLANE LOADS

Nomenclature

a	depth of equivalent rectangular stress block	in
$A_{\max}$	maximum area of reinforcement that will satisfy MSJC Sec. 3.3.3.5.1	in^2
A_{se}	effective area of reinforcing steel	in^2
b	effective thickness of wall	in
c	depth of neutral axis	in
C_m	force in masonry stress block	kips
d	effective depth	in
d_v	depth of wall in direction of shear	in
e	eccentricity of applied axial load	in
e_u	eccentricity of applied factored load	in
E_s	modulus of elasticity of reinforcement	lbf/in^2
E_m	modulus of elasticity of masonry	lbf/in^2
f'_m	specified masonry compressive strength	lbf/in^2
f_r	modulus of rupture of masonry	lbf/in^2
f_s	stress in reinforcement	lbf/in^2
f_y	yield strength of reinforcement	lbf/in^2
h	wall height	in
I_{cr}	moment of inertia of cracked transformed section about the neutral axis	in^4
I_g	moment of inertia of gross wall section	in^4
L_w	length of wall	in
M_{cr}	cracking moment	ft-kips
M_n	nominal bending moment strength	ft-kips
M_{ser}	service moment	ft-kips
M_u	factored bending moment	ft-kips
n	modular ratio	–
P	unfactored axial load	kips
P_f	service load level from tributary floor and roof loads	kips
P_u	sum of P_{uw} and P_{uf}	kips
P_{uf}	factored load from tributary floor or roof loads	kips
P_{uw}	factored weight of wall tributary to section considered	kips
P_w	service level load tributary to the section considered	kips
S_n	section modulus of net wall section	in^3
V_u	factored shear force	lbf
w_u	factored lateral load	lbf/ft

Symbols

δ_u	deflection at midheight of wall caused by factored loads and including P-delta effects	in
ϵ_{mu}	maximum usable compressive strain of masonry, 0.0035 for clay masonry and 0.0025 for concrete masonry	–
ϵ_s	strain in reinforcement	–
ϵ_y	strain at yield in tension reinforcement	–
$\rho_{\max}$	maximum reinforcement ratio that will satisfy MSJC Sec. 3.3.3.5.1	–
ϕ	strength reduction factor	–

Flexural Strength

For the design of slender masonry walls, NCEES requires the use of the strength design method.

The design flexural strength for a wall with out-of-plane loading is given by MSJC Eq. (3-27) as

$$M_u \leq \phi M_n$$

An allowance is made for the axial load on the wall, and the effective reinforcement area is calculated as

$$A_{se} = \frac{P_u + A_s f_y}{f_y}$$

With the reinforcing steel placed in the center of the wall as shown in Fig. 6.7, and using the design assumptions of MSJC Sec. 3.3.2, the nominal moment is given by MSJC Eq. (3-28) as

$$M_n = A_{se} f_y \left(d - \frac{a}{2} \right)$$

$$a = \frac{P_u + A_s f_y}{0.80 f'_m L_w} \quad \text{[MSJC 3-29]}$$

Figure 6.7 Nominal Moment for Out-of-Plane Loading of Concrete Masonry

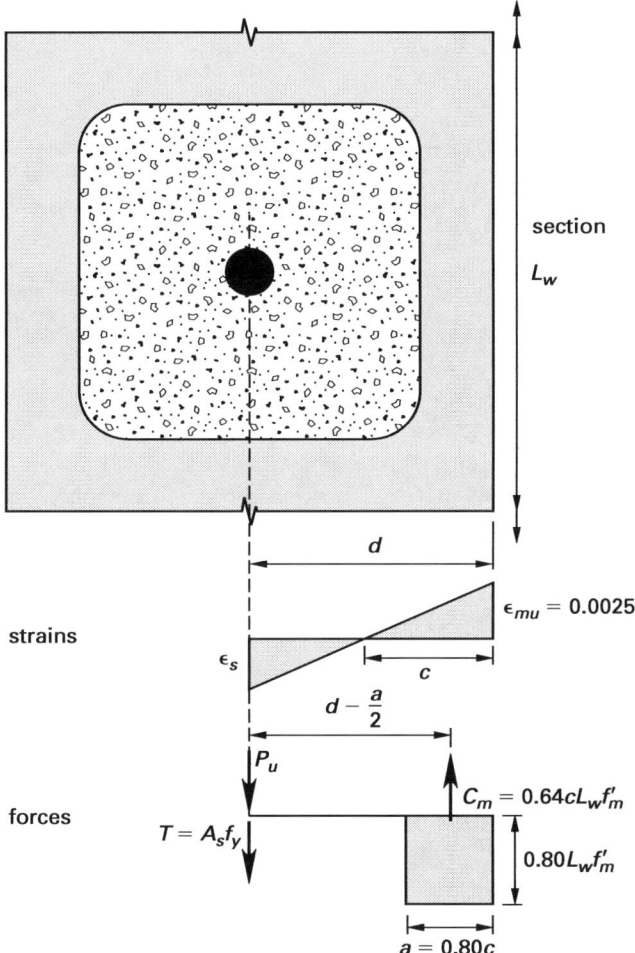

Illustration for Example 6.10

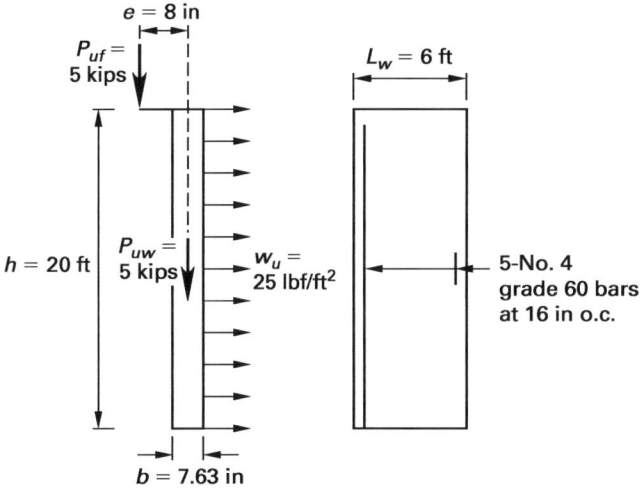

Example 6.10

The nominal 8 in, solid grouted concrete block masonry wall shown in the illustration has a specified strength of 3000 lbf/in². It is reinforced longitudinally with five no. 4 grade 60 bars placed centrally in the wall. The wall has a height of 20 ft, and is simply supported at the top and bottom. The factored applied loads are shown in the illustration. The lateral load is due to wind. Determine the design flexural strength of the wall.

Solution

The total factored axial load at the midheight of the wall is

$$P_u = P_{uf} + P_{uw}$$
$$= 5 \text{ kips} + 5 \text{ kips}$$
$$= 10 \text{ kips}$$

The reinforcement area in the wall is

$$A_s = (5)(0.20 \text{ in}^2) = 1.00 \text{ in}^2$$

For strength level loads, the equivalent reinforcement area is

$$A_{se} = \frac{P_u + A_s f_y}{f_y} = \frac{10 \text{ kips} + (1.00 \text{ in}^2)\left(60 \frac{\text{kips}}{\text{in}^2}\right)}{60 \frac{\text{kips}}{\text{in}^2}}$$
$$= 1.17 \text{ in}^2$$

The depth of the rectangular stress block is given by MSJC Eq. (3-29) as

$$a = \frac{A_{se} f_y}{0.80 f'_m L_w} = \frac{(1.17 \text{ in}^2)\left(60 \frac{\text{kips}}{\text{in}^2}\right)}{(0.80)\left(3 \frac{\text{kips}}{\text{in}^2}\right)(6 \text{ ft})\left(12 \frac{\text{in}}{\text{ft}}\right)}$$
$$= 0.41 \text{ in}$$

The nominal moment strength is given by MSJC Eq. (3-28) as

$$M_n = A_{se} f_y \left(d - \frac{a}{2}\right)$$
$$= \frac{(1.17 \text{ in}^2)\left(60 \frac{\text{kips}}{\text{in}^2}\right)\left(3.82 \text{ in} - \frac{0.41 \text{ in}}{2}\right)}{12 \frac{\text{in}}{\text{ft}}}$$
$$= 21.15 \text{ ft-kips}$$

The design moment strength is

$$\phi M_n = (0.9)(21.15 \text{ ft-kips}) = 19.04 \text{ ft-kips}$$

Maximum Reinforcement Ratio for Walls

The reinforcement ratio in a wall subject to out-of-plane loading must not exceed the value necessary to satisfy the requirements of MSJC Sec. 3.3.3.5.1. When $M_u/V_u d_v \geq 1$, the maximum reinforcement ratio is determined using the following design assumptions.

- Strain in the extreme tension reinforcement is 1.5 times the strain associated with the reinforcement yield stress, f_y.
- Unfactored gravity axial loads are included in the analysis using the combination

$$P = D + 0.75L + 0.525 Q_E$$

From the strain distribution shown in Fig. 6.8, the neutral axis depth and the depth of the equivalent rectangular stress block are

$$c = \frac{0.0025 d}{0.00560} = 0.446 d$$
$$a = 0.80 c = 0.357 d$$

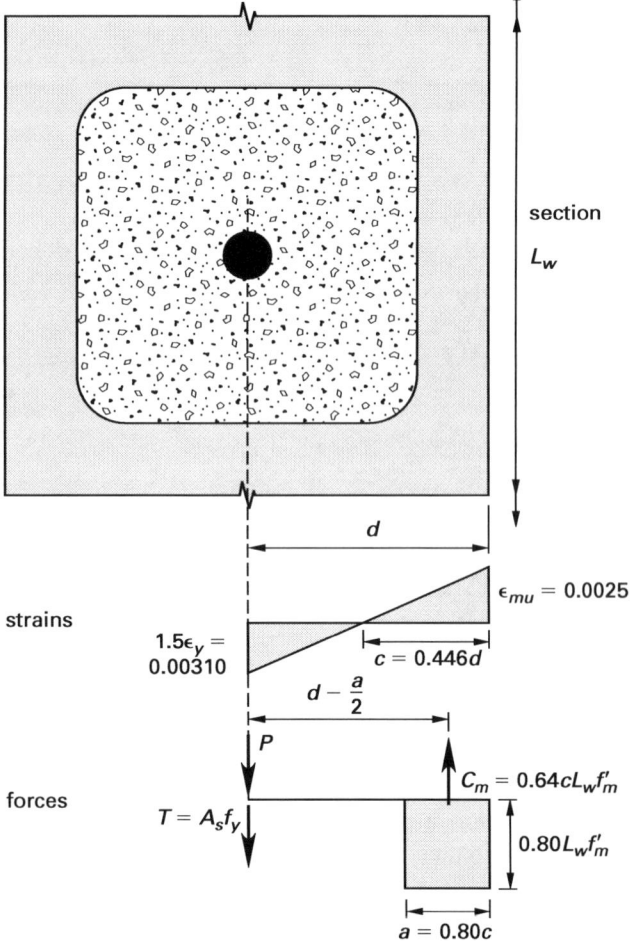

Figure 6.8 Maximum Reinforcement Requirements for Concrete Masonry Walls

The force in the equivalent rectangular stress block is

$$C_m = 0.80 a L_w f'_m$$
$$= 0.286 L_w d f'_m$$

The force in the reinforcing bars is given by MSJC Sec. 3.3.3.5.1 as

$$T = A_{\max} f_y$$

Equating compressive and tensile forces acting on the section gives

$$P = C_m - T$$
$$= 0.286 L_w d f'_m - A_{\max} f_y$$

The maximum area of the tension reinforcement that will satisfy MSJC Sec. 3.3.3.5.1 is

$$A_{\max} = \frac{0.286 L_w d f'_m - P}{f_y}$$

Example 6.11

The nominal 8 in, solid grouted concrete block masonry wall described in Ex. 6.10 has a specified strength of 3000 lbf/in². It is reinforced with five no. 4 grade 60 bars. Determine whether the reinforcement area provided satisfies MSJC Sec. 3.3.3.5.1.

Solution

From Ex. 6.10, the factored tributary roof load is

$$P_{uf} = 5 \text{ kips}$$

The service level tributary roof load is

$$P_f = \frac{5 \text{ kips}}{1.2}$$
$$= 4.17 \text{ kips}$$

From Ex. 6.10, the factored wall dead load at the mid-height of the wall is

$$P_{uw} = 5 \text{ kips}$$

The service level roof dead load is

$$P_w = \frac{5 \text{ kips}}{1.2}$$
$$= 4.17 \text{ kips}$$

The axial load combination specified in MSJC Sec. 3.3.3.5.1 for determining the maximum reinforcement limit is

$$P = D + 0.75L + 0.525Q_E$$
$$\approx P_f + P_w$$
$$= 4.17 \text{ kips} + 4.17 \text{ kips}$$
$$= 8.34 \text{ kips}$$

The relevant dimensions are obtained from Ex. 6.10 as

$$b = 7.63 \text{ in}$$
$$d = 3.82 \text{ in}$$

The maximum area of the tension reinforcement that will satisfy MSJC Sec. 3.3.3.5.1 is

$$A_{\max} = \frac{0.286 L_w d f'_m - P}{f y}$$

$$= \frac{(0.286)(3.82)(6 \text{ ft}) \times \left(12 \frac{\text{in}}{\text{ft}}\right)\left(3 \frac{\text{kips}}{\text{in}^2}\right) - 8.34 \text{ kips}}{60 \frac{\text{kips}}{\text{in}^2}}$$

$$= 3.79 \text{ in}^2$$
$$> 1.0 \text{ in}^2 \text{ provided} \quad [\text{satisfactory}]$$

Design Loads

Slender masonry walls are designed for factored applied loads, taking into consideration the P-delta effects caused by the vertical loads and the lateral deflection of the wall. The design method of MSJC Sec. 3.3.5.4 assumes the wall is simply supported, and uniformly laterally loaded, with the critical section at the midheight of the wall. In addition, MSJC Eq. (3-24) limits the factored axial load stress at the midheight of the wall to a maximum value of

$$\frac{P_u}{A_g} = 0.20 f'_m$$

P_u is the factored load caused by applied loads and wall self-weight.

When the slenderness ratio exceeds 30,

$$\frac{P_u}{A_g} \leq 0.05 f'_m$$

As shown in Fig. 6.9, the factored moment at the midheight of the wall is given by MSJC Eq. (3-25) as

$$M_u = \frac{w_u h^2}{8} + \frac{P_{uf} e_u}{2} + P_u \delta_u$$

Figure 6.9 Wall with Out-of-Plane Loading

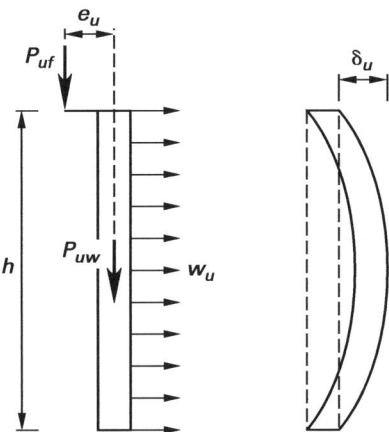

The deflection at the midheight of the wall, δ_u, caused by factored loads and including P-delta effects, is derived from MSJC Eqs. (3-32) and (3-31) as

$$\delta_u = \frac{5 M_{cr} h^2}{48 E_m I_g} + \frac{5 h^2 (M_u - M_{cr})}{48 E_m I_{cr}} \quad [M_{cr} < M_u < M_n]$$

$$\delta_u = \frac{5 M_u h^2}{48 E_m I_g} \quad [M_u \leq M_{cr}]$$

The cracking moment, M_{cr}, is

$$M_{cr} = S_n f_r$$

The moment of inertia of a cracked transformed section about the neutral axis is

$$I_{cr} = \frac{L_w c^3}{3} + n A_{se}(d - c)^2$$

The modular ratio is

$$n = \frac{E_s}{E_m}$$

The distance of the reinforcing bar from the neutral axis is $d - c$. The effective area of reinforcing steel is

$$A_{se} = \frac{P_u + A_s f_y}{f_y}$$

The moment of inertia of the gross wall section is

$$I_g = \frac{L_w b^3}{12}$$

The section modulus of the net wall section is

$$S_n = \frac{L_w b^2}{6}$$

The modulus of rupture of masonry, f_r, is given in MSJC Table 3.1.8.2. The modulus of elasticity of reinforcement is

$$E_s = 29{,}000 \text{ kips/in}^2 \quad \text{[MSJC Sec. 1.8.2.1]}$$

The modulus of elasticity for concrete masonry is

$$E_m = 900 f'_m \quad \text{[MSJC Sec. 1.8.2.2.1]}$$

An iterative process is required[5] until the values for M_u and δ_u converge.

Example 6.12

The nominal 8 in, solid grouted concrete block masonry wall described in Ex. 6.10 has a specified strength of 3000 lbf/in². It is reinforced with five no. 4 grade 60 bars. The factored loads acting on the wall are indicated in the illustration for Ex. 6.10. Determine the factored design moment at the midheight of the wall.

Solution

Assume a deflection at midheight caused by factored loads of

$$\delta_{u1} = 0.10 \text{ in}$$

The modulus of rupture for out-of-plane forces on a fully grouted masonry wall is given by MSJC Table 3.1.8.2 as

$$f_r = 158 \frac{\text{lbf}}{\text{in}^2} \quad \text{[type N Portland cement/lime mortar]}$$

The section modulus of the net wall section is

$$S_n = \frac{b^2 L_w}{6}$$
$$= \frac{(7.63 \text{ in})^2 (72 \text{ in})}{6}$$
$$= 699 \text{ in}^3$$

Ignoring the effects of axial load, the nominal cracking moment strength is

$$M_{cr} = S_n f_r$$
$$= \frac{(699 \text{ in}^3)\left(158 \frac{\text{lbf}}{\text{in}^2}\right)}{\left(12 \frac{\text{in}}{\text{ft}}\right)\left(1000 \frac{\text{lbf}}{\text{kip}}\right)}$$
$$= 9.20 \text{ ft-kips}$$

The nominal wall moment is

$$M_n = 21.15 \text{ ft-kips} \quad \text{[from Ex. 6-10]}$$
$$\frac{M_n}{M_{cr}} = \frac{21.15 \text{ ft-kips}}{9.20 \text{ ft-kips}}$$
$$= 2.3$$
$$> 1.3 \quad \text{[complies with MSJC Sec. 3.3.4.2.2.2]}$$

The applied strength level moment at the midheight of the wall is given by MSJC Eq. (3-25) as

$$M_{u1} = \frac{w_u h^2}{8} + \frac{P_{uf} e_u}{2} + P_u \delta_{u1}$$
$$= \frac{\left(0.025 \frac{\text{kips}}{\text{ft}^2}\right)(6 \text{ ft})(20 \text{ ft})^2 \left(12 \frac{\text{in}}{\text{ft}}\right)}{8}$$
$$+ \frac{(5 \text{ kips})(8 \text{ in})}{2} + (10 \text{ kips})(0.10 \text{ in})$$
$$= \frac{111 \text{ in-kips}}{12 \frac{\text{in}}{\text{ft}}}$$
$$= 9.25 \text{ ft-kips}$$

The deflection corresponding to the factored moment is determined in accordance with MSJC Eq. (3-32). The moment of inertia of the gross wall section is

$$I_g = \frac{L_w b^3}{12}$$
$$= \frac{(72 \text{ in})(7.63 \text{ in})^3}{12}$$
$$= 2665 \text{ in}^4$$

From Ex. 6.10, the depth of the equivalent rectangular stress block is

$$a = 0.41 \text{ in}$$

The depth of the neutral axis is

$$c = \frac{a}{0.80}$$
$$= \frac{0.41 \text{ in}}{0.80}$$
$$= 0.51 \text{ in}$$

The modulus of elasticity of reinforcement is given by MSJC Sec. 1.8.2.1.1 as

$$E_s = 29{,}000 \frac{\text{kips}}{\text{in}^2}$$

---

[5]Ekwueme, C.G., 2003 (See References and Codes)

The modulus of elasticity of concrete masonry is given by MSJC Sec. 1.8.2.2 as

$$E_m = 900 f'_m$$
$$= (900)\left(3 \ \frac{\text{kips}}{\text{in}^2}\right)$$
$$= 2700 \ \text{kips/in}^2$$

The modular ratio is

$$n = \frac{E_s}{E_m}$$
$$= \frac{29{,}000 \ \dfrac{\text{kips}}{\text{in}^2}}{2700 \ \dfrac{\text{kips}}{\text{in}^2}}$$
$$= 10.74$$

The moment of inertia of the cracked transformed section about the neutral axis is

$$I_{cr} = \frac{L_w c^3}{3} + nA_{se}(d-c)^2$$
$$= \frac{(72 \ \text{in})(0.51 \ \text{in})^3}{3}$$
$$\quad + (10.74)(1.17 \ \text{in}^2)(3.815 \ \text{in} - 0.51 \ \text{in})^2$$
$$= 140 \ \text{in}^4$$

Because M_{u1} is greater than M_{cr}, the midheight deflection corresponding to the factored moment is derived from MSJC Eq. (3-32) as

$$\delta_u = \frac{5 M_{cr} h^2}{48 E_m I_g} + \frac{5 h^2 (M_{u1} - M_{cr})}{48 E_m I_{cr}}$$
$$= \frac{(5)(110.40 \ \text{in-kips})(240 \ \text{in})^2}{(48)\left(2700 \ \dfrac{\text{kips}}{\text{in}^2}\right)(2665 \ \text{in}^4)}$$
$$\quad + \frac{(5)(0.60 \ \text{in-kips})(240 \ \text{in})^2}{(48)\left(2700 \ \dfrac{\text{kips}}{\text{in}^2}\right)(140 \ \text{in}^4)}$$
$$= 0.102 \ \text{in}$$
$$\approx \delta_{u1} \quad [\text{satisfactory}]$$

The original assumptions are correct, and the factored applied moment is

$$M_{u1} = 9.25 \ \text{ft-kips}$$
$$< \phi M_n \quad [\text{satisfactory}]$$

The flexural capacity is adequate.

Service Load Deflections

The maximum permissible deflection at the midheight of the wall, caused by service level vertical and lateral loads, and including P-delta effects, is given by MSJC Eq. (3-30) as

$$\delta_s = 0.007 h$$

When the applied service moment, M_{ser}, exceeds the cracking moment, M_{cr}, the service deflection is given by MSJC Eq. (3-32) as

$$\delta_s = \frac{5 M_{cr} h^2}{48 E_m I_g} + \frac{5 h^2 (M_{\text{ser}} - M_{cr})}{48 E_m I_{cr}}$$

The service moment at the midheight of the wall, including P-delta effects, is

$$M_{\text{ser}} = \frac{w h^2}{8} + \frac{P_f e}{2} + P \delta_s$$

The service level lateral load is w. The unfactored axial load is

$$P = P_f + P_w$$

The cracked moment of inertia of the wall section, assuming the stress in the masonry is essentially elastic, is

$$I_{cr} = \frac{L_w c^3}{3} + n A'_{se}(d-c)^2$$

The depth to the neutral axis is

$$c = kd$$
$$k = \sqrt{2 n \rho_e + (n \rho_e)^2} - n \rho_e$$
$$n = \frac{E_s}{E_m}$$
$$\rho_e = \frac{A'_{se}}{L_w d}$$

The equivalent reinforcement area at working load is

$$A'_{se} = \frac{P + A_s f_y}{f_y}$$

When the applied service moment is less than the cracking moment, the service deflection is given by MSJC Eq. (3-31) as

$$\delta_s = \frac{5 M_{\text{ser}} h^2}{48 E_m I_g}$$

An iterative process is required until the values for δ_s and the values for M_{ser} converge.

Example 6.13

Determine whether the midheight deflection of the slender wall of Ex. 6.10 under service level loads is within the permissible limits.

Solution

The maximum permissible deflection at midheight of the wall caused by service level vertical and lateral loads, including P-delta effects, is given by MSJC Eq. (3-30) as

$$\delta_s = 0.007h$$
$$= (0.007)(20 \text{ ft})\left(12 \ \frac{\text{in}}{\text{ft}}\right)$$
$$= 1.68 \text{ in}$$

From Ex. 6.12, the midheight deflection produced by the factored loads is

$$\delta_u = 0.10 \text{ in}$$
$$< \delta_s$$

The deflection under service loads is within the permissible limit.

7. DESIGN OF ANCHOR BOLTS[6]

Nomenclature

A_b	nominal cross-sectional area of the bolt	in^2
A_o	overlap of projected areas	in^2
A_{pt}	projected area of tensile breakout surface	in^2
b_a	tensile force on an anchor bolt	kips
b_v	shear force on an anchor bolt	kips
B_{ab}	axial capacity in tension of an anchor bolt when governed by breakout	kips
B_{as}	design capacity in steel tensile yield	kips
B_{vb}	design capacity in shear of an anchor bolt when governed by breakout	kips
B_{vc}	capacity in shear of an anchor bolt when governed by crushing	kips
B_{vpry}	design capacity in shear pryout	kips
B_{vs}	design capacity in shear of an anchor bolt when governed by bolt steel	kips
d_b	bolt diameter	in
f_y	bolt yield stress	–
l_b	effective embedment depth of anchor bolt	in
l_{be}	anchor bolt edge distance	in
r	radius of projected area	in
s	bolt spacing	in

Symbols

θ	half the angle subtended by the chord at the intersection of overlapping projected areas	deg

---
[6]Ekwueme, C. G., 2010 (See References and Codes)

Headed Anchor Bolts in Tension

In accordance with MSJC Commentary Sec. 1.16.2, anchors that are solidly grouted in masonry fail in tension by the pullout of a conically shaped section of masonry (see Fig. 6.10). The failure surface slopes at 45°. The bolt's projected area, given by MSJC Eq. (1-2), is a circle of radius equal to the embedment length of the bolt. The projected area is

$$A_{pt} = \pi l_b^2$$

The effective embedment depth of the anchor bolt, measured from the surface of the masonry to the bearing surface of the bolt head, is l_b.

The anchor bolt edge distance, measured from the edge of the masonry to the center of the bolt is l_{be}.

Figure 6.10 Masonry Failure in Tension

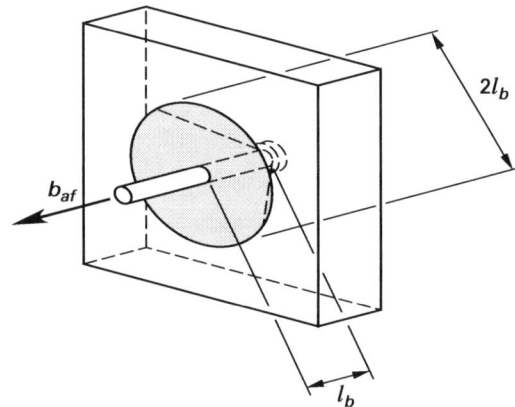

For tensile strength governed by masonry breakout, the design capacity in tension is given by MSJC Eq. (2-1) as

$$B_{ab} = 1.25 A_{pt}\sqrt{f'_m}$$

The design capacity in steel tensile yield is

$$B_{as} = 0.6 A_b f_y$$

The minimum effective embedment length is specified by MSJC Sec. 1.16.6 as the greater of four bolt diameters or 2 in.

When bolts are spaced at less than twice the embedment length apart, the projected areas of the bolts overlap, and MSJC Sec. 1.16.2 requires the combined projected area to be reduced by the overlapping area. As shown in Fig. 6.11, the overlapping area is

$$A_o = \left(\frac{\pi\theta}{90°} - \sin 2\theta\right) r^2$$

Figure 6.11 *Overlap of Projected Areas*

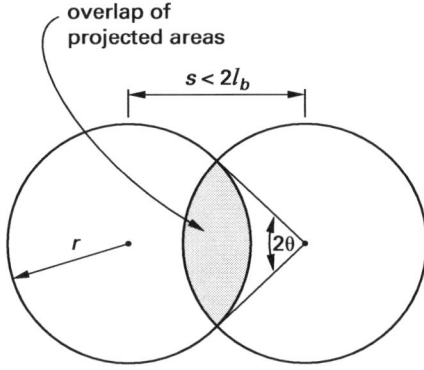

The angle subtended at the center of the projected area by the chord of the intersecting circles is 2θ, where

$$\theta = \cos^{-1}\left(\frac{s}{2r}\right)$$
$$r = l_b$$

Similarly, when a bolt is located less than the embedment length from the edge of a member, that portion of the projected area falling in an open cell shall be deducted from the calculated area.

For tensile strength governed by the bolt steel, the design capacity in tension is given by MSJC Eq. (2-2) as

$$B_{as} = 0.6 A_b f_y$$

Example 6.14

The anchorage for a bracket to a nominal 8 in, solid grouted concrete block masonry wall is shown in the illustration. The masonry compressive strength is 3000 lbf/in². Anchor bolts are $7/8$ in diameter, ASTM A307 type C, with a minimum specified yield strength of 36 kips/in², and an effective cross-sectional area at the root of the threads of 0.46 in². For the loads indicated in the illustration, determine whether the bolts are adequate for the tensile forces on the bracket.

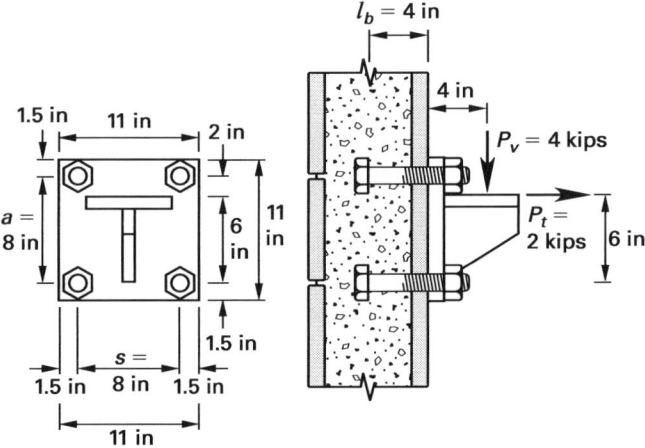

Solution

By taking moments about the bottom bolts, the tensile force in each top bolt is obtained as

$$b_a = \frac{4P_v + 6P_t}{2a}$$
$$= \frac{(4 \text{ in})(4 \text{ kips}) + (6 \text{ in})(2 \text{ kips})}{(2)(8 \text{ in})}$$
$$= 1.75 \text{ kips}$$

For tensile strength governed by the $7/8$ in bolts, the design capacity of one bolt in tension is given by MSJC Eq. (2-2) as

$$B_{as} = 0.6 A_b f_y$$
$$= (0.6)(0.46 \text{ in}^2)\left(36 \frac{\text{kips}}{\text{in}^2}\right)$$
$$= 9.94 \text{ kips}$$
$$> b_a \quad [\text{satisfactory}]$$

The effective embedment length of an anchor bolt, measured from the surface of the masonry to the bearing surface of the bolt head, is

$$l_b = 4 \text{ in}$$
$$> 4d_b \quad [\text{satisfies MSJC Sec. 1.16.6}]$$
$$> 2 \text{ in} \quad [\text{satisfies MSJC Sec. 1.16.6}]$$

The spacing of the bolts is

$$s = 8 \text{ in}$$
$$= 2l_b$$

The projected areas of the bolts do not overlap. MSJC Sec. 1.16.2 gives the projected area of one bolt as

$$A_{pt} = \pi l_b^2$$
$$= (3.14)(4 \text{ in})^2$$
$$= 50.24 \text{ in}^2$$

For tensile strength governed by masonry breakout, the design capacity in tension is given by MSJC Eq. (2-1) as

$$B_{ab} = 1.25 A_{pt} \sqrt{f'_m}$$
$$= \frac{(1.25)(50.24 \text{ in}^2)\sqrt{3000 \frac{\text{lbf}}{\text{in}^2}}}{1000 \frac{\text{lbf}}{\text{kip}}}$$
$$= 3.44 \text{ kips} \quad [\text{governs}]$$
$$> b_a \quad [\text{satisfactory}]$$

Headed Anchor Bolts in Shear

The projected area in shear for bolts is given by MSJC Eq. (1-3) as

$$A_{pv} = \frac{\pi l_{be}^2}{2}$$

The masonry design capacity in shear crushing is given by MSJC Eq. (2-7) as

$$B_{vc} = 350\sqrt[4]{f'_m A_b}$$

The design capacity in shear breakout is given by MSJC Eq. (2-6) as

$$B_{vb} = 1.25 A_{pv} \sqrt{f'_m}$$

For shear strength governed by the bolt steel yielding, the design capacity in shear is given by MSJC Eq. (2-9) as

$$B_{vs} = 0.36 A_b f_y$$

The design capacity in shear pryout is given by MSJC Eq. (2-8) as

$$B_{vpry} = 2.0 B_{ab}$$
$$= 2.5 A_{pt} \sqrt{f'_m}$$

Example 6.15

For the bracket anchored to a nominal 8 in, solid grouted concrete block masonry wall described in Ex. 6.14, the masonry compressive strength is 3000 lbf/in². The threaded portions of the anchor bolts are located inside the shear plane. The bolts are ⅞ in diameter, ASTM A307 type C, with minimum specified yield strength of 36 kips/in². For the loads indicated in the illustration for Ex. 6.14, determine whether the bolts are adequate for the shear forces on the bracket.

Solution

The shear force on each bolt is

$$b_v = \frac{P_v}{4}$$
$$= \frac{4 \text{ kips}}{4}$$
$$= 1.0 \text{ kip}$$

The bolts are far from a free edge of the wall and masonry breakout in shear does not govern. The shear strength is governed by shear crushing. The design capacity in shear crushing is given by MSJC Eq. (2-7) as

$$B_{vc} = 0.350 \sqrt[4]{f'_m A_b}$$
$$= 0.350 \sqrt[4]{\left(3000 \frac{\text{lbf}}{\text{in}^2}\right)(0.46 \text{ in}^2)}$$
$$= 2.13 \text{ kips}$$
$$> b_v \quad [\text{satisfactory}]$$

Headed Anchor Bolts in Combined Tension and Shear

For combined tension and shear, MSJC Eq. (2-10) shall be satisfied

$$\frac{b_a}{B_a} + \frac{b_v}{B_v} \leq 1$$

The governing design axial capacity of an anchor bolt is B_a. The governing design shear capacity of an anchor bolt is B_v. In addition, the design capacity in shear and tension shall each exceed the applied loads.

Example 6.16

For the bracket anchored to a nominal 8 in, solid grouted concrete block masonry wall described in Ex. 6.14, the masonry compressive strength is 3000 lbf/in². Anchor bolts are ⅞ in diameter, ASTM A307 type C, with a minimum specified yield strength of 36 kips/in². For the loads indicated in the illustration for Ex. 6.14, determine whether the bolts are adequate for the combined tension and shear forces.

Solution

For combined tension and shear, MSJC Eq. (2-10) must be satisfied

$$\frac{b_a}{B_a} + \frac{b_v}{B_v} \leq 1$$

The left-hand side of the expression is

$$\frac{1.75 \text{ kips}}{3.44 \text{ kips}} + \frac{1.0 \text{ kip}}{2.13 \text{ kips}} = 0.51 + 0.47$$
$$= 0.98$$
$$< 1.00 \quad [\text{satisfactory}]$$

8. DESIGN OF PRESTRESSED MASONRY

Nomenclature

A_n	net cross-sectional area of masonry	in²
A_{ps}	area of prestressing steel	in²
A_s	area of reinforcement	in²
E_m	modulus of elasticity of masonry	kips/in²
f_a	calculated compressive stress in masonry caused by axial load	lbf/in²

Symbol	Description	Units
f_b	calculated compressive stress in masonry caused by flexure	lbf/in²
f'_m	specified compressive strength of masonry	lbf/in²
f'_{mi}	specified compressive strength of masonry at time of prestress transfer	lbf/in²
f_{ps}	stress in prestressed tendon at nominal strength	kips/in²
f_{psi}	initial stress in prestressed tendon	kips/in²
f_{pu}	specified tensile strength of prestressing tendons	kips/in²
f_{py}	specified yield strength of prestressing tendons	kips/in²
f_s	stress in prestressing tendons	kips/in²
f_{se}	effective stress in prestressed tendon after allowance for all prestress losses	kips/in²
f_y	specified yield stress of non-prestressed reinforcement	kips/in²
F_a	allowable compressive stress in masonry caused by axial load	lbf/in²
F_b	allowable compressive stress in masonry caused by flexure	lbf/in²
F_t	allowable tensile stress in masonry caused by flexure	lbf/in²
h	effective height of wall	ft
I_n	moment of inertia of net cross-sectional area	in⁴
l_p	clear span of the prestressed member in the direction of the prestressing tendon	ft
M	maximum moment at the section under consideration	ft-kips
M_n	nominal flexural strength	ft-kips
M_u	factored moment	ft-kips
P	sum of P_w and P_f	kips
P_f	unfactored load from tributary floor or roof loads	kips
P_{ps}	prestressing tendon force at time and location relevant for design	kips
P_u	factored axial load	kips
P_w	unfactored weight of wall tributary to section considered	kips
r	radius of gyration	in
S_n	section modulus of net cross-sectional area	in³
w_u	factored lateral load	lbf/ft

Symbols

ϕ	strength reduction factor	—

General Considerations

A masonry member is prestressed by tensioning a prestressing tendon that is anchored within the member. This produces a compression in the member, increasing its flexural tensile capacity, and augmenting its resistance to lateral loading.[7] Compared with conventional reinforced masonry construction, prestressed masonry reduces construction time and costs by eliminating most of the grout and conventional reinforcement. This also reduces the weight of the masonry, resulting in smaller seismic forces and cost savings in foundations. The technique is particularly well suited to walls, and design consists of an initial allowable stress approach followed by a strength check. The addition of prestressing to an existing masonry structure is a relatively simple technique, and can be used to strengthen deficient structures.

Prestressed masonry members must be designed for the three design stages: transfer, service, and ultimate. At the transfer design stage, a prestressing force is applied to the member. Immediate prestress losses result from the elastic deformation of the masonry and anchor set. At the serviceability design stage, all time-dependent prestress losses have occurred as a result of creep, shrinkage of the masonry, and relaxation of the tendon stress. At the strength design stage, a rectangular stress block is assumed in the masonry, with a maximum strain in concrete masonry of 0.0025.

Construction Technique

Figure 6.12 shows a section through a typical prestressed masonry wall. Prestressing tendons are anchored at the foot and top of the wall. Tendons are normally unbonded, as shown, and require corrosion protection in walls that are subject to a moist and corrosive environment. Corrosion protection may be provided by coating the tendon with a corrosion-inhibiting material, and enclosing the tendon in a continuous plastic sheath.

Figure 6.12 *Prestressed Masonry Construction*

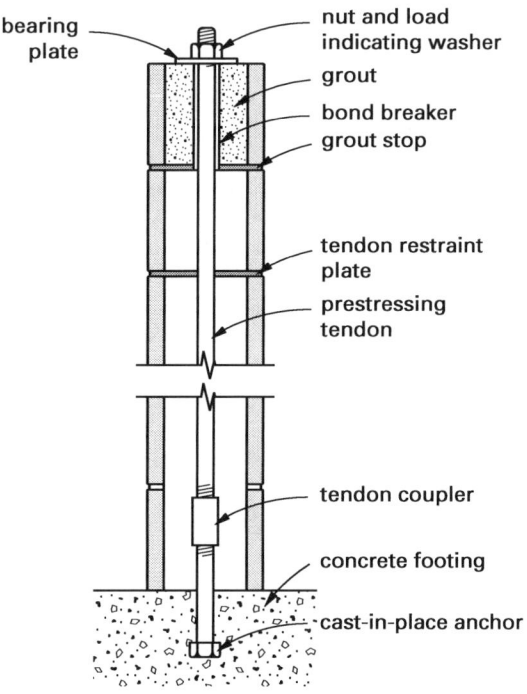

---

[7]Durning, T.A., 2000 (See References and Codes)

Alternatively, tendons may be galvanized threaded high-tensile steel rods.

Tendons are attached by a threaded coupler at the foot of the wall to an anchor bolt embedded in the concrete foundation. An inspection port is provided at the location of the coupler to enable the tendon to be connected to the coupler after the block wall has been constructed to full height. Blocks are usually laid in face-shell mortar bedding. In accordance with MSJC Sec. 4.8.2, couplers shall develop 95% of the specified tensile strength of the prestressing tendons. When a moment connection is not required at the base of the wall, anchor bolts may be located in the first block course of masonry.

The tendons are normally laterally restrained in order to conform to the lateral deformation of the wall. This has the advantage that the prestressing force does not contribute to the elastic instability of the wall, as lateral displacement of the wall is counterbalanced by an equal and opposite restraint from the prestressing tendons. In addition, the ultimate strength of the wall is increased by ensuring that the wall and the tendon deform together. MSJC Commentary Sec. 4.4.2 stipulates that three restraints along the length of a tendon provide adequate restraint. Restraint is provided by a steel plate with a central hole. The sides of the plate are embedded in the joint between courses. Alternatively, a cell can be filled with grout and a bond breaker applied to the tendon to allow it to move freely within the cell.

Similarly, the topmost cell is filled with grout and a steel bearing plate bedded on top. The size of bearing plate is determined from the requirement of MSJC Sec. 4.8.4.2 that bearing stresses shall not exceed 50% of the masonry compressive strength at the time of transfer. The tendon may be stressed by means of an hydraulic jack, or by tightening a nut with a standard wrench. In the former method, the load is measured by a load cell or a calibrated gauge on the jack. The force produced by tightening a nut may be measured with a direct tension indicator washer. This is a washer with dimples formed on its top face. A hardened steel washer is placed on top of the indicator washer, and as the nut is tightened, the dimples are compressed. The required force in the tendon has been produced when the gap between the two washers reaches a specified amount as measured by a feeler gauge. MSJC Specification Sec. 3.6B requires that the elongation of the tendons be measured and compared with the elongation anticipated for the applied prestressing force. If the discrepancy between the two methods exceeds 7%, the cause must be determined and corrected.

Serviceability Design Stage

For laterally restrained tendons, the permissible stresses under service loads after all prestressing losses have occurred are specified in MSJC Secs. 4.3, 4.4, and 2.2.3. The allowable compressive stress in a member that is subjected to axial load and flexure, having a height-to-radius of gyration ratio not greater than 99, is given by MSJC Eq. (2-15) as

$$F_a = 0.25 f'_m \left(1 - \left(\frac{h}{140r}\right)^2\right)$$

For members having a height-to-radius of gyration ratio greater than 99, the allowable compressive stress is given by MSJC Eq. (2-16) as

$$F_a = (0.25 f'_m)\left(\frac{70r}{h}\right)^2$$

The allowable compressive stress in a member caused by flexure is given by MSJC Eq. (2-17) as

$$F_b = \frac{f'_m}{3}$$

The compressive stress produced in the wall by floor and roof loads, wall self-weight, and the effective prestressing force after all losses is

$$f_a = \frac{P + P_{ps}}{A_n}$$

The flexural stress produced in the wall by applied lateral loads and eccentric axial load is

$$f_b = \frac{M}{S_n}$$

The resulting combined stress caused by compression and flexure shall satisfy the interaction expression of MSJC Eq. (2-13), which is

$$\frac{f_a}{F_a} + \frac{f_b}{F_b} \leq 1.0$$

The allowable tensile stress, F_t, caused by flexure is given by MSJC Table 2.2.3.2. In accordance with MSJC Sec. 2.2.3.2,

$$f_b - f_a \leq F_t$$

For laterally restrained tendons, the member cannot buckle under its own prestressing force. The Euler critical load is given by MSJC Eq. (2-18) as

$$P_e = \frac{\pi^2 E_m I_n \left(1 - \frac{0.577e}{r}\right)^3}{h^2}$$

To ensure elastic stability, the axial load on a wall is limited by MSJC Eq. (2-14) to a value of

$$P \leq \frac{P_e}{4}$$

For laterally restrained tendons, the effective prestressing force, P_{ps}, is not considered in the calculation of the axial force, P.

At the time the prestress is applied, the permissible stress in prestressing tendons at anchorages and couplers is given by MSJC Sec. 4.3.3 as

$$f_s \leq 0.78 f_{py}$$
$$\leq 0.70 f_{pu}$$

The total loss of prestress for concrete masonry after long-term service load conditions is given by MSJC Commentary Sec. 4.3.4 as 30–35%.

Example 6.17

A nominal 8 in, concrete block masonry wall, laid in face-shell type N Portland cement/lime mortar bedding, has a specified strength of 2000 lbf/in^2. The wall is 16 ft high, is pinned at the top and bottom, and weighs 45 lbf/ft^2. The wind load on the wall, which may act in either direction, is 20 lbf/ft^2. The wall supports an axial load of 500 lbf/ft. Assuming the total loss of prestress after long-term service is 35%, determine the spacing required for $^7/_{16}$ in diameter steel rod tendons. The tendons are laterally restrained.

Solution

The design is based on a 1 ft length of wall. The relevant details[8] are

$$A_{ps} = 0.142 \text{ in}^2 \quad \text{[for one rod]}$$
$$f_{py} = 100 \text{ kips/in}^2$$
$$f_{pu} = 122 \text{ kips/in}^2$$

$A_n = 30 \text{ in}^2$ [face-shell mortar bedding]
$I_n = 309 \text{ in}^4$ [face-shell mortar bedding]
$S_n = 81 \text{ in}^3$ [face-shell mortar bedding]
$r = 3.21 \text{ in}$ [face-shell mortar bedding]

$$E_m = 900 f'_m$$
$$= 1800 \text{ kips/in}^2$$

The weight of the wall above midheight is

$$P_w = \left(45 \frac{\text{lbf}}{\text{ft}}\right)(8 \text{ ft})$$
$$= 360 \text{ lbf}$$

The axial load at the top of the wall is

$$P_f = 500 \text{ lbf}$$

The total axial load at midheight of the wall is

$$P = P_w + P_f$$
$$= 360 \text{ lbf} + 500 \text{ lbf}$$
$$= 860 \text{ lbf}$$

The compressive stress produced in the wall by floor and roof loads, wall self-weight, and the effective prestressing force after all losses is

$$f_a = \frac{P + P_{ps}}{A_n}$$
$$= \frac{860 \text{ lbf} + P_{ps}}{30 \text{ in}^2}$$
$$= 28.67 \frac{\text{lbf}}{\text{in}^2} + \frac{P_{ps}}{30 \text{ in}^2}$$

The bending moment acting on the wall is

$$M = \frac{wh^2}{8}$$
$$= \frac{\left(20 \frac{\text{lbf}}{\text{ft}}\right)(16 \text{ ft})^2}{8}$$
$$= 640 \text{ ft-lbf}$$

The flexural stress produced in the wall by applied lateral load is

$$f_b = \frac{M}{S_n}$$
$$= \frac{(640 \text{ ft-lbf})\left(12 \frac{\text{in}}{\text{ft}}\right)}{81 \text{ in}^3}$$
$$= 94.81 \text{ lbf/in}^2$$

The slenderness ratio of the wall is

$$\frac{h}{r} = \frac{(16 \text{ ft})\left(12 \frac{\text{in}}{\text{ft}}\right)}{3.21 \text{ in}}$$
$$= 59.81$$

The allowable compressive stress in a member having a height-to-radius of gyration ratio not greater than 99, and subjected to axial load and flexure, is given by MSJC Eq. (2-15) as

$$F_a = 0.25 f'_m \left(1 - \left(\frac{h}{140r}\right)^2\right)$$
$$= (0.25)\left(2000 \frac{\text{lbf}}{\text{in}^2}\right)\left(1 - \left(\frac{59.81}{140}\right)^2\right)$$
$$= 408.74 \text{ lbf/in}^2$$

---

[8]Masonry Society, 2010 (See References and Codes)

The allowable compressive stress in a member caused by flexure is given by MSJC Eq. (2-17) as

$$F_b = \frac{f'_m}{3} = \frac{2000 \frac{\text{lbf}}{\text{in}^2}}{3}$$
$$= 666.67 \text{ lbf/in}^2$$

The combined stress caused by compression and flexure shall satisfy MSJC Eq. (2-13), which is

$$\frac{f_a}{F_a} + \frac{f_b}{F_b} \leq 1.0$$

$$\frac{28.67 \frac{\text{lbf}}{\text{in}^2} + \frac{P_{ps}}{30 \text{ in}^2}}{408.74 \frac{\text{lbf}}{\text{in}^2}} + \frac{94.81 \frac{\text{lbf}}{\text{in}^2}}{666.67 \frac{\text{lbf}}{\text{in}^2}} \leq 1.0$$

$$0.070 + \frac{P_{ps}}{12{,}262} + 0.14 \leq 1.0$$

$$P_{ps} \leq \frac{9687 \text{ lbf}}{1000 \frac{\text{lbf}}{\text{kip}}}$$
$$\leq 9.69 \text{ kips}$$

The allowable tensile stress caused by flexure for type N Portland cement/lime mortar is given by MSJC Table 2.2.3.2 as

$$F_t = 19 \frac{\text{lbf}}{\text{in}^2}$$

The difference between calculated compressive stress caused by axial load and flexure is

$$f_b - f_a \leq 19 \frac{\text{lbf}}{\text{in}^2}$$

$$94.82 \frac{\text{lbf}}{\text{in}^2} - \left(28.67 \frac{\text{lbf}}{\text{in}^2} + \frac{P_{ps}}{30 \text{ in}^2}\right) \leq 19 \frac{\text{lbf}}{\text{in}^2}$$

$$P_{ps} \geq \frac{1415 \text{ lbf}}{1000 \frac{\text{lbf}}{\text{kip}}}$$
$$\geq 1.41 \text{ kips}$$

At the time of application of the prestress, the maximum permissible stress in prestressing tendons at anchorages and couplers is given by MSJC Sec. 4.3.3 as the lesser of

$$f_{psi} = 0.70 f_{pu}$$
$$= (0.70)\left(122 \frac{\text{kips}}{\text{in}^2}\right)$$
$$= 85.4 \text{ kips/in}^2$$

$$f_{psi} = 0.78 f_{py}$$
$$= (0.78)\left(100 \frac{\text{kips}}{\text{in}^2}\right)$$
$$= 78 \text{ kips/in}^2 \quad \text{[governs]}$$

Allowing for a total long-term loss of prestress of 35%, the final tendon force is

$$P_{ps} = 0.65 f_{psi} A_{ps}$$
$$= (0.65)\left(78 \frac{\text{kips}}{\text{in}^2}\right)(0.142 \text{ in}^2)$$
$$= 7.2 \text{ kips}$$

Providing a $7/16$ in diameter tendon at 2 ft on center gives an effective final prestressing force per foot of

$$P_{ps} = 3.60 \text{ kips}$$
$$> 1.41 \text{ kips} \quad \text{[satisfactory]}$$
$$< 9.66 \text{ kips} \quad \text{[satisfactory]}$$

For laterally restrained tendons, the Euler critical load is given by MSJC Eq. (2-18) as

$$P_e = \frac{\pi^2 E_m I_n \left(1 - \frac{0.577 e}{r}\right)^3}{h^2}$$

$$= \frac{\pi^2 \left(1800 \frac{\text{kips}}{\text{in}^2}\right)(309 \text{ in}^4)(1 - 0)^3}{(192 \text{ in})^2}$$

$$= 148.91 \text{ kips}$$

To ensure elastic stability, the axial load on a wall is limited by MSJC Eq. (2-14) to a maximum value of

$$P = \frac{P_e}{4}$$
$$= \frac{148.91 \text{ kips}}{4}$$
$$= 37.23 \text{ kips}$$
$$> 860 \text{ lbf} \quad \text{[satisfactory]}$$

The wall is stable.

Transfer Design Stage

The allowable stresses in the masonry at transfer are given by MSJC Sec. 4.4.1.2 as

$$F_{ai} = 1.2 F_a$$
$$F_{bi} = 1.2 F_b$$

MSJC Eq. (2-13) for combined compression and flexure becomes

$$\frac{f_{ai}}{F_{ai}} + \frac{f_{bi}}{F_{bi}} \leq 1.0$$

The allowable tensile stress F_t caused by flexure is given by MSJC Table 2.2.3.2. Then, in accordance with MSJC Sec. 2.2.3.2,

$$f_{bi} - f_{ai} \leq F_t$$

Immediately after transfer, the permissible stress in prestressing tendons is given by MSJC Sec. 4.3.2 as

$$f_{psi} \leq 0.82 f_{py}$$
$$\leq 0.74 f_{pu}$$

However, for post-tensioned tendons with couplers, the permissible stress is governed by the maximum allowed at the time the prestress is applied, given by MSJC Sec. 4.3.3 as

$$f_{psi} \leq 0.78 f_{py}$$
$$\leq 0.70 f_{pu}$$

MSJC Commentary Sec. 4.3.4 gives the initial loss of prestress for concrete masonry at transfer as 5–10%.

Example 6.18

The nominal 8 in, concrete block masonry wall described in Ex. 6.17 is post-tensioned with $7/16$ in diameter, steel rod tendons at 2 ft centers. Assuming the initial loss of prestress at transfer is 5%, determine whether the stresses at transfer are satisfactory. The specified masonry strength at transfer is 1500 lbf/in^2.

Solution

At the time the prestress is applied, the allowable stress is governed by MSJC Sec. 4.3.3. The maximum permissible stress was obtained in Ex. 6.17 as

$$f_{psi} = 78.0 \ \frac{\text{kips}}{\text{in}^2}$$

Allowing for an initial loss of prestress of 5%, the tendon force immediately after transfer is

$$P_{psi} = 0.95 f_{psi} A_{ps}$$
$$= (0.95)\left(78.0 \ \frac{\text{kips}}{\text{in}^2}\right)(0.142 \ \text{in}^2)$$
$$= 10.52 \ \text{kips}$$

For tendons spaced at 2 ft on center, the initial force per foot of wall is

$$P_{psi} = 5.26 \ \text{kips}$$

The total axial load at the midheight of the wall is

$$P = 860 \ \text{lbf}$$

The compressive stress produced in the wall by floor and roof loads, wall self-weight, and the effective prestressing force after initial losses is

$$f_{ai} = \frac{P + P_{psi}}{A_n}$$
$$= \frac{860 \ \text{lbf} + 5260 \ \text{lbf}}{30 \ \text{in}^2}$$
$$= 204 \ \text{lbf/in}^2$$

The flexural stress produced in the wall by applied lateral load at transfer is

$$f_{bi} = 94.82 \ \text{lbf/in}^2$$

The allowable compressive stress in the wall at transfer, caused by combined compression and flexure, is given by MSJC Sec. 4.4.1.2 and MSJC Eq. (2-15) as

$$F_{ai} = \frac{1.2 F_a f'_{mi}}{f'_m}$$
$$= \frac{(1.2)\left(408.74 \ \frac{\text{lbf}}{\text{in}^2}\right)\left(1500 \ \frac{\text{lbf}}{\text{in}^2}\right)}{2000 \ \frac{\text{lbf}}{\text{in}^2}}$$
$$= 367.87 \ \text{lbf/in}^2$$

The allowable compressive stress in the wall at transfer, caused by flexure, is given by MSJC Sec. 4.4.1.2 and MSJC Eq. (2-17) as

$$F_{bi} = \frac{1.2 f'_{mi}}{3}$$
$$= \frac{(1.2)\left(1500 \ \frac{\text{lbf}}{\text{in}^2}\right)}{3}$$
$$= 600 \ \text{lbf/in}^2$$

The interaction equation for combined compression and flexure is

$$\frac{f_{ai}}{F_{ai}} + \frac{f_{bi}}{F_{bi}} \leq 1.0$$

The left-hand side of the expression is evaluated as

$$\frac{204 \ \frac{\text{lbf}}{\text{in}^2}}{367.87 \ \frac{\text{lbf}}{\text{in}^2}} + \frac{94.82 \ \frac{\text{lbf}}{\text{in}^2}}{600 \ \frac{\text{lbf}}{\text{in}^2}} = 0.713$$

$$< 1.0 \quad \text{[satisfactory]}$$

The allowable tensile stress caused by flexure, F_t, is given by MSJC Table 2.2.3.2, and in accordance with MSJC Sec. 2.2.3.2,

$$f_{bi} - f_{ai} \leq F_t$$

$$94.82 \, \frac{\text{lbf}}{\text{in}^2} - 204 \, \frac{\text{lbf}}{\text{in}^2} \leq 19 \, \frac{\text{lbf}}{\text{in}^2}$$

$$-109.18 \, \text{lbf/in}^2 < 19 \, \text{lbf/in}^2$$

[satisfactory]

The stresses at transfer are satisfactory.

Strength Design Stage

The design flexural strength for a prestressed member is given by MSJC Sec. 4.4.3.3 as

$$\phi M_n = 0.8 M_n$$

The nominal strength of a masonry member is determined as detailed in MSJC Sec. 4.4.3.

For walls with laterally restrained, unbonded tendons, the stress in the tendons at nominal load is given by MSJC Eq. (4-3) as

$$f_{ps} = f_{se} + (1{,}000{,}000)\left(\frac{d}{l_p}\right)\sqrt{1 - (1.4)\left(\frac{f_{pu}A_{ps}}{bdf'_m}\right)}$$

$$\leq f_{ps}$$

$$\geq f_{se}$$

An equivalent rectangular stress block is assumed in the masonry, with a stress of $0.80f'_m$. The depth of the stress block is given by MSJC Eq. (4-1) as

$$a = \frac{f_{ps}A_{ps} + f_y A_s + P_u}{0.80 f'_m b}$$

The nominal moment is given by MSJC Eq. (4-2) as

$$M_n = (f_{ps}A_{ps} + f_y A_s + P_u)\left(d - \frac{a}{2}\right)$$

To ensure a ductile failure, MSJC Sec. 4.4.3.6 specifies a maximum depth for the stress block of

$$a = 0.425 d$$

Also,

$$a \leq \text{face shell thickness}$$

Example 6.19

The nominal 8 in, concrete block masonry wall described in Ex. 6.17 is post-tensioned with $7/16$ in diameter, steel rod tendons at 2 ft centers. Determine whether the design moment strength is adequate. The factored axial dead load at midheight is $P_u = 900$ lbf/ft, and the factored lateral force is $w_u = 32$ lbf/ft^2.

Solution

Consider one foot length of wall.

The factored moment is

$$M_u = \frac{w_u h^2}{8}$$

$$= \frac{\left(32 \, \frac{\text{lbf}}{\text{ft}}\right)(16 \, \text{ft})^2}{8}$$

$$= 1024 \, \text{ft-lbf}$$

From Ex. 6.17, the effective stress in the tendons after all losses is

$$f_{se} = 0.65 f_{psi}$$

$$= (0.65)\left(78 \, \frac{\text{kips}}{\text{in}^2}\right)$$

$$= 50.70 \, \text{kips/in}^2$$

The area of the tendons per foot of wall is

$$A_{ps} = \frac{0.142 \, \text{in}^2}{2 \, \text{ft}}$$

$$= 0.071 \, \text{in}^2$$

For walls with laterally restrained, unbonded tendons, the stress in the tendons at nominal load per foot of wall is given by MSJC Eq. (4-3) as

$$f_{ps} = f_{se} + (1{,}000{,}000)\left(\frac{d}{l_p}\right)\sqrt{1 - (1.4)\left(\frac{f_{pu}A_{ps}}{bdf'_m}\right)}$$

$$= 50.70 \, \frac{\text{kips}}{\text{in}^2} + (1000)\left(\frac{3.82 \, \text{in}}{(16 \, \text{ft})\left(12 \, \frac{\text{in}}{\text{ft}}\right)}\right)$$

$$\times \sqrt{1 - (1.4)\left(\frac{\left(122 \, \frac{\text{kips}}{\text{in}^2}\right)(0.071 \, \text{in}^2)}{(12 \, \text{in})(3.82 \, \text{in})\left(2 \, \frac{\text{kips}}{\text{in}^2}\right)}\right)}$$

$$= 69.23 \, \frac{\text{kips}}{\text{in}^2} \quad \text{[governs]}$$

$$\leq f_{py}$$

$$= 100 \, \text{kips/in}^2$$

The force in the tendons at nominal load per foot of wall is

$$f_{ps}A_{ps} = \left(69{,}240 \, \frac{\text{lbf}}{\text{in}^2}\right)(0.071 \, \text{in}^2)$$

$$= 4916 \, \text{lbf}$$

The depth of the stress block is given by MSJC Eq. (4-1) as

$$a = \frac{f_{ps}A_{ps} + f_y A_s + P_u}{0.80 f'_m b}$$

$$= \frac{4916 \text{ lbf} + 0 + 900 \text{ lbf}}{(0.80)\left(2000 \, \dfrac{\text{lbf}}{\text{in}^2}\right)(12 \text{ in})}$$

$$= 0.303 \text{ in}$$

$$< \text{face shell thickness} \quad [\text{satisfactory}]$$

$$a = \left(\frac{0.303 \text{ in}}{3.82 \text{ in}}\right)d$$

$$= 0.08 d$$

$$< 0.425 d \quad [\text{satisfies MSJC Sec. 4.4.3.6}]$$

The nominal moment is given by MSJC Eq. (4-2) as

$$M_n = (f_{ps}A_{ps} + f_y A_s + P_u)\left(d - \frac{a}{2}\right)$$

$$= \left(\frac{4916 \text{ lbf} + 0 + 900 \text{ lbf}}{1000 \, \dfrac{\text{lbf}}{\text{kips}}}\right)$$

$$\times \left(3.82 \text{ in} - \frac{0.303 \text{ in}}{2}\right)$$

$$= 21.34 \text{ in-kips}$$

The design flexural strength for a prestressed member is given by MSJC Sec. 4.4.3.3 as

$$\phi M_n = 0.8 M_n$$

$$= (0.8)(21.34 \text{ in-kips})\left(\frac{1 \text{ ft}}{12 \text{ in}}\right)\left(1000 \, \frac{\text{lbf}}{\text{kip}}\right)$$

$$= 1423 \text{ ft-lbf}$$

$$> M_u \quad [\text{satisfactory}]$$

The design moment strength is adequate.

PRACTICE PROBLEMS

Problems 1–3 refer to the nominal 8 in, solid grouted concrete block masonry beam shown. The beam is simply supported over a clear span of 12 ft. The overall depth of the beam is 48 in, and its effective depth, d, is 45 in. The unit weight may be assumed to be 80 lbf/ft². The masonry has a compressive strength of 1500 lbf/in², and the reinforcement consists of two no. 7 grade 60 bars.

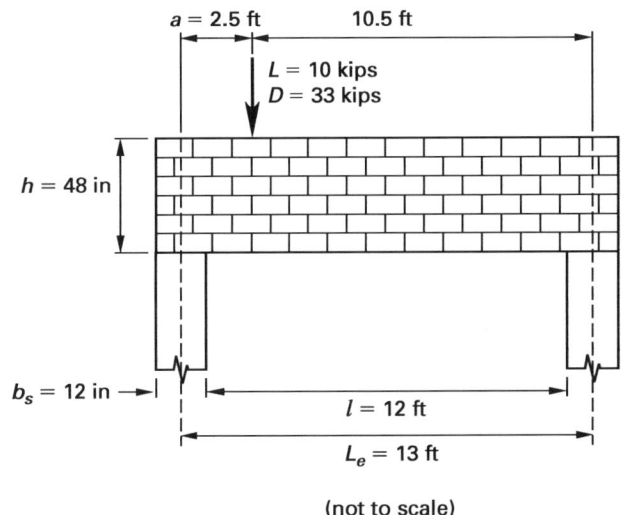

(not to scale)

1. What is most nearly the maximum nonfactored bending moment on the beam caused by the applied loads shown?

(A) 20 ft-kips

(B) 66 ft-kips

(C) 91 ft-kips

(D) 110 ft-kips

2. What is most nearly the moment capacity of the section?

(A) 92.0 ft-kips

(B) 96.0 ft-kips

(C) 100 ft-kips

(D) 104 ft-kips

3. What is most nearly the spacing required for no. 5 stirrups at the critical section?

(A) 2 in

(B) 4 in

(C) 6 in

(D) 8 in

Problems 4 and 5 refer to a nominal 8 in square, solid grouted, concrete block masonry column with a specified strength of $f'_m = 1500$ lbf/in^2 and reinforced with four no. 4 grade 60 bars. The column supports an axial load of $P = 25$ kips, has a height of $h = 12$ ft, and may be considered pinned at each end. Neglect accidental eccentricity.

4. Which of the given statements is/are true?

I. The allowable axial load is approximately 24 kips.

II. The allowable axial load is approximately 27 kips.

III. The column is adequate.

IV. The column is inadequate.

(A) I and III only

(B) I and IV only

(C) II and III only

(D) II and IV only

5. What is most nearly the minimum required size and spacing of lateral ties?

(A) lateral ties $= 0.25$ in diameter, spacing not more than 8 in

(B) lateral ties $= 0.25$ in diameter, spacing not more than 10 in

(C) lateral ties < 0.25 in diameter, spacing not more than 8 in

(D) lateral ties < 0.25 in diameter, spacing not more than 10 in

SOLUTIONS

1. The effective span length is given by MSJC Sec. 1.13 as

$$L_e = L_s$$
$$= l + b_s$$
$$= 12 \text{ ft} + 1 \text{ ft}$$
$$= 13 \text{ ft}$$

The maximum moment occurs at point P, the point of application of the concentrated load, a distance $a = 2.5$ ft from the center of the left-hand support.

The beam self-weight is

$$w = \left(80 \, \frac{\text{lbf}}{\text{ft}^2}\right)(4 \text{ ft})$$
$$= 320 \text{ lbf/ft}$$

The bending moment produced by this self-weight at point P is

$$M_s = \frac{wa(L_e - a)}{2}$$
$$= \frac{\left(320 \, \frac{\text{lbf}}{\text{ft}}\right)(2.5 \text{ ft})(13 \text{ ft} - 2.5 \text{ ft})}{(2)\left(1000 \, \frac{\text{lbf}}{\text{kip}}\right)}$$
$$= 4.2 \text{ ft-kips}$$

The bending moment produced by the concentrated dead load at point P is

$$M_D = \frac{Da(L_e - a)}{L_e}$$
$$= \frac{(33 \text{ kips})(2.5 \text{ ft})(13 \text{ ft} - 2.5 \text{ ft})}{13 \text{ ft}}$$
$$= 66.63 \text{ ft-kips}$$

The bending moment produced by the concentrated live load at point P is

$$M_L = \frac{La(L_e - a)}{L_e}$$
$$= \frac{(10 \text{ kips})(2.5 \text{ ft})(13 \text{ ft} - 2.5 \text{ ft})}{13 \text{ ft}}$$
$$= 20.19 \text{ ft-kips}$$

The total moment at point P is

$$\begin{aligned} M &= M_s + M_D + M_L \\ &= 4.2 \text{ ft-kips} + 66.63 \text{ ft-kips} \\ &\quad + 20.19 \text{ ft-kips} \\ &= 91.02 \text{ ft-kips} \quad (91 \text{ ft-kips}) \end{aligned}$$

The answer is (C).

2. The total moment at point P is derived in Prob. 1 as

$$M = 91.02 \text{ ft-kips}$$

The relevant parameters of the beam are

$$\begin{aligned} b_w &= 7.63 \text{ in} \\ d &= 45 \text{ in} \\ f'_m &= 1500 \ \frac{\text{lbf}}{\text{in}^2} \\ f_y &= 60{,}000 \ \frac{\text{lbf}}{\text{in}^2} \\ L_e &= 13 \text{ ft} \end{aligned}$$

The allowable stresses are

$$\begin{aligned} F_b &= \frac{f'_m}{3} \\ &= \frac{1500 \ \frac{\text{lbf}}{\text{in}^2}}{3} \\ &= 500 \ \frac{\text{lbf}}{\text{in}^2} \\ F_s &= 24{,}000 \text{ lbf/in}^2 \end{aligned}$$

The modular ratio is

$$\begin{aligned} n &= \frac{E_s}{E_m} \\ &= \frac{29{,}000{,}000 \ \frac{\text{lbf}}{\text{in}^2}}{900 f'_m} \\ &= \frac{29{,}000{,}000 \ \frac{\text{lbf}}{\text{in}^2}}{(900)\left(1500 \ \frac{\text{lbf}}{\text{in}^2}\right)} \\ &= 21.48 \end{aligned}$$

Assuming the reinforcement consists of two no. 7 bars, the tension reinforcement ratio is

$$\begin{aligned} \rho &= \frac{A_s}{b_w d} \\ &= \frac{1.20 \text{ in}^2}{(7.63 \text{ in})(45 \text{ in})} \\ &= 0.00349 \\ \rho n &= (0.00349)(21.48) \\ &= 0.0750 \end{aligned}$$

Using this value of ρn, the neutral axis depth factor is obtained from Table 6.1 as

$$k = 0.3197$$

The lever-arm factor is

$$\begin{aligned} j &= 1 - \frac{k}{3} \\ &= 1 - \frac{0.3197}{3} \\ &= 0.893 \end{aligned}$$

The moment capacity of the section is the lesser of

$$\begin{aligned} M_s &= F_s j \rho b_w d^2 \\ &= \frac{\left(24{,}000 \ \frac{\text{lbf}}{\text{in}^2}\right)(0.893)(0.00349)(7.63 \text{ in})(45 \text{ in})^2}{\left(12 \ \frac{\text{in}}{\text{ft}}\right)\left(1000 \ \frac{\text{lbf}}{\text{kip}}\right)} \\ &= 96.31 \text{ ft-kips} \\ M_m &= \frac{F_b j k b_w d^2}{2} \\ &= \frac{\left(500 \ \frac{\text{lbf}}{\text{in}^2}\right)(0.893)(0.3197)(7.63 \text{ in})(45 \text{ in})^2}{(2)\left(12 \ \frac{\text{in}}{\text{ft}}\right)\left(1000 \ \frac{\text{lbf}}{\text{kip}}\right)} \\ &= 91.90 \text{ ft-kips} \quad (92 \text{ ft-kips}) \quad [\text{governs}] \\ &> M \quad [\text{satisfactory}] \end{aligned}$$

Two no. 7 bars are adequate.

The answer is (A).

3. The critical section for shear occurs at a distance of $d/2$ from the face of the support. This is a distance from the left-hand effective support given by

$$x = \frac{b_s + d}{2}$$
$$= \frac{1 \text{ ft} + 3.75 \text{ ft}}{2}$$
$$= 2.38 \text{ ft}$$

The shear force produced by the beam self-weight at a distance x from the left-hand effective support is

$$V_s = w\left(\frac{L_e}{2} - x\right)$$
$$= \left(0.320 \ \frac{\text{kips}}{\text{ft}}\right)\left(\frac{13 \text{ ft}}{2} - 2.38 \text{ ft}\right)$$
$$= 1.32 \text{ kips}$$

The shear force produced by the concentrated dead load at a distance x from the left-hand effective support is

$$V_D = \frac{D(L_e - a)}{L_e}$$
$$= \frac{(33 \text{ kips})(13 \text{ ft} - 2.5 \text{ ft})}{13 \text{ ft}}$$
$$= 26.65 \text{ kips}$$

The shear force produced by the concentrated live load at a distance x from the left-hand effective support is

$$V_L = \frac{L(L_e - a)}{L_e}$$
$$= \frac{(10 \text{ kips})(13 \text{ ft} - 2.5 \text{ ft})}{13 \text{ ft}}$$
$$= 8.08 \text{ kips}$$

The total shear force at a distance x from the left-hand effective support is

$$V = V_s + V_D + V_L$$
$$= 1.32 \text{ kips} + 26.65 \text{ kips} + 8.08 \text{ kips}$$
$$= 36.05 \text{ kips}$$

The shear stress at the critical section is given by MSJC Eq. (2-23) as

$$f_v = \frac{V}{b_w d} = \frac{(36.05 \text{ kips})\left(1000 \ \frac{\text{lbf}}{\text{kip}}\right)}{(7.63 \text{ in})(45 \text{ in})}$$
$$= 105 \text{ lbf/in}^2$$

The allowable shear stress without shear reinforcement is given by MSJC Eq. (2-24) as

$$F_v = \sqrt{f'_m}$$
$$= \sqrt{1500 \ \frac{\text{lbf}}{\text{in}^2}}$$
$$= 38.7 \ \frac{\text{lbf}}{\text{in}^2}$$
$$< f_v \quad \text{[shear reinforcement is required]}$$

The shear stress with shear reinforcement provided to carry the total shear force is limited by MSJC Eq. (2-27) to

$$F_v = 3\sqrt{f'_m}$$
$$= 3\sqrt{1500 \ \frac{\text{lbf}}{\text{in}^2}}$$
$$= 116 \ \frac{\text{lbf}}{\text{in}^2}$$
$$\leq 150 \ \frac{\text{lbf}}{\text{in}^2} \quad \text{[satisfies MSJC Sec. 2.3.5.2.3]}$$
$$> f_v \quad \text{[satisfactory]}$$

The minimum area of shear reinforcement required is given by MSJC Eq. (2-30) as

$$\frac{A_v}{s} = \frac{V}{F_s d}$$
$$= \frac{(36.05 \text{ kips})\left(12 \ \frac{\text{in}}{\text{ft}}\right)}{\left(24 \ \frac{\text{kips}}{\text{in}^2}\right)(45 \text{ in})}$$
$$= 0.401 \text{ in}^2/\text{ft}$$

Providing no. 5 grade 60 stirrups at 8 in centers supplies a value of

$$\frac{A_v}{s} = 0.470 \ \frac{\text{in}^2}{\text{ft}}$$
$$> 0.401 \text{ in}^2/\text{ft} \quad \text{[satisfactory]}$$

The answer is (D).

4. The effective column width is

$$b = 7.63 \text{ in}$$

The effective column height is

$$b = 12 \text{ ft}$$

The reinforcement area is

$$A_s = 0.80 \text{ in}^2$$

The effective column area is

$$\begin{aligned} A_s &= b^2 \\ &= (7.63 \text{ in})^2 \\ &= 58 \text{ in}^2 \end{aligned}$$

The reinforcement ratio is

$$\begin{aligned} \rho &= \frac{A_s}{A_n} \\ &= \frac{0.80 \text{ in}^2}{58 \text{ in}^2} \\ &= 0.014 \\ &< 0.04 \\ &> 0.0025 \quad \text{[satisfies MSJC Sec. 1.14.1.2]} \end{aligned}$$

The allowable reinforcement stress is

$$\begin{aligned} F_s &= 0.4 f_y \\ &= (0.4)\left(60 \ \frac{\text{kips}}{\text{in}^2}\right) \\ &= 24 \text{ kips/in}^2 \end{aligned}$$

The radius of gyration of the column is

$$\begin{aligned} r &= \sqrt{\frac{I_n}{A_n}} = \sqrt{\frac{b^2}{12}} \\ &= (0.289)(7.63 \text{ in}) \\ &= 2.21 \text{ in} \end{aligned}$$

The slenderness ratio of the column is

$$\begin{aligned} \frac{h}{r} &= \frac{(12 \text{ ft})\left(12 \ \frac{\text{in}}{\text{ft}}\right)}{2.21 \text{ in}} \\ &= 65.16 \\ &< 99 \quad \text{[MSJC Eq. (2-20) is applicable]} \end{aligned}$$

The allowable column load is given by

$$\begin{aligned} P_a &= (0.25 f'_m A_n + 0.65 A_{st} F_s)\left(1.0 - \left(\frac{h}{140r}\right)^2\right) \\ &= \begin{pmatrix} (0.25)\left(1.5 \ \frac{\text{kips}}{\text{in}^2}\right)(58 \text{ in}^2) \\ + (0.65)(0.80 \text{ in}^2)\left(24 \ \frac{\text{kips}}{\text{in}^2}\right) \end{pmatrix} \\ &\quad \times \left(1.0 - \left(\frac{65.16}{140}\right)^2\right) \\ &= 26.81 \text{ kips} \quad (27 \text{ kips}) \\ &> P \quad \text{[satisfactory]} \end{aligned}$$

The column is adequate.

The answer is (C).

5. As specified by MSJC Sec. 1.14.1.3, lateral ties for the confinement of longitudinal reinforcement cannot be less than $1/4$ in diameter. The spacing must not exceed the lesser of

$$\begin{aligned} s &= 48 \text{ lateral tie diameters} \\ &= (48)(0.375 \text{ in}) \\ &= 18 \text{ in} \end{aligned}$$

$$\begin{aligned} s &= \text{the least cross-sectional dimension of the column} \\ &= 8 \text{ in} \end{aligned}$$

$$\begin{aligned} s &= 16 \text{ longitudinal bar diameters} \\ &= (16)(0.5 \text{ in}) \\ &= 8 \text{ in} \quad \text{[governs]} \end{aligned}$$

The answer is (A).

Index

A

Action
  composite, 3-20
  factor, group, 5-23 (tbl), 5-28
  full composite, 3-22, 4-28, 4-31
  partial composite, 4-31
Adjustment factor, 5-23 (tbl)
  glued laminated timber, 5-3
  sawn lumber, 5-2
Ambient relative humidity, 3-19
Analogous truss model, 1-1
Analysis
  method, plastic, 4-2
  of masonry beam, 6-9 (fig)
Anchor
  bar, 1-6 (fig)
  bolt, 6-24
  bolt, headed, 6-24
  bolt in shear, 6-25, 6-26
  bolt in tension, 6-24
  seating loss, 3-17
  set, 6-27
Anchorage
  length, 1-3, 2-6 (fig)
  slip, 3-17
Applied
  load, 3-24
  torque, 3-12
  torsion, 3-11
Area
  bearing factor, 5-2 (tbl), 5-5
  bonded reinforcement, 3-5 (fig)
  effective reinforcement, 6-18
  for shear, stirrup, 3-11
  for torsion, stirrup, 3-11, 3-12
  maximum reinforcement, 6-19, 6-20
  maximum tensile reinforcement, 6-20
  method, elastic unit, 4-12
  minimum for reinforcement
    at interface, 2-9
  minimum shear reinforcement, 6-12
  overlap, 6-24
  overlap, projected, 6-24, 6-25 (fig)
  reinforcing bar, 6-4
  shear reinforcement,
    minimum, 6-14, 6-15
  transformed flange, 3-20
  unbonded tendons, 3-5 (fig)
Arm, lever, 6-9
ASD method
  bolt group loaded normal to faying
    surface, 4-17 (fig)
  composite beam design, 4-30
  eccentrically loaded bolt group, 4-13 (fig)
  eccentrically loaded weld
    group, 4-21 (fig)
  instantaneous center of
    rotation, 4-15 (fig), 4-24
  weld group loaded normal to faying
    surface, 4-17 (fig)
Auxiliary reinforcement, 3-2, 3-4, 3-6
Axial
  capacity, bolt design, 6-26
  compression, 6-13
  compressive strength, allowable, 6-13
  load, design for, 5-11
  load, plastic, 4-10
  strength, design, 4-11

  tension, design for, 5-16
Axially loaded column, 2-4
Axis
  depth, neutral, 6-4, 6-6, 6-9, 6-10, 6-20
  neutral, 6-4, 6-6, 6-9, 6-10, 6-20
  plastic neutral, 4-28, 4-29 (fig)

B

B region, 1-1, 1-2 (fig)
Balancing load, 3-24
Band width, 2-8 (fig), 2-15
  transverse reinforcement, 2-8 (fig)
Bar
  anchor, 1-6 (fig)
  reinforcing, diameter, 6-4
  reinforcing, size, 6-4
Base of column, force transfer, 2-9
Beam
  analysis of masonry, 6-9 (fig)
  bracing, 4-6
  composite, 3-20, 3-23, 4-27, 4-28, 4-30
  continuous, 4-4
  deep, 1-1, 1-3
  design, ASD method composite, 4-30
  design, LRFD method composite, 4-28
  dimensional limitation, 6-4
  fully composite, at ultimate
    load, 4-29 (fig)
  mechanism, 4-7, 4-8 (fig)
  nonuniform, 4-5 (fig)
  notched, 5-19, 5-20 (fig)
  notched, glued laminated, 5-20 (fig)
  notched, lumber, 5-20
  region, 1-1
  reinforcement requirement, 6-4
  stability factor, 5-2 (tbl), 5-6
  strap, 2-16, 2-17 (fig), 2-19, 2-20
  theory, 1-1
  torsion in rectangular, 1-8 (fig)
Bearing
  area factor, 5-2 (tbl), 5-5
  capacity, 2-9
  plate, 6-27 (fig), 6-28
  pressure, soil, 2-17 (fig)
  pressure, uniform, 2-11 (fig)
  strength, 2-9
  type bolt, 4-15, 4-16, 4-17
Behavior, ductile, 6-15, 6-32
Bending moment
  diagram, free, 4-4, 4-5 (fig)
  column, 2-4
  primary, 3-26
  secondary, 3-27
Block
  equivalent rectangular stress, 6-32
  equivalent stress, 6-32, 6-33
  rectangular stress, 6-27
  stress, 3-1, 3-2, 4-28, 4-30, 6-19, 6-32
Bolt
  anchor, 6-24
  anchor, in shear, 6-25, 6-26
  anchor, in tension, 6-24
  bearing type, 4-15, 4-16, 4-17
  design, axial capacity, 6-26
  design, shear capacity, 6-25, 6-26
  edge distance, 5-25

  group, eccentrically loaded, 4-11,
    4-13, 4-16
  group, eccentrically loaded, ASD
    method, 4-13 (fig)
  group, eccentrically loaded, LRFD
    method, 4-12 (fig)
  group loaded normal to faying surface,
    ASD method, 4-17 (fig)
  group loaded normal to faying surface,
    LRFD method, 4-16 (fig)
  hole size, 5-24
  spacing, 5-26 (fig)
Bolted connection, 5-21 (fig), 5-25
Bond
  breaker, 6-27 (fig), 6-28
  stress, 6-4
Bonded
  reinforcement area, 3-5 (fig)
  reinforcement, auxiliary, 3-5
  tendons, 3-3, 3-4
Bottle-shaped strut, 1-2, 1-3 (tbl)
Bound, upper, 4-9
Brace, lateral, 5-7 (fig), 5-12 (fig)
Braced
  column, 4-10
  frame, 4-11
Bracing, beam, 4-6
Bracket, 1-5
Breaker bond, 6-27 (fig), 6-28
Breakout masonry, 6-24, 6-25
Brittle
  failure, 6-11, 6-13
  shear failure, 6-11
Buckling
  factor, 5-2 (tbl)
  length coefficient, 5-11 (fig)
β_1, 3-2
β_n, 1-3
  factor, 1-3
  value, 1-3 (tbl)
β_s, 1-2, 1-3
  factor, 1-2, 1-3 (tbl)

C

C-C
  -C node, 1-2 (fig), 1-3
  -T node, 1-2 (fig), 1-3
C_1, 4-23
Cable profile, 3-25 (fig), 3-26
Cantilever bracket, 1-5
Capacity
  bearing, 2-9
  bolt design axial, 6-24
  bolt design shear, 6-25, 6-26
  nominal flexural, 4-29, 4-30
  shear, 3-8, 3-10
Category seismic design, 6-15
Cell load, 6-28
Center
  instantaneous, 4-14, 4-15, 4-25, 4-26
  of rotation, instantaneous, ASD
    method, 4-15 (fig), 4-24 (fig)
  of rotation, instantaneous, LRFD
    method, 4-14 (fig)
Central band width, 2-8
Characteristic, elastic-plastic, 4-2
Check, static equilibrium, 4-9

Clear span, 6-2
Closed
    maximum spacing, 3-13
    stirrup, 1-9, 3-11, 3-13
    ties, 1-6
Coefficient
    buckling length, 5-11 (fig)
    friction, 1-6, 3-22
Collapse mechanism, 4-5, 4-7
    independent, 4-7, 4-8 (fig)
    mode, 4-9
Column
    axial load, 2-4
    base, force transfer to, 2-9
    bending moment, 2-4
    braced, 4-10
    depth, minimum, 6-12
    design requirements, 4-10
    detail, 6-13 (fig), 6-37
    dimensional limitation, 6-12
    reinforcement, maximum, 6-12
    reinforcement, minimum, 6-12
    reinforcement requirement, 6-12
    slenderness parameter, 4-10
    slenderness ratio, 4-11
    stability factor, 5-2 (tbl), 5-11
    unbraced, 4-10
    width, minimum, 6-12
Combination load, 4-4
Combined
    compression and flexure, 5-13, 6-28
    footing, 2-10
    lateral and withdrawal loads,
      5-27 (fig), 5-30, 5-31
    mechanism, 4-7, 4-8 (fig)
    tension and flexure, 5-16
    tension and shear, 6-26
Compatibility torsion, 1-9, 3-12
Composite
    action, 3-20
    action, full, 3-22, 4-28, 4-31
    action, partial, 4-31
    beam, 3-20, 3-23, 4-27, 4-28, 4-30
    beam design, ASD method, 4-30
    beam design, LRFD method, 4-28
    construction, 3-19
    section, 3-20 (fig), 3-23, 4-29 (fig)
    section properties, 3-20
Compression
    and flexure, combined, 5-13
    concrete strut, 3-12
    controlled, 3-2, 3-3 (fig)
    diagonal, concrete, 1-10, 3-13
    fiber, extreme, 3-8
    reinforcement, 3-2, 3-4
    strut, 1-2 (fig)
    zone factor, 3-2
Compressive
    strain, masonry, 6-6, 6-26, 6-28
    strain, maximum useable, 3-2
    strength, 1-2
Concentrated load, 3-8, 4-31
    shear from, 5-18 (fig)
Concordant
    cable, 3-27
    cable profile, 3-26
    profile, 3-27 (fig)
Concrete
    compression diagonal, 1-10, 3-13
    compression strut, 1-9, 3-12
    flange, 3-20
    girder, precast prestressed, 3-20
    strut, 1-1, 1-2 (fig)
Condition, governing load, 6-1
Confining reinforcement, 1-2, 1-5
Connection, 5-22
    bolted, 5-21 (fig)
    design of, 5-22
    lag screw, 5-26
    shear, 3-22, 5-21
    shear, ASD method, 4-32
    shear, LRFD method, 4-31
    shear plate, 5-28

split ring, 5-28
toe-nail, 5-31 (fig)
wood screw, 5-29
Connector, 4-31
    shear, 4-28, 4-31
    shear stud, 4-31, 4-32
Construction
    composite, 3-19
    load, 5-4 (tbl)
    prestressed masonry, 6-26, 6-27 (fig)
    shored, 3-23, 4-29, 4-30
    unshored, 3-23, 4-29, 4-30
Continuous beam, 4-4
Control
    compression, 3-2, 3-3 (fig)
    tension, 3-2, 3-3 (fig)
Corbel, 1-5, 1-6
Correction factor, 3-22, 4-23 (tbl)
    electrode strength, 4-23 (tbl), 4-24
Corrosion
    -inhibiting material, 6-27
    protection, 6-27
Coupler, 6-27 (fig), 6-30
    threaded, 6-28
Crack
    diagonal, 3-12, 6-11
    flexural, 3-3
    shear, 6-10, 6-11
    torsion, 3-12
Cracked
    moment, 6-21, 6-23
    moment, inertia, 6-21
    transformed section, 6-21, 6-23
Cracking, 3-12
    moment, 3-3
    torsion, 1-8, 3-11, 3-12
    torsional, 1-8, 3-11
Creep
    loss, 3-18
    rate of, 3-18
Critical
    perimeter, 2-4 (fig), 2-12 (fig)
    section, flexural shear, 2-6 (fig), 2-14
    section for flexure, 2-6 (fig)
    section for shear, 3-7 (fig)
    section for torsion, 3-11
    shear section, 6-11
Curvature factor, 5-2 (tbl), 5-3

D

D region, 1-1, 1-2 (fig)
Dead load, 3-5, 3-6, 5-4 (tbl)
    superimposed, 3-18, 3-23
Decking, 5-1
Deep beam, 1-1, 1-3
Deflection, 3-25 (fig)
    elastic, deformation, 6-28
    maximum permissible, 6-23
    service, 6-23
    service load, 6-23, 6-24
Deformation, elastic, 6-30
Depth
    effective, 1-6
    effective embedment, 6-24, 6-25
    factor, penetration, 5-23 (tbl)
    minimum column, 6-12
    neutral axis, 6-6, 6-20, 6-21, 6-35
Design
    axial capacity, 6-26
    axial strength, 4-11
    composite beam, ASD method, 4-30
    composite beam, LRFD method, 4-28
    flexural capacity, 4-29
    flexural strength, 3-3
    for axial tension, 5-16
    for flexure, 3-23, 5-6
    for shear, 3-7, 5-21, 6-10, 6-12
    for torsion, 3-7, 3-11
    mechanism, 4-4, 4-9
    method, statical, 4-4
    of connections, 5-22

plastic, 4-1
requirements, column, 4-10
service load, 3-20
shear bolt capacity, 6-26
shear, maximum, 6-11
shear strength, 3-8
stage, serviceability, 6-27, 6-28
stage, strength, 6-28, 6-32
stage, transfer, 6-27, 6-30
statical, 4-4, 4-5 (fig)
strength, 6-1, 6-9, 6-18
strength, flexural, 6-19
value, lag screw lateral, 5-26
value, lag screw withdrawal, 5-27
value, nail lateral, 5-29
value, reference, 5-2, 5-23
value, wood screw lateral, 5-29
value, wood screw withdrawal, 5-30
Detail
    column, 6-12 (fig)
    shear wall reinforcement, 6-15 (fig)
Detailed plain masonry shear walls, 6-14
Determinate, statically, 4-4
Determination, shape factor, 4-2 (fig)
Development length, 1-3, 2-6
    reinforcing bar, 6-5
    straight, 1-3
Diagonal
    concrete compression, 1-10, 3-13
    crack, 3-12, 6-11
Diagram free-body, 3-24 (fig)
Diameter
    minimum longitudinal
      reinforcement, 3-13
    reinforcing bar, 6-4, 6-5
Diaphragm factor, 5-23 (tbl), 5-24
Dimension, lumber, 5-1
Dimensional limitation
    beam, 6-4
    column, 6-12
Direct tension indicator washer, 6-28
Discontinuity region, 1-1, 1-2 (fig)
Displacement
    imaginary, 4-8
    virtual, 4-8
Distance
    bolt edge, 5-25
    shear plate, 5-28
    split ring, 5-28
Distributed load, shear from, 5-18 (fig)
Distribution
    elastic stress, 4-17 (fig)
    equivalent rectangular stress, 6-32
    linear strain, 3-1
    plastic stress, 4-16 (fig)
    reinforcement, 2-6
    strain, 3-1, 6-6 (fig), 6-20
    stress, 6-6 (fig)
Dowel, 2-9
Dressed size, 5-1
Ductile
    behavior, 6-15, 6-32
    failure, 6-32
Ducts out-of-straightness, 3-16
Duration
    factor, impact load, 5-23
    factor, load, 5-1, 5-2, 5-4 (tbl), 5-23
    impact load, 5-23

E

Earthquake load, 5-4
Eccentric
    load, footing with, 2-2 (fig)
    shear stress, 2-4
Eccentrically loaded
    bolt group, 4-11, 4-13, 4-16
    bolt group, ASD method, 4-13 (fig)
    bolt group, LRFD method, 4-12 (fig)
    weld group, 4-18, 4-19, 4-21
    weld group, ASD method, 4-21 (fig)
    weld group, LRFD method, 4-19 (fig)

Eccentricity, 2-1
   effecitve tendon, 3-27
   initial, 3-27
Edge
   distance, bolt, 5-25
   loaded, 5-26 (fig)
   unloaded, 5-21 (fig), 5-26 (fig)
Effect
   P-delta, 4-10, 6-18, 6-21, 6-23
   second-order, 4-10
   secondary, 3-27 (fig)
   slenderness, 6-29, 6-37
Effective
   -length factor, 4-11
   depth, 1-6
   embedment depth, 6-24
   embedment length, 6-24
   flange width, 3-21
   length, 5-7 (fig), 5-8
   prestressing force, 3-8
   reinforcement area, 6-18
   slab width, 4-28 (fig)
   span, 6-2
   span length, 5-6
   stress, 3-3, 3-4
   tendon eccentricity, 3-27
   width, 3-20
Elastic
   -plastic characteristics, 4-2
   -plastic material, 4-2 (fig)
   deformation, 6-30
   deformation deflection, 6-28
   shortening, 3-17
   shortening losses, 3-17
   stability, 6-30
   stress distribution, 4-17 (fig)
   unit area method, 4-12
   vector analysis technique, 4-19, 4-21, 4-24, 4-26
Elasticity, modulus of, 5-6, 6-21, 6-22
Electrode strength, correction factor, 4-23 (tbl), 4-24
Embedment
   depth, effective, 6-24, 6-25
   length, 6-24
End
   distance, shear plate, 5-28
   distance, split ring, 5-28
   grain factor, 5-23 (tbl), 5-24
   plate, 1-3
Energy, internal strain, 4-8
Envelope, moment, 4-4
Equilibrium
   check, static, 4-9
   method, static, 4-9
   relationships, 4-8
   torsion, 1-9, 3-12
Equivalent
   expression, interaction, 6-28
   rectangular stress distribution, 6-32
   reinforcement area, 6-23
   retangular stress block, 6-32
Expression, interaction, 4-10
   equivalent, 6-28
Extended nodal zone, 1-3 (fig)
External
   restraint, 4-8
   work, 4-8
Extreme
   compression fiber, 3-8
   tension fiber, 3-5

F

Factor
   adjustment, 5-2, 5-23 (tbl)
   adjustment for glued laminated timber, 5-3
   beam stability, 5-2 (tbl), 5-6
   bearing area, 5-2 (tbl), 5-5
   buckling, 5-2 (tbl)
   column stability, 5-2 (tbl), 5-11
   compression zone, 3-2
   correction, 3-22
   correction for electrode strength, 4-23 (tbl), 4-24 (tbl)
   curvature, 5-2 (tbl), 5-3
   determination, shape, 4-2 (fig)
   diaphragm, 5-23 (tbl), 5-24
   effective length, 4-11
   end grain, 5-23 (tbl), 5-24
   flat use, 5-2 (tbl), 5-5
   geometry, 5-23 (tbl)
   group action, 5-23 (tbl)
   incising, 5-2 (tbl), 5-3
   load, 4-3
   load duration, 5-1, 5-2, 5-4, 5-23 (tbl)
   metal side plate, 5-23 (tbl), 5-24
   penetration depth, 5-23 (tbl), 5-24
   repetitive member, 5-2 (tbl), 5-3
   sawn lumber adjustment, 5-2
   shape, 4-2
   size, 5-2 (tbl), 5-6
   stability, 5-6
   strength reduction, 1-6, 3-2
   temperature, 5-2 (tbl), 5-5, 5-23 (tbl)
   toe-nail, 5-23 (tbl), 5-24
   volume, 5-2 (tbl), 5-3, 5-9
   wet service, 5-2 (tbl), 5-5, 5-23 (tbl)
   β_n, 1-3
   β_s, 1-2, 1-3
Factored
   forces, 2-19
   load, 2-12 (fig), 2-18 (fig)
   moment, 2-20
   pressure, 2-12 (fig)
   shear force, 3-7, 3-22
   soil pressure, 2-3, 2-12, 2-18
   torque, 3-11
   torsional moment, 3-12
Failure
   brittle, 6-11
   brittle shear, 6-11
   ductile, 6-32
   masonry, 6-24 (fig)
   surface, spiral, 3-12
   tension, masonry, 6-24
Fastener, staggered, 5-23 (fig)
Faying surface
   bolt group loaded normal to, ASD method, 4-17 (fig)
   bolt group normal to, LRFD method, 4-16 (fig)
   bolt group, eccentrically loaded, 4-12 (fig)
   weld group, eccentrically loaded, 4-19
   weld group eccentrically loaded normal to, ASD method, 4-26 (fig)
   weld group eccentrically loaded normal to, LRFD method, 4-24, 4-25 (fig)
Feeler gauge, 6-28
Fixing moment line, 4-5 (fig)
Flange
   concrete, 3-20
   transformed area, 3-20
   width, effective, 3-21
   width, overhanging, 3-11 (fig)
Flanged section, 3-11
Flat use factor, 5-2 (tbl), 5-5
Flexural
   capacity, nominal, 4-29, 4-30
   crack, 3-3
   design strength, 3-3
   reinforcement, 3-8
   shear, 2-3, 2-5, 2-6, 2-12, 2-14, 2-18
   shear, critical section, 2-6 (fig), 2-14
   strength, 3-3, 3-4
   strength, design, 3-3
   strength, nominal, 3-2, 4-32, 6-28
   strength, wall, 6-18
Flexure, 2-3, 2-5, 2-6, 2-12, 2-14, 2-18, 2-20
   and compression, combined, 5-13
   and tension, combined, 5-16
   critical section for, 2-6 (fig)
   design for, 3-23, 5-6
Flow, shear, 1-8 (fig), 1-9, 3-11 (fig), 3-12
Footing, 2-1
   combined, 2-10
   pad, 2-16 (fig)
   rectangular, 2-1, 2-2
   reinforced concrete, 2-3, 2-12
   strap, 2-16, 2-17 (fig)
   transfer load to, 2-9 (fig)
   with eccentric load, 2-1, 2-2 (fig)
Force
   effective prestressing, 3-8
   factored, 2-19
   factored shear, 3-7, 3-22
   horizontal, 1-5, 1-6
   horizontal shear, 3-22
   jacking, 3-17
   redistribution of internal, 3-12
   shear, 1-6, 3-8, 3-22
   strut-and-tie, 1-3
   tensile, 1-5, 1-6
   transfer, column base, 2-9
   vertical, 1-6
Formation, plastic hinge, 4-3
Formula, Hankinson, 5-27, 5-30
Frame
   braced, 4-11
   sway, 4-10
Free
   -body diagram, 3-24 (fig)
   bending moment diagram, 4-4, 4-5 (fig)
Friction, 3-16
   coefficient, 1-6, 3-22
   loss, 3-16
   method, shear-, 3-22
   shear reinforcement, 1-6
Full composite action, 3-22, 4-28, 4-31
Fully composite section, 4-29 (fig)

G

Gable mechanism, 4-7, 4-8 (fig)
Gauge feeler, 6-28
Geometry factor, 5-23 (tbl)
Girder
   precast, 3-20
   precast prestressed concrete, 3-20
   propped, 3-23
Glued laminated timber, 5-2
   adjustment factor, 5-3
Governing condition, load, 6-1
Grade, 5-1
Graded lumber, 5-2
Grain factor, end, 5-23 (tbl), 5-24
Gravity load, unfactored, 6-20
Group action factor, 5-23 (tbl), 5-28
Gyration, radius of, 4-10, 6-13

H

Hankinson formula, 5-27, 5-30
Hardened steel
   threaded nail, 5-30
   washer, 6-28
Headed anchor bolt, 6-24
Hinge
   formation, plastic, 4-3, 4-4 (fig)
   last to form, 4-6, 4-10
   locations, 4-7 (fig)
   plastic, 4-3, 4-4 (fig), 4-7
   rotation, 4-6
Hole
   bolt, size, 5-24
   prebored, 5-30
Hook, 1-3
   standard, 6-11 (fig), 6-15
Horizontal
   closed stirrups, 1-6
   force, 1-5, 1-6
   reinforcement, 1-3
   shear, 3-20, 4-31, 4-32
   shear force, 3-22
   shear, reinforced, 6-15

shear strength, nominal, 3-22
tensile force, 1-5, 1-6
Hydrostatic nodal zone, 1-3

I

Imaginary displacement, 4-8
Impact load, 5-4 (tbl)
    duration, 5-23
Incising factor, 5-2 (tbl), 5-3
Independent
    collapse mechanism, 4-7, 4-8 (fig)
    mechanisms, number of, 4-7
Index, reinforcement, 3-4
Indicator washer, direct tension, 6-27 (fig)
Inertia
    cracked moment, 6-21, 6-23
    inspection port, 6-28
    instability, lateral, 6-12
    moment, 4-13, 4-17
    polar moment, 2-4, 4-12, 4-13, 4-19
Initial eccentricity, 3-27
Inspection port, inertia, 6-28
Instability, lateral, 6-12
    inertia, 6-12
Instantaneous center, 4-14, 4-15, 4-25, 4-26
    of rotation, 4-14
    of rotation, ASD method,
        4-15 (fig), 4-24 (fig)
    of rotation, LRFD method, 4-14 (fig)
Intentionally roughened interface, 3-22
Interaction
    equation, 5-13
    expression, 4-10
Interface
    intentionally roughened, 3-22
    minimum reinforcement area, 2-9
    smooth, 3-22
Intermediate reinforced masonry shear
        walls, 6-15
Internal
    force, redistribution of, 3-12
    strain energy, 4-8
    work, 4-8
Intertia, moment, 4-12

J

Jacking force, 3-17
Joint mechanism, 4-7, 4-8 (fig)
Joist, 5-1

L

Lag screw connection, 5-26
    lateral design value, 5-26
    withdrawal design value, 5-27
Laminated
    beam, notched glued, 5-20 (fig)
    timber, glued, adjustment factor for, 5-3
    timber, structural glued, 5-2
Lap splice, 6-5
Last hinge to form, 4-6, 4-10
Lateral
    and withdrawal load, nail, 5-30
    and withdrawal loads, combined,
        5-27 (fig), 5-30, 5-31
    brace, 5-7 (fig), 5-12 (fig)
    design value, lag screw, 5-26
    design value, nail, 5-29
    design value, wood screw, 5-29
    instability, 6-12
    support, 6-3, 6-12
    tie, 6-12, 6-13 (fig)
    tie, spacing, 6-13 (fig)
Laterally restrained, tendon, 6-28
Length
    anchorage, 1-3, 2-6 (fig)
    coefficient, buckling, 5-11 (fig)
    design span, 5-6
    development, 1-3, 2-6

effective, 5-7 (fig), 5-8
effective embedment, 6-24
straight development, 1-3
unbraced, 4-6, 4-10
unbraced, maximum, 4-6
Lever arm, 6-35
Limitation, dimensional
    beam, 6-4
    column, 6-12
Line
    fixing moment, 4-5 (fig)
    property, 2-11
Linear
    strain distribution, 3-1
    transformation, 3-27
Live load, 3-6
Load
    applied, 3-24
    axial, design for, 5-11
    balancing, 3-24 (fig)
    cell, 6-34
    combination, 4-4
    combined lateral and withdrawal,
        5-27 (fig), 5-30, 5-31
    concentrated, 3-8, 4-31
    concentrated, shear from, 5-18 (fig)
    condition governing, 6-1
    construction, 5-4 (tbl)
    dead, 3-5, 3-6, 5-4 (tbl)
    deflections, service, 6-23
    design for service, 3-20
    duration factor, 5-1, 5-2, 5-4, 5-23 (tbl)
    duration, impact, 5-23
    earthquake, 5-4 (tbl)
    eccentric, footing with, 2-1
    factor, 4-3
    factored, 2-12 (fig), 2-18 (fig)
    impact, 5-4 (tbl)
    live, 3-6
    nail lateral and withdrawal load, 5-30
    occupancy live, 5-4 (tbl)
    out-of-balance, 3-25
    service, 2-11, 2-16, 2-17 (fig)
    shear from concentrated, 5-18 (fig)
    shear from distributed, 5-18 (fig)
    snow, 5-4 (tbl)
    soil pressure for service, 2-11 (fig)
    superimposed dead, 3-18, 3-23
    transfer of, to footing, 2-9 (fig)
    ultimate, 3-2, 4-4
    ultimate, fully composite beam, 4-29 (fig)
    unfactored, 6-23
    unfactored gravity, 6-20
    wind, 5-4 (tbl)
Loaded
    eccentrically, bolt group,
        ASD method, 4-13 (fig)
    eccentrically, bolt group,
        LRFD method, 4-12 (fig)
    eccentrically, weld group,
        ASD method, 4-21 (fig)
    eccentrically, weld group,
        LRFD method, 4-19 (fig)
    edge, 5-26 (fig)
    normal to faying surface, bolt group,
        ASD method, 4-17, 4-17 (fig)
    normal to faying surface, bolt group,
        LRFD method, 4-16 (fig)
    normal to faying surface, weld group,
        ASD method, 4-26 (fig)
    normal to faying surface, weld group,
        LRFD method, 4-24, 4-25 (fig)
Loading, out-of-plane, 6-18, 6-19 (fig),
        6-20, 6-21 (fig), 6-22
Locations, hinge, 4-7 (fig)
Longitudinal reinforcement, 1-9, 3-12
    maximum spacing, 1-9, 3-14
    minimum area, 3-13
    minimum diameter, 1-9, 3-13
    torsional, 3-11
Loss
    anchor seating, 3-17
    creep, 3-18

elastic shortening, 3-17
friction, 3-16
prestress, 3-3, 3-15, 6-27, 6-28
relaxation, 3-19, 6-27
shrinkage, 3-18, 3-19, 6-27
LRFD method
    bolt group loaded normal to faying
        surface, 4-16 (fig)
    composite beam design, 4-28
    eccentrically loaded bolt group, 4-12 (fig)
    eccentrically loaded
        weld group, 4-19 (fig)
    instantaneous center of rotation,
        4-14, 4-14 (fig), 4-22, 4-23
    shear connection, 4-31
    weld group loaded normal to faying
        surface, 4-25 (fig)
Lumber, 5-2
    dimension, 5-1
    mechanically graded, 5-2
    sawn, adjustment factor, 5-2
    visually stress-graded, 5-2

M

Masonry
    beam, analysis, 6-9 (fig)
    breakout, 6-24, 6-25
    compressive, strain, 6-26
    construction, prestressed, 6-27 (fig)
    failure, 6-24 (fig)
    member, prestressed, 6-27
    shear walls, special reinforced, 6-14, 6-15
    strain, maximum, 6-27
    wall, slender, 6-18
Material
    corrosion-inhibiting, 6-27
    elastic-plastic, 4-2 (fig)
Maximum
    area, tensile reinforcement, 6-20
    column reinforcement, 6-12
    design shear, 6-11
    factored torsional moment, 3-12
    masonry strain, 6-27
    permissible deflection, 6-23, 6-24
    reinforcement area, 6-19, 6-20
    reinforcement for walls, 6-19, 6-20 (fig)
    spacing, longitudinal reinforcement, 1-9
    spacing, reinforcement, 2-6, 6-15
    spacing, stirrup, 3-8, 3-10, 3-13
    spacing, tie, 3-22
    unbraced length, 4-6
    useable compressive strain, 3-2
Mechanical splice, 6-5
Mechanically graded lumber, 5-2
Mechanism
    beam, 4-7, 4-8 (fig)
    collapse, 4-5, 4-7
    combined, 4-7, 4-8 (fig)
    design method, 4-4, 4-7, 4-9
    gable, 4-7, 4-8 (fig)
    independent, 4-8
    joint, 4-7, 4-8 (fig)
    sway, 4-7, 4-8 (fig)
Member
    factor, repetitive, 5-2 (tbl), 5-3
    post-tensioned, 3-16, 3-17
    prestressed masonry, 6-27
    pretensioned, 3-17
Metal
    side plate factor, 5-23 (tbl), 5-24
    washer, 5-25
Method
    ASD, bolt group loaded normal to faying
        surface, 4-17 (fig)
    ASD, composite beam design, 4-30
    ASD, eccentrically loaded bolt
        group, 4-13 (fig)
    ASD, eccentrically loaded weld
        group, 4-21 (fig)
    ASD, instantaneous center of rotation,
        4-15 (fig), 4-24 (fig)

ASD, shear connection, 4-32
ASD, weld group loaded normal to faying
    surface, 4-26 (fig)
elastic unit area, 4-12
LRFD, bolt group loaded normal to
    faying surface, 4-16 (fig)
LRFD, composite beam design, 4-28
LRFD, eccentrically loaded bolt
    group, 4-12 (fig)
LRFD, eccentrically loaded weld
    group, 4-19 (fig)
LRFD, instantaneous center of
    rotation, 4-14, 4-22, 4-23
LRFD, shear connection, 4-31
LRFD, weld group loaded normal to
    faying surface, 4-25 (fig)
mechanism design, 4-4, 4-9
plastic, 4-2
shear-friction, 3-22
static equilibrium, 4-9
Minimum
    area, shear reinforcement, 6-14, 6-15
    column depth, 6-12
    column reinforcement, 6-12
    column width, 6-12
    diameter, longitudinal
        reinforcement, 3-13
    diamter, longitudinal reinforcement, 1-9
    reinforcement area at interface, 2-9
Mode, collapse, 4-9
Model
    analogous truss, 1-1
    strut-and-tie, 1-1, 1-2, 1-4 (fig)
Modular ratio, 3-20, 6-21
Modulus
    of elasticity, 5-6, 6-21, 6-22
    plastic, 4-2
    rupture, 3-2, 3-3, 6-22
Moment
    cracked, 6-21, 6-23
    cracking, 3-3
    diagram, free bending, 4-4, 4-5 (fig)
    envelope, 4-4
    factored, 2-20, 3-12
    free bending, 4-5 (fig)
    inertia, 4-12
    line, fixing, 4-5 (fig)
    nominal, 6-32
    of inertia, 4-13, 4-17
    of inertia, cracked, 6-21
    of inertia, polar, 2-4, 4-12, 4-13, 4-19
    of resistance, plastic, 4-2, 4-3, 4-5
    out-of-balance, 3-25
    plastic, 4-2
    primary bending, 3-26
    resisting, 4-2
    resultant, 3-27
    secondary bending, 3-27
    section, 6-23
    service, 6-23
    threshold torsional, 3-13
    yield, 4-2

N

Nail, 5-30
Neutral axis, 6-6
    depth, 6-6, 6-20, 6-21, 6-35
    plastic, 4-28, 4-29 (fig)
Nodal zone, 1-2 (fig), 1-3 (fig), 1-4
    extended, 1-3 (fig)
    hydrostatic, 1-3
Node, 1-3
    C-C-C, 1-2 (fig), 1-3
    C-C-T, 1-2 (fig), 1-3
Nominal
    flexural capacity, 4-29, 4-30
    flexural strength, 3-2, 4-32, 6-28, 6-32
    horizontal shear strength, 3-22
    moment, 6-32
    shear capacity, 3-8, 3-10
    shear strength, 3-8, 3-10

size, 5-2
strength, 3-4
Nonprestressed reinforcement, 3-4
Nonuniform beam, 4-5 (fig)
Normal to faying surface,
    bolt group loaded,
        ASD method, 4-17 (fig)
    bolt group loaded,
        LRFD method, 4-16 (fig)
    weld group loaded,
        ASD method, 4-25 (fig)
Notched beam, 5-19, 5-20 (fig)
    glued laminated, 5-20 (fig)
    lumber, 5-20
Number of independent mechanisms, 4-7

O

Occupancy live load, 5-4 (tbl)
One piece stirrup, 6-11 (fig)
Ordinary
    plain masonry shear walls, 6-14
    reinforced masonry shear walls, 6-15
Out-of
    -balance load, 3-25
    -balance moment, 3-25
    -plane loading, 6-18, 6-19 (fig),
        6-20, 6-21 (fig), 6-22
    -straightness ducts, 3-16
Outside perimeter, 3-11
Overhanging flange width, 3-11 (fig)
Overlap, projected areas, 6-25 (fig)

P

P-delta effect, 4-10, 6-18, 6-21, 6-23
Pad footing, 2-16 (fig)
Panel, wood structural, 5-2
Parabolic profile, 3-24
Parameter, column slenderness, 4-11
Partial composite action, 4-31
Penetration depth factor, 5-23 (tbl), 5-24
Perimeter
    critical, 2-4 (fig), 2-12 (fig)
    outside, 3-11
Permissible deflection, maximum, 6-24
Plastic
    analysis method, 4-2
    design, 4-1
    hinge, 4-3, 4-4 (fig), 4-7
    hinge formation, 4-3, 4-4 (fig)
    method, 4-2
    modulus, 4-2
    moment, 4-2
    moment of resistance, 4-2, 4-3, 4-5
    neutral axis, 4-28, 4-29 (fig)
    stress distribution, 4-16 (fig), 4-29 (fig)
    yielding, 4-2 (fig)
Plate, 1-3
    bearing, 6-27 (fig), 6-28
    end, 1-3
    shear connection, 5-28
Polar moment of
    inertia, 2-4, 4-12, 4-13, 4-19
Port, inspection, 6-28
Post-tensioned
    members, 3-16, 3-17
    slab, 3-3
Prebored hole, 5-30
Precast
    girder, 3-20
    prestressed concrete girder, 3-20
Pressure
    factored, 2-12 (fig)
    maximum value, soil, 2-2
    minimum value, soil, 2-2
    soil, 2-1, 2-11 (fig)
    soil bearing, 2-17 (fig)
    tendon, 3-24 (fig)
    uniform bearing, 2-11 (fig)
Prestress loss, 3-3, 3-15

Prestressed
    loss, 6-27, 6-28
    masonry construction, 6-27 (fig)
    masonry member, 6-27
    precast concrete girder, 3-20
    reinforcement, 3-4
    technique, masonry construction, 6-27
Pretensioned member, 3-17
Primary
    bending moment, 3-26
    reinforcement, 1-6 (fig)
    steel, 1-6
    tension reinforcement, 1-6
Principal tensile stress, 1-8, 3-11
Principle of virtual work, 4-8
Prism strut, 1-2 (fig)
Profile
    cable, 3-25 (fig), 3-26
    concordant, 3-27 (fig)
    concordant cable, 3-26
    parabolic, 3-24
    transformed, 3-27 (fig)
Projected area, overlap, 6-24, 6-25 (fig)
Property
    composite section, 3-20
    line, 2-11, 2-17 (fig)
Propped girder, 3-23
Protection, corrosion, 6-27
Pullout, 6-24
Punching shear, 2-3, 2-4 (fig), 2-5, 2-12 (fig)
Pure torsion, 3-12

R

Radius of gyration, 4-10, 6-13
Rate, creep, 3-18
Ratio
    maximum reinforcement, 6-20
    modular, 3-20, 6-21
    reinforcement, 1-2, 2-6, 6-6
    slenderness, 4-11, 5-11, 5-12 (fig)
    span-to-depth, 3-5
Reaction, support, 3-8, 3-26
Rectangular
    beam, torsion in, 1-8 (fig)
    footing, 2-1, 2-2
    stress block, 3-1, 3-2, 4-28,
        4-30, 6-27, 6-32
Redistribution, internal force, 3-12
Reduction factor, strength, 1-6, 3-2
Reference design value, 5-2, 5-23
Region
    B, 1-1, 1-2 (fig)
    beam, 1-1
    D, 1-1, 1-2 (fig)
    discontinuity, 1-1, 1-2 (fig)
Reinforced
    concrete footing, 2-3, 2-12
    concrete strap, 2-18
    horizontal shear, 6-15
    masonry, special shear walls, 6-15, 6-20
Reinforcement, 6-7
    area, bonded, 3-5 (fig)
    area, effective, 6-18
    area, maximum, 6-20
    auxiliary, 3-2, 3-4, 3-6
    auxiliary bonded, 3-5
    band width, 2-8
    band width, transverse, 2-8 (fig)
    compression, 3-2, 3-4
    confining, 1-2, 1-5
    details, shear wall, 6-15 (fig)
    distribution, 2-6
    equivalent, area, 6-23
    flexural, 3-8
    horizontal, 1-3
    index, 3-4
    longitudinal, 1-9, 3-12
    longitudinal, minimum diameter, 3-13
    longitudinal torsional, 3-11
    maximum column, 6-12
    maximum spacing, 2-6, 6-15

maximum spacing, shear, 6-15
minimum area at interface, 2-9
minimum spacing of, 1-9
nonprestressed, 3-4
prestressed, 3-4
primary, 1-6 (fig)
primary tension, 1-6
ratio, 1-2, 2-6, 6-6
ratio, maximum, 6-19, 6-20
requirement, beam, 6-4
requirement, column, 6-13
shear, 1-9, 3-8, 3-10, 6-11 (fig), 6-13
shear friction, 1-6
shear, minimum area, 6-15
shear, wall, 6-16
spacing, 1-9
tensile, maximum area, 6-19, 6-20
tension, 3-4
torsion, 1-9, 3-11
transverse, 3-12
vertical, 1-3
wall, maximum, 6-20 (fig)
Reinforcing bar
    development length, 6-5
    diameter, 6-5
    spacing, 6-4, 6-5
    splice length, 6-5
Relationship, equilibrium, 4-8
Relative humidity, 3-19
Relaxation loss, 3-19, 6-27
Repetitive member factor, 5-2 (tbl), 5-3
Requirement
    beam reinforcement, 6-4
    bolt spacing, 5-26 (fig)
    column design, 4-10
    column reinforcement, 6-13
Resistance, plastic moment of, 4-2, 4-3, 4-5
Resisting moment, 4-2
Restrained tendon, laterally, 6-28
Restraint, external, 4-8
Resultant moment, 3-27
Ring, split
    connection, 5-28
    distance, 5-28
    spacing, 5-28
Rotation
    hinge, 4-6
    instantaneous center of, 4-14, 4-15
    instantaneous center of,
        ASD method, 4-15 (fig), 4-24 (fig)
    instantaneous center of,
        LRFD method, 4-14 (fig)
Roughening of surface, 3-22
Rupture, modulus, 3-2, 3-3, 6-22

S

Sawn lumber
    adjustment factor, 5-2
    beam, notched, 5-20 (fig)
Screw
    connection, lag, 5-26
    connection, wood, 5-29
    lag, lateral design value, 5-26
    lag, withdrawal design value, 5-27
    wood, lateral design value, 5-29
    wood, withdrawal design value, 5-30
Second-order effect, 4-10
Secondary
    bending moment, 3-27
    effect, 3-27 (fig), 4-11
Section
    composite, 3-20 (fig), 3-23, 4-29 (fig)
    composite, properties, 3-20
    cracked, transformed, 6-23
    critical, flexural shear, 2-6 (fig)
    critical, for flexure, 2-6 (fig)
    critical, for shear, 6-11
    critical, for torsion, 3-11
    flanged, 3-11
    for shear, critical, 3-7 (fig)
    transformed, 3-20 (fig)

transformed composite, 4-28
Seismic design category, 6-17
Service
    factor, wet, 5-5, 5-23 (tbl)
    load, 2-11, 2-16, 2-17 (fig)
    load deflection, 6-23, 6-24
    load design, 3-20
    load, soil pressure for, 2-11 (fig)
    moment, 6-23
Set, 3-17
    anchor, 6-27
Shape factor, 4-2
    determination, 4-2 (fig)
Shear, 3-7
    -friction method, 3-22
    anchor bolt in, 6-25, 6-26
    capacity, 3-8, 3-10
    capacity, bolt design, 6-26
    capacity, nominal, 3-8, 3-10
    capacity, wall, 6-15
    combined tension and, 6-26
    connection, 3-22, 4-31, 5-21
    connection, ASD method, 4-32
    connection, LRFD method, 4-31
    connector, 4-28, 4-31
    crack, 6-10, 6-11
    critical section for, 3-7 (fig)
    design, 6-10, 6-12
    design for, 3-7, 5-21
    detail, reinforcement wall, 6-15
    flexural, 2-3, 2-5, 2-6, 2-12, 2-14, 2-18
    flow, 1-8 (fig), 1-9, 3-11 (fig), 3-12
    force, 1-6, 3-7, 3-8, 3-22
    force, horizontal, 3-22
    friction reinforcement, 1-6
    from concentrated loads, 5-18 (fig)
    from distributed loads, 5-18 (fig)
    horizontal, 3-20, 4-31, 4-32
    maximum design, 6-11
    minimum area, reinforcement, 6-15
    plate connection, 5-28
    plate distance, 5-28
    plate end distance, 5-28
    plate spacing, 5-28
    punching, 2-3, 2-4 (fig), 2-5, 2-12 (fig)
    reinforcement, 1-9, 3-8, 3-10, 6-11, 6-13
    reinforcement, maximum spacing, 6-15
    span, 1-5
    stirrup area for, 3-11
    strength, design, 3-8
    strength, nominal, 3-8, 3-10
    strength, nominal horizontal, 3-22
    stress, 1-8, 3-11
    stress, eccentric, 2-4
    stud connector, 4-31, 4-32
    walls, 6-16
    walls, detailed plain, 6-14
    walls, intermediate reinforced
        masonry, 6-15
    walls, ordinary plain masonry, 6-14
    walls, ordinary reinforced masonry, 6-15
    walls, reinforcement, 6-15 (fig), 6-16
    walls, special reinforced masonry, 6-15
    walls, unreinforced masonry, 6-14
Shored construction, 3-23, 4-29, 4-30
Shortening, elastic, 3-17
Shrinkage
    loss, 3-18, 3-19, 6-27
    strain, 3-18, 3-19
Side plate factor, metal, 5-23 (fig), 5-24
Size
    bolt, hole, 5-24
    dressed, 5-1
    factor, 5-2 (tbl), 5-6
    nominal, 5-2
    reinforcing bar, 6-4
Slab
    two-way, 3-5, 3-6
    unbonded post-tensioned, 3-3
    width, effective, 4-28 (fig)
Slender wall, masonry, 6-18
Slenderness
    effect, 6-29, 6-37

ratio, 4-11, 5-11, 5-12 (fig)
Slip, 3-17
    -critical bolt, 4-15
Snow load, 5-4 (tbl)
Soil
    bearing pressure, 2-17 (fig)
    pressure, 2-1, 2-11 (fig), 2-16
    pressure distribution, 2-1, 2-11, 2-16
    pressure, factored, 2-3, 2-12, 2-18
    pressure, factored loads for, 2-12 (fig)
    pressure, maximum value, 2-1
    pressure, maximum, minimum, value, 2-2
    pressure, minimum value, 2-1
    pressure, service loads for, 2-11 (fig)
    reaction, 2-16
Space truss, tubular, 1-9, 3-12
Spacing
    bolt, 5-26 (fig)
    lateral tie, 6-13 (fig)
    maximum for reinforcement, 2-6
    maximum tie, 3-22
    reinforcing bar, 6-4, 6-5
    shear plate, 5-28
    split ring, 5-28
Span
    -to-depth ratio, 3-5
    clear, 6-2
    effective, 6-2
    length, effective, 5-6
    shear, 1-5
Special reinforced masonry shear walls, 6-15
Spikes and nails, 5-30
Spiral failure surface, 3-12
Splice
    lap, 6-5
    length, reinforcing bar, 6-5
    mechanical, 6-5
    welded, 6-5
Split ring
    connection, 5-28
    distance, 5-28
    spacing, 5-28
Stability
    beam, factor, 5-2 (tbl)
    column, 5-11
    column, factor, 5-2 (tbl)
    elastic, 6-30
    factor, 5-6
    factor, beam, 5-6
Stage
    serviceability design, 6-27, 6-28
    strength design, 6-27, 6-28, 6-32
    transfer, 6-27, 6-30
    transfer design, 6-27, 6-30
Staggered fasteners, 5-23 (fig)
Standard hook, 6-11, 6-15
Static equilibrium
    check, 4-9
    method, 4-9
Statical design, 4-4, 4-5 (fig)
Statically determinate, 4-4
Steel
    hardened threaded nail, 5-30
    hardened washer, 6-28
    primary, 1-6
    tie, 1-1, 1-2 (fig)
Stirrup, 6-11
    area for shear, 3-11
    area for torsion, 3-11, 3-12
    area, minimum combined, 3-13
    closed, 1-9, 3-11, 3-13
    horizontal closed, 1-6
    one piece, 6-11 (fig)
    spacing, maximum, 3-8, 3-10
Strain
    distribution, 3-1, 6-6 (fig), 6-20
    internal energy, 4-8
    masonry compressive, 6-6, 6-28
    maximum masonry, 6-27
    maximum useable compressive, 3-2
    shrinkage, 3-18, 3-19
    sustained tensile, 3-19
    tensile, 3-2

Strap
  beam, 2-16, 2-17 (fig), 2-19, 2-20
  footing, 2-16, 2-17 (fig)
  reinforced, 2-18
Strength
  axial, design, 4-11
  bearing, 2-9
  compressive, 1-2
  correction factor for electrode,
      4-23 (tbl), 4-24 (tbl)
  design, 3-1, 6-1, 6-9, 6-18
  design, flexural, 3-3, 6-19
  design, stage, 6-27
  flexural, 3-3, 3-4
  flexural, nominal, 6-32
  flexural, wall, 6-18
  nominal, 3-4
  nominal flexural, 3-2, 4-32
  nominal horizontal shear, 3-22
  reduction factor, 1-6, 3-2
  tensile, 3-4, 3-8
Stress
  -graded lumber, 5-2
  block, 3-1, 3-2, 4-28, 4-30, 6-32
  block, equivalent, 6-32, 6-33
  block, rectangular, 4-28, 4-30, 6-27
  distribution, 6-6 (fig)
  distribution, elastic, 4-17 (fig)
  distribution, plastic, 4-16 (fig), 4-29 (fig)
  eccentric shear, 2-4
  effective, 3-3, 3-4
  principal tensile, 1-8, 3-11
  shear, 1-8, 3-11
  torsional, 1-9
  yield, 4-2, 6-20
Structural
  glued laminated timber, 5-2
  wood panel, 5-2
Strut, 1-1, 1-2 (fig), 1-3 (tbl)
  -and-tie, 1-1, 1-2
  -and-tie force, 1-3
  -and-tie model, 1-1, 1-2, 1-4 (fig)
  bottle-shaped, 1-2, 1-3 (tbl)
  compression, 1-2 (fig)
  concrete, 1-1, 1-2 (fig)
  concrete compression, 1-9, 3-12
  prism, 1-2 (fig)
Stud connector, shear, 4-31
Styrofoam, 2-16, 2-17 (fig)
Superimposed dead load, 3-23
Support
  lateral, 6-3, 6-12
  reaction, 3-8, 3-26
Surface
  roughening, 3-22
  spiral, failure, 3-12
Sustained tensile strain, 3-19
Sway
  frame, 4-10
  mechanism, 4-7, 4-8 (fig)

T

Technique
  elastic vector analysis, 4-19, 4-21,
      4-24, 4-26
  prestressed masonry construction, 6-27
Temperature factor, 5-2 (tbl), 5-5, 5-23 (tbl)
Tendon
  area, unbonded, 3-5 (fig)
  bonded, 3-3, 3-4
  effective eccentricity, 3-27
  laterally restrained, 6-28
  pressure, 3-24 (fig)
  restraint, 6-28
  unbonded, 3-5 (fig), 6-27
Tensile
  force, 1-5, 1-6
  force, horizontal, 1-5, 1-6
  reinforcement, maximum area, 6-19, 6-20
  reinforcement, primary, 1-6
  strain, 3-2

strain, sustained, 3-19
strength, 3-4, 3-8
stress, principal, 1-8
Tension
  anchor bolt in, 6-24
  and flexure, combined, 5-16
  and shear, combined, 6-26
  axial, design for, 5-16
  controlled, 3-2, 3-3 (fig)
  extreme, 3-5
  indicator washer, 6-28
  masonry failure, 6-24
  reinforcement, 3-4
  reinforcement only, 6-9
Theory, beam, 1-1
Thin-walled tube, 1-8 (fig), 3-11 (fig)
Threaded
  coupler, 6-28
  hardened-steel nail, 5-30
Threshold
  torsion, 1-8, 3-11
  torsional moment, 3-13
Tie, 1-1, 1-2 (fig), 3-22
  closed, 1-6
  lateral, 6-12, 6-13 (fig)
  lateral, spacing, 6-13 (fig)
  spacing, maximum, 3-22
  steel, 1-1, 1-2 (fig)
Timber
  adjustment factor for glued
      laminated, 5-3
  structural glued laminated, 5-2
Toe-nail, 5-31
  connection, 5-31 (fig)
  factor, 5-23 (tbl), 5-24
Torque
  applied, 3-12
  factored, 3-11
Torsion, 1-8, 3-7, 3-11
  applied, 3-11
  compatibility, 1-9, 3-12
  concentrated, 3-11
  crack, 3-12
  cracking, 1-8, 3-11
  critical section, 3-11
  design for, 3-11
  equilibrium, 1-9, 3-12
  pure, 3-12
  rectangular beam, 1-8 (fig)
  reinforcement, 1-9, 3-11
  stirrup area for, 3-11, 3-12
  threshold, 1-8, 3-11
Torsional
  cracking, 1-8, 3-11, 3-12
  effects, 3-11
  moment, maximum factored, 3-12
  reinforcement, 3-11
  reinforcement, longitudinal, 3-11
  stress, 1-9
Transfer, 3-17
  design stage, 6-27
  load to footing, 2-9 (fig)
  stage, 6-30
Transformation, linear, 3-27
Transformed
  flange area, 3-20
  profile, 3-27 (fig)
  section, 3-20 (fig)
  section, cracked, 6-21
Transition zone, 3-2, 3-3 (fig)
Transverse reinforcement, 3-12
  band width, 2-8 (fig)
Truss
  analogous model, 1-1
  tubular space, 3-12
Tube, thin-walled, 1-8 (fig), 3-11 (fig)
Tubular space truss, 1-9, 3-12
Two-way
  action, 2-5
  slab, 3-5, 3-6

U

Ultimate load, 3-2, 4-4
  fully composite beam, 4-29 (fig)
Unbonded
  post-tensioned slab, 3-3
  tendon, 3-5 (fig), 6-27
Unbraced
  column, 4-10
  length, 4-6, 4-10
Unfactored
  gravity load, 6-20
  load, 6-23
Uniform
  bearing pressure, 2-11 (fig)
  shear flow, 3-11
Unit area method, elastic, 4-12
Unloaded edge, 5-21 (fig), 5-26 (fig)
Unreinforced masonry shear walls, 6-14
Unshored construction, 3-23, 4-29, 4-30
Upper bound, 4-9

V

Value
  design, lag screw lateral, 5-26
  design, lag screw withdrawal, 5-27
  design, reference, 5-23
  design, wood screw lateral, 5-29
  design, wood screw withdrawal, 5-30
  nail lateral design, 5-29
  nail widthdrawal design, 5-30
  reference design, 5-2
Vertical
  force, 1-6
  reinforcement, 1-3
Virtual
  displacement, 4-8
  work principle, 4-8
Visually stress-graded lumber, 5-2
Volume factor, 5-2 (tbl), 5-3, 5-9

W

Wall
  maximum reinforcement for, 6-20 (fig)
  shear, 6-16
  shear, reinforcement, 6-15 (fig)
  shear, reinforcement detail, 6-15 (fig)
  slender masonry, 6-21
  with out-of-plane loading, 6-21 (fig)
Washer
  direct tension indicator, 6-28
  metal, 5-25
Weld group, eccentrically loaded, 4-18
      4-19, 4-21
  normal to faying surface, ASD
      method, 4-26 (fig)
  normal to faying surface, LRFD
      method, 4-25 (fig)
  ASD method, 4-21 (fig)
  LRFD method, 4-19 (fig)
Welded splice, 6-5
Wet service factor, 5-2 (tbl), 5-5, 5-23 (tbl)
Width
  band, 2-8 (fig), 2-15
  central band, 2-8
  effective slab, 4-28
  minimum column, 6-12
  overhanging flange, 3-11 (fig)
Wind load, 5-4 (tbl)
Withdrawal and lateral
  load, 5-27 (fig), 5-30, 5-31
  nail, 5-30
Wood
  screw connection, 5-29
  screw, lateral design value, 5-29
  screw, withdrawal design value, 5-30
  structural panel, 5-2
Work
  external, 4-8
  internal, 4-8

Y

Yield
  moment, 4-2
  stress, 4-2, 6-20
Yielding, plastic, 4-2 (fig)

Z

Zone
  factor, compression, 3-2
  hydrostatic nodal, 1-3
  nodal, 1-2 (fig), 1-3 (fig), 1-4
  nodal extended, 1-3 (fig)
  transition, 3-2, 3-3 (fig)

Index of Codes

A

ACI
 Eq. (9-10), 3-3
 Eq. (11-3), 2-6
 Eq. (11-18), 3-15
 Eq. (11-21), 1-10, 3-14
 Eq. (11-22), 1-11, 3-14
 Eq. (11-23), 1-10
 Eq. (11-24), 1-9
 Eq. (11-33), 2-5
 Eq. (18-1), 3-16
 Eq. (18-3), 3-5
 Eq. (A-3), 1-5
 Eq. (A-8), 1-5
 Sec. 7.12.2, 2-6, 2-7, 2-15
 Sec. 8.5, 3-17, 3-21
 Sec. 8.12, 3-20
 Sec. 9.3.2, 3-2
 Sec. 9.3.2.3, 3-9
 Sec. 9.3.2.6, 1-4
 Sec. 10.2, 1-7, 2-6, 3-1
 Sec. 10.2.7.1, 3-2
 Sec. 10.2.7.3, 2-7, 2-15, 3-2, 3-4
 Sec. 10.3.3, 3-2
 Sec. 10.3.4, 2-7, 2-15, 3-2, 3-5, 3-6
 Sec. 10.5.1, 2-20
 Sec. 10.5.3, 2-20
 Sec. 10.5.4, 2-6
 Sec. 10.14.1, 2-9, 2-10
 Sec. 11.1.1, 3-10
 Sec. 11.1.3, 3-8
 Sec. 11.1.3.1, 2-6, 2-14
 Sec. 11.2.1.1, 2-6, 2-14
 Sec. 11.3.1, 3-8
 Sec. 11.3.2, 3-8, 3-9
 Sec. 11.4.3, 3-8
 Sec. 11.4.5, 1-9, 3-8
 Sec. 11.4.5.1, 3-10
 Sec. 11.4.6.1, 1-10, 3-8, 3-9
 Sec. 11.4.6.3, 3-8, 3-10, 3-22
 Sec. 11.4.6.4, 3-8, 3-10
 Sec. 11.4.7.2, 3-10
 Sec. 11.4.7.9, 3-10
 Sec. 11.5.1, 1-8, 1-10, 3-11, 3-12
 Sec. 11.5.1.1, 3-11
 Sec. 11.5.2.2, 1-9
 Sec. 11.5.2.4, 1-8
 Sec. 11.5.2.5, 3-11
 Sec. 11.5.3.1, 1-10, 3-13, 3-15
 Sec. 11.5.3.6, 1-9, 3-12, 3-13, 3-14
 Sec. 11.5.3.7, 1-9, 3-13
 Sec. 11.5.3.8, 3-11
 Sec. 11.5.5.2, 1-9, 3-13, 3-14
 Sec. 11.5.5.3, 1-9, 3-13
 Sec. 11.5.6, 1-9
 Sec. 11.5.6.1, 1-11, 3-13, 3-14
 Sec. 11.5.6.2, 1-9, 3-13, 3-15
 Sec. 11.6.4, 3-22
 Sec. 11.6.4.3, 1-6, 3-22
 Sec. 11.7.1, 1-1
 Sec. 11.7.4, 1-3
 Sec. 11.7.5, 1-3
 Sec. 11.8, 1-5
 Sec. 11.8.2, 1-6
 Sec. 11.8.3.1, 1-6
 Sec. 11.8.3.4, 1-6
 Sec. 11.8.3.5, 1-6, 1-7
 Sec. 11.8.4, 1-6, 1-7
 Sec. 11.11.1.2, 2-4, 2-12
 Sec. 11.11.2.1, 2-5, 2-13
 Sec. 11.11.7.1, 2-4
 Sec. 12.2.2, 2-7
 Sec. 12.2.4, 2-7
 Sec. 12.3.2, 2-10
 Sec. 12.5.2, 1-5
 Sec. 12.5.3, 1-5
 Sec. 13.2.4, 3-11
 Sec. 13.5.3.2, 2-4
 Sec. 15.4.2, 2-6
 Sec. 15.4.4.2, 2-8, 2-15
 Sec. 15.5.2, 2-4, 2-6, 2-12, 2-14
 Sec. 15.8.2.1, 2-9, 2-10
 Sec. 17.2.4, 3-23
 Sec. 17.5.2, 3-22
 Sec. 17.5.3, 3-22
 Sec. 17.5.3.1, 3-22
 Sec. 17.5.3.2, 3-22
 Sec. 17.5.3.3, 3-22, 3-23
 Sec. 17.5.3.4, 3-22, 3-23
 Sec. 17.6.1, 3-22
 Sec. 18.6.2.1, 3-16
 Sec. 18.7.1, 3-1
 Sec. 18.7.2, 3-4, 3-5, 3-6
 Sec. 18.8.2, 3-3, 3-4, 3-6
 Sec. 18.9, 3-5
 Sec. 18.9.2, 3-5, 3-6
 Sec. 18.9.3.2, 3-6
 Sec. A.1, 1-3
 Sec. A.3.1, 1-2
 Sec. A.3.2, 1-2
 Sec. A.3.3, 1-2, 1-4
 Sec. A.4.1, 1-3
 Sec. A.4.3.2, 1-3
 Sec. A.5.1, 1-3
 Sec. R11.3.2, 3-8, 3-9
 Sec. R11.5, 1-8, 3-11
 Sec. R11.5.1, 1-8
 Sec. R11.5.3.10, 1-9, 3-13
 Sec. R11.5.3.8, 1-9
 Sec. R11.6.4.1, 1-7
 Sec. R11.11.7.2, 2-4, 2-5
 Sec. R15.8.1.1, 2-9

AISC
 App. 1.1, 4-4
 App. 1.2, 4-3
 App. 1.3, 4-3
 App. 1.5, 4-10, 4-11
 App. 1.7, 4-7
 Eq. (A-1-7), 4-6
 Eq. (F1-1), 4-7
 Eq. (H1-1b), 4-10
 Eq. (J3-2), 4-11
 Eq. (J3-3a), 4-17
 Eq. (J3-3b), 4-18
 Sec. B2, 4-4
 Sec. F2.2, 4-6
 Sec. F2.2b, 4-7
 Sec. I1.1a, 4-28, 4-30
 Sec. I3.1, 4-28
 Sec. I3.2d, 4-31, 4-32
 Sec. J2.2b, 4-20, 4-21, 4-22
 Sec. J3.7, 4-16, 4-17
 Table 1-1, 4-3
 Table 3-6, 4-11
 Table 3-19, 4-29, 4-30
 Table 3-21, 4-31, 4-32
 Table 4-22, 4-11
 Table 7-1, 4-13, 4-14, 4-15, 4-16
 Table 7-9, 4-16
 Table 8-4, 4-25, 4-26, 4-27
 Table 8-8, 4-20, 4-21, 4-23, 4-24
 Table C-C2.2, 4-11
 Table J2.4, 4-20, 4-22, 4-23, 4-26, 4-27
 Table J3.2, 4-16, 4-17, 4-18

AISC Commentary
 Fig. C-C2.4, 4-11
 Sec. 1.5, 4-10
 Sec. C2, 4-10, 4-11

ASCE Sec. 2.4.1, 6-1, 6-3

I

IBC
 Eq. (16-4), 4-4
 Eq. (16-5), 4-4
 Sec. 1605.2, 4-4
 Sec. 2107.1, 6-1
 Sec. 2107.3, 6-5
 Sec. 2107.5, 6-4

M

MSJC
 Eq. (1-2), 6-24
 Eq. (1-3), 6-26
 Eq. (2-1), 6-24, 6-25
 Eq. (2-2), 6-25
 Eq. (2-6), 6-26
 Eq. (2-7), 6-26
 Eq. (2-8), 6-26
 Eq. (2-10), 6-26
 Eq. (2-12), 6-5
 Eq. (2-13), 6-28, 6-30, 6-31
 Eq. (2-14), 6-28, 6-30
 Eq. (2-15), 6-28, 6-29, 6-31
 Eq. (2-16), 6-28
 Eq. (2-17), 6-28, 6-30, 6-31
 Eq. (2-18), 6-28, 6-30
 Eq. (2-23), 6-11, 6-16, 6-17
 Eq. (2-24), 6-11
 Eq. (2-26), 6-16
 Eq. (2-27), 6-12
 Eq. (2-29), 6-17
 Eq. (2-30), 6-12, 6-17
 Eq. (3-24), 6-21
 Eq. (3-25), 6-21, 6-22
 Eq. (3-27), 6-18
 Eq. (3-28), 6-18
 Eq. (3-29), 6-19
 Eq. (3-30), 6-23, 6-24
 Eq. (3-31), 6-21, 6-23
 Eq. (3-32), 6-21, 6-22, 6-23
 Eq. (4-1), 6-32, 6-33
 Eq. (4-2), 6-32, 6-33
 Eq. (4-3), 6-32
 Sec. 1.6, 6-12, 6-14
 Sec. 1.8.2.1.1, 6-22
 Sec. 1.8.2.2, 6-22, 6-23
 Sec. 1.13, 6-2

Sec. 1.13.1.2, 6-3
Sec. 1.13.2, 6-4
Sec. 1.14, 6-12
Sec. 1.14.1.2, 6-12
Sec. 1.14.1.3, 6-13, 6-14
Sec. 1.15.2, 6-4
Sec. 1.15.3.1, 6-4, 6-5
Sec. 1.15.3.4, 6-4
Sec. 1.15.5.5, 6-5
Sec. 1.16.2, 6-24, 6-25
Sec. 1.16.6, 6-24
Sec. 1.17.3.2.3.1, 6-14, 6-15, 6-16
Sec. 1.17.3.2.6, 6-15
Sec. 1.17.3.2.6(c), 6-18
Sec. 1.17.3.2.6.1.2, 6-15
Sec. 2.1.2, 6-1
Sec. 2.1.6.1, 6-12
Sec. 2.1.9.3, 6-5
Sec. 2.1.9.5.1, 6-5
Sec. 2.1.9.7, 6-1
Sec. 2.1.9.7.1.1, 6-1
Sec. 2.2.3, 6-28
Sec. 2.2.3.2, 6-28, 6-31, 6-32
Sec. 2.3.2.1, 6-6
Sec. 2.3.3.2.1, 6-13
Sec. 2.3.3.2.2, 6-6
Sec. 2.3.3.3, 6-4
Sec. 2.3.5.2.1, 6-11, 6-16
Sec. 2.3.5.2.2, 6-11, 6-16
Sec. 2.3.5.2.2(b), 6-15
Sec. 2.3.5.2.3, 6-11
Sec. 2.3.5.2.3(b), 6-16
Sec. 2.3.5.3, 6-11, 6-16, 6-17
Sec. 2.3.5.5, 6-11
Sec. 3.3.2, 6-18
Sec. 3.3.3.5.1, 6-20, 6-21
Sec. 3.3.4.2.3, 6-10
Sec. 3.3.5.4, 6-21
Sec. 4.3, 6-28
Sec. 4.3.2, 6-31
Sec. 4.3.3, 6-30, 6-31
Sec. 4.4, 6-28
Sec. 4.4.1.2, 6-30, 6-31
Sec. 4.4.3, 6-32
Sec. 4.4.3.3, 6-32, 6-33
Sec. 4.4.3.6, 6-32
Sec. 4.8.2, 6-28
Sec. 4.8.4.2, 6-28
Table 2.2.3.2, 6-28, 6-30, 6-31, 6-32
Table 3.1.8.2, 6-21, 6-22

MSJC Commentary
Sec. 1.9.3, 6-14
Sec. 1.16.2, 6-24
Sec. 2.3.5, 6-10
Sec. 2.3.5.3, 6-16, 6-17
Sec. 4.3.4, 6-29, 6-31
Sec. 4.4.2, 6-28

N
NDS
Eq. (3.4-6), 5-21
Sec. 2.3, 5-2
Sec. 2.3.3, 5-5
Sec. 3.2.1, 5-6, 5-19
Sec. 3.3.3, 5-6, 5-8, 5-9,
  5-10, 5-14, 5-15, 5-17
Sec. 3.4.2, 5-18, 5-19
Sec. 3.4.3.1, 5-18, 5-19
Sec. 3.4.3.1(a), 5-19
Sec. 3.4.3.2(a), 5-20
Sec. 3.4.3.2(e), 5-20
Sec. 3.4.3.3, 5-21
Sec. 3.7.1, 5-11
Sec. 3.7.1.2, 5-11
Sec. 3.7.1.3, 5-11
Sec. 3.7.1.5, 5-13
Sec. 3.8.2, 5-16
Sec. 3.9.1, 5-16, 5-18

Sec. 3.9.2, 5-13, 5-15
Sec. 3.10.4, 5-5
Sec. 4.3.6, 5-2
Sec. 4.3.7, 5-5
Sec. 4.3.8, 5-3
Sec. 4.4.1, 5-6
Sec. 4.4.3, 5-19
Sec. 5.3.6, 5-3
Sec. 5.3.8, 5-3
Sec. 5.4.4, 5-19
Sec. 10.3, 5-23
Sec. 11.1.2, 5-25
Sec. 11.1.3.6, 5-26
Sec. 11.1.4.2, 5-29
Sec. 11.1.4.3, 5-29
Sec. 11.1.4.6, 5-30
Sec. 11.1.5, 5-31
Sec. 11.1.5.5, 5-31
Sec. 11.2.3.2, 5-31
Sec. 11.3, 5-25, 5-27, 5-30
Sec. 11.3.7, 5-31
Sec. 11.4.1, 5-30
Sec. 11.4.2, 5-31
Sec. 11.5, 5-25
Sec. 11.5.1, 5-23, 5-24, 5-25,
  5-26, 5-27, 5-28
Sec. 11.5.2, 5-24
Sec. 11.5.3, 5-24, 5-31
Sec. 11.5.4, 5-24, 5-31
Sec. 12.2.4, 5-24
Sec. 12.2.5, 5-28
Sec. 12.3.2, 5-23
Sec. 12.3.4, 5-29
Sec. 12.3.5, 5-29
Sec. C11.5.4, 5-31
Table App. G1, 5-11, 5-12
Table 2.3.2, 5-4, 5-12, 5-16,
  5-25, 5-28, 5-30, 5-32
Table 2.3.3, 5-5
Table 4.3.1, 5-2
Table 4.3.8, 5-3
Table 5.3.1, 5-2
Table 10.3.1, 5-23
Table 10.3.3, 5-23, 5-25, 5-28, 5-29
Table 10.3.4, 5-23
Table 10.3.6B, 5-28, 5-29
Table 10.3.6C, 5-25, 5-28
Table 11.1.3, 5-26
Table 11.2A, 5-27
Table 11.2B, 5-30
Table 11.2C, 5-31
Table 11.4.1, 5-27
Table 11.5.1A, 5-25
Table 11.5.1B, 5-23, 5-24, 5-25, 5-27, 5-28
Table 11.5.1C, 5-23, 5-24, 5-27
Table 11.5.1E, 5-27
Table 11A, 5-25
Table 11B, 5-24, 5-25
Table 11C, 5-25
Table 11D, 5-25
Table 11E, 5-25
Table 11F, 5-25
Table 11G, 5-25
Table 11H, 5-25
Table 11I, 5-25
Table 11J, 5-24, 5-26, 5-27
Table 11K, 5-24, 5-27, 5-28
Table 11L, 5-24, 5-29, 5-30
Table 11M, 5-24, 5-30
Table 11N, 5-24, 5-31, 5-32
Table 11O, 5-24
Table 11P, 5-31, 5-32
Table 11Q, 5-24
Table 11R, 5-24
Table 12.2.3, 5-28
Table 12.2.4, 5-28
Table 12.2A, 5-28
Table 12.2B, 5-28, 5-29
Table 12.3, 5-23, 5-28, 5-29

NDS Commentary
Table C11.1.4.7, 5-29
Table C11.1.5.6, 5-31

NDS Supplement, 5-1
Table 4A, 5-1, 5-2, 5-5, 5-8, 5-16, 5-17
Table 4B, 5-1, 5-2, 5-5
Table 4C, 5-1
Table 4D, 5-1, 5-4, 5-12, 5-14
Table 4E, 5-1, 5-3
Table 4F, 5-1, 5-2
Table 5A, 5-1, 5-3, 5-5, 5-6,
  5-9, 5-19, 5-20, 5-21
Table 5B, 5-1
Table 5C, 5-1
Table 5D, 5-1

Trust PPI for Your Civil PE Exam Review

Visit **www.ppi2pass.com** to view all your choices.

Comprehensive Reference and Practice Materials

Civil Engineering Reference Manual
Michael R. Lindeburg, PE

- More than 500 example problems
- Over 400 defined engineering terms
- References to over 3,300 equations, 760 figures, and 500 tables
- Example problems use both SI and U.S. Customary units

Practice Problems for the Civil Engineering PE Exam
Michael R. Lindeburg, PE

- Over 750 practice problems
- Quantitative and nonquantitative problems presented
- Complete step-by-step solutions
- Over 130 tables and 530 figures

Civil Engineering Solved Problems
Michael R. Lindeburg, PE

- Scenario-based practice for the Civil PE exam
- Over 370 practice problems arranged in order of increasing complexity
- Complete step-by-step solutions

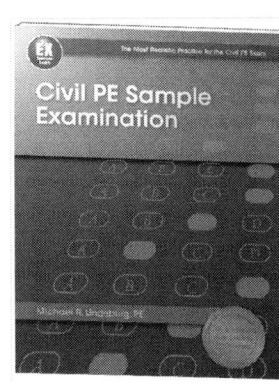

Civil PE Sample Examination
Michael R. Lindeburg, PE

- Similar format, level of difficulty, and problem distribution to the exam
- A 40-problem sample exam for the morning session and each of the 5 afternoon sections (240 problems in total)
- Complete step-by-step solutions

Quick Reference for the Civil Engineering PE Exam
Michael R. Lindeburg, PE

- Puts the most frequently-used equations and formulas at your fingertips
- Includes a comprehensive index for rapid retrieval
- Cross-references additional information found in the *Civil Engineering Reference Manual*

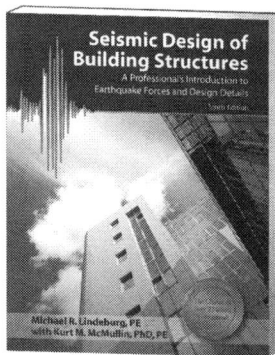

Seismic Design of Building Structures
Michael R. Lindeburg, PE, with Kurt M. McMullin, PhD, PE

- 129 practice problems
- 34 example problems
- Detailed illustrations and definitions of seismic terminology
- Up-to-date building code information

Find out about all of PPI's Civil PE exam preparation products, including Review Courses, Passing Zones, Exam Cafe, and the exam forum at **www.ppi2pass.com**.

The Power to Pass®
www.ppi2pass.com

Coverage of the Civil PE Exam Depth Sections
For more information, visit www.ppi2pass.com.

Construction

Construction Depth Reference Manual for the Civil PE Exam
Thomas M. Korman, PhD, PE, PLS

- Clear, easy-to-understand explanations of construction engineering concepts
- 35 example problems
- 178 equations, 29 tables, 74 figures, and 5 appendices
- An easy-to-use index

Six-Minute Solutions for Civil PE Exam Construction Problems
Elaine Huang, PE

- 100 challenging multiple-choice problems
- 20 morning and 80 afternoon session problems
- Coverage of exam-adopted design standards
- Step-by-step solutions

Geotechnical

Soil Mechanics and Foundation Design: 201 Solved Problems
Liiban A. Affi, PE

- A comprehensive review for the Civil PE exam's geotechnical depth section and the California GE exam
- More than 200 solved problems
- Step-by-step solutions

Six-Minute Solutions for Civil PE Exam Geotechnical Problems
Bruce A. Wolle, MSE, PE

- 100 challenging multiple-choice problems
- 20 morning and 80 afternoon session problems
- Coverage of exam-adopted codes
- Step-by-step solutions

Structural

Structural Depth Practice Exams for the Civil PE Exam
James Giancaspro, PhD, PE

- Two, 40-problem multiple-choice practice exams
- Problems that closely match the topics, level of difficulty, and reference standards on the exam
- Detailed solutions

Six-Minute Solutions for Civil PE Exam Structural Problems
Christine A. Subasic, PE

- 100 challenging multiple-choice problems
- 20 morning and 80 afternoon session problems
- Coverage of exam-adopted codes
- Step-by-step solutions

Coverage of the Civil PE Exam Depth Sections

For more information, visit www.ppi2pass.com.

Transportation

Transportation Depth Reference Manual for the Civil PE Exam
Norman R. Voigt, PE, PLS

- Clear, easy-to-understand explanations of transportation engineering concepts
- 45 example problems
- 86 exam-like, end-of-chapter problems with complete solutions
- 242 equations, 90 tables, 88 figures, and 35 appendices

Six-Minute Solutions for Civil PE Exam Transportation Problems
Norman R. Voigt, PE, PLS

- 86 challenging multiple-choice problems
- 15 morning and 71 afternoon session problems
- Coverage of exam-adopted standards
- Step-by-step solutions

Water Resources and Environmental

Water Resources and Environmental Depth Reference Manual for the Civil PE Exam
Jonathan Brant, PhD;
Gerald J. Kauffman, MPA, PE

- Clear, easy-to-understand explanations of water resources and environmental engineering concepts
- An overview of the *Ten States Standards*
- 115 example problems
- 101 exam-like, end-of-chapter problems with complete solutions
- 230 equations, 65 tables, 102 figures, and 8 appendices

Six-Minute Solutions for Civil PE Exam Water Resources and Environmental Problems
R. Wane Schneiter, PhD, PE

- 100 challenging multiple-choice problems
- 31 morning problems and 69 afternoon problems
- Step-by-step solutions
- Explanations of how to avoid common errors

Don't miss all the Civil PE exam news, the latest exam advice, the exam FAQs, and the unique community of the Exam Forum at **www.ppi2pass.com.**

The Power to Pass®
www.ppi2pass.com

Supplement Your Review With Design Principles

For more information, visit www.ppi2pass.com.

Comprehensive Reference and Practice Materials

Bridge Design for the Civil and Structural PE Exams
Robert H. Kim, MSCE, PE; Jai B. Kim, PhD, PE

- An overview of key bridge design principles
- 5 design examples
- 2 practice problems
- Step-by-step solutions

Concrete Design for the Civil and Structural PE Exams
C. Dale Buckner, PhD, PE, SECB

- Comprehensive concrete design review
- A complete overview of the relevant codes and standards
- 37 practice problems
- Easy-to-use tables, figures, and concrete design nomenclature

Steel Design for the Civil PE and Structural SE Exams
Frederick S. Roland, PE, SECB, RA, CFEI, CFII

- Comprehensive overview of the key elements of structural steel design and analysis
- Side-by-side LRFD and ASD solutions
- More than 50 examples and 35 practice problems
- Solutions cite specific *Steel Manual* sections, equations, or table numbers

Timber Design for the Civil and Structural PE Exams
Robert H. Kim, MSCE, PE; Jai B. Kim, PhD, PE

- A complete overview of the relevant codes and standards
- 40 design examples
- 6 scenario-based practice problems
- Easy-to-use tables, figures, and timber design nomenclature

Don't miss all the Civil PE exam news, the latest exam advice, the exam FAQs, and the Exam Forum at **www.ppi2pass.com.**

Move Your Civil PE Exam Review into the Passing Zone
For more information, visit www.ppi2pass.com/passingzone.

Interactive, Online Review for the Civil PE Exam

Get Organization, Support, and Practice to Help You Succeed
- Complete the weekly assignments at your own pace.
- Post questions to your online instructor, who is an experienced engineer familiar with the Civil PE exam.
- Get answers to your questions and clarify problems as you review.
- Use the weekly study schedule.
- Prepare with additional online practice problems.
- Receive weekly email reminders.

Join the thousands of exam candidates who found the focus to help them pass their Civil PE exams through the Passing Zone.

Civil PE Exam Cafe
Get Immediate Results with
Online Civil PE Sample Exams and Practice Problems

www.ppi2pass.com/examcafe

PPI's Exam Cafe is an online collection of over 1,000 nonquantitative and 300 quantitative problems similar in format and level of difficulty to the ones found on the Civil PE exam. Easily create realistic timed exams, or work through problems one at a time, going at your own pace. Since Exam Cafe is online, it is available to you 24 hours a day, seven days a week, so you can practice for your exam anytime, anywhere.

Over 1,300 Civil PE Exam-Like Problems

- Over 1,000 nonquantitative and 300 quantitative practice problems
- Morning session problems covered
- Afternoon problems covered for all sections: construction, geotechnical, structural, transportation, and water resources and environmental
- Problems can be solved individually or in an exam format

Realistic, Timed Civil PE Exams

- Create 4-hour, 40-problem morning session exams
- Create 4-hour, 40-problem afternoon section exams
- Pause and return to exams at any time

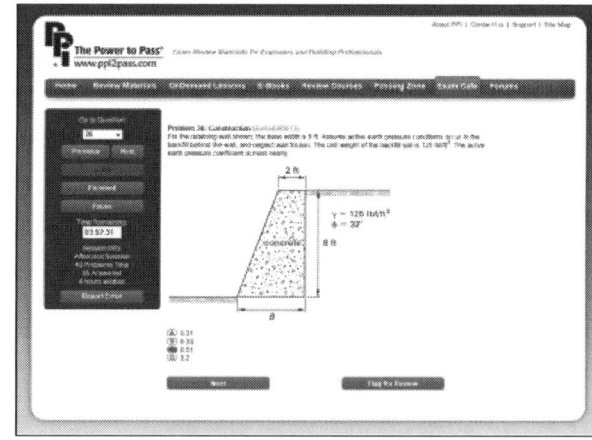

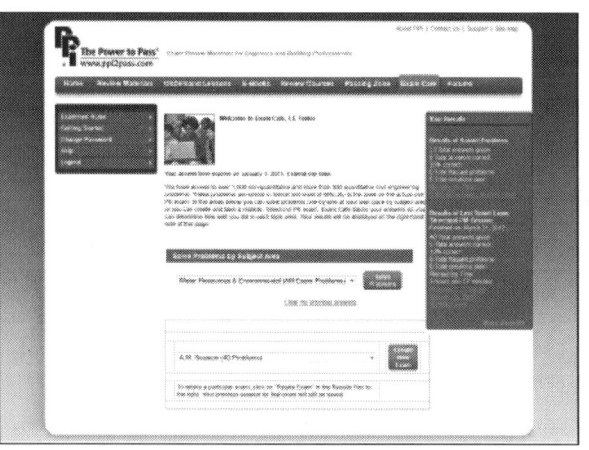

Immediate Exam Results and Problem Solutions

- Real-time results
- Detailed solutions shown for every exam problem
- Results organized by topic or by problem

Visit Exam Cafe at
www.ppi2pass.com/examcafe.